BASIC ELECTRICITY FOR INDUSTRY:

CIRCUITS AND MACHINES

BASIC ELECTRICITY FOR INDUSTRY:

CIRCUITS AND MACHINES

Theodore Wildi

President, Sperika Enterprises Ltd.

Professor Emeritus of Laval University

Department of Electrical Engineering

JOHN WILEY & SONS

New York   Chichester   Brisbane   Toronto   Singapore

To Rachel, Suzanne and Pauline

Library of Congress Cataloging in Publication Data:

Wildi, Théodore.
   Basic electricity for industry.

   Includes index.
   1. Electric engineering.   I. Title.
TK146.W457   1985   621.3   84-27145
ISBN 0–471–88489–8

Printed in the United States of America

10 9 8 7 6 5 4 3 2 1

Preface

As its title implies, this is a textbook on electrical circuits and machines, adapted to industrial needs. However, this book also covers the generation, transmission, and distribution of electrical energy and presents all the fundamental laws, such as Ohm's law, Faraday's law, Lenz's law, and the origins of electromagnetic forces. An introductory chapter also gives a brief overview of units, mechanics and heat, along with a short review of trigonometry.

The solution of dc and ac circuits is developed from basic principles and gradually evolves into the solution of three-phase circuits, using phasor diagrams. The notation used to describe voltages and currents is fully explained, with particular regard to the meaning of positive and negative values. The rate of change of a voltage or current is stressed because it has direct applications in the study of inductors, capacitors, and electromagnetic induction.

The properties of electrical materials such as conductors and insulators, as well as magnetic materials, are examined to better understand their use in electrical machines.

After the most important types of industrial motors, generators and transformers are discussed, the control of rotating machinery is explained with the aid of control diagrams.

Throughout this book, theory is supported by many worked-out examples. Furthermore, multi-choice questions at the end of each chapter are supplemented by review questions and problems. The book therefore gives a broad overview of electrical circuits and machines.

In the sense that it reflects a set of motivating beliefs, concepts, and principles, this book may be said to embody a philosophy. Most textbooks on electrical machines assume some familiarity with circuits; however, this book starts at the very beginning so that the student is presented with a complete, coherent picture. The same notation and methods of explanation are used throughout. For example, the basic theory used to describe magnetism and circuits is again used in discussing rotating machines and transformers.

The presentation is simple and assumes only an elementary knowledge of algebra and trigonometry. Nevertheless, simplicity has been achieved while retaining technical rigor. Thus an electrical engineer familiar with differential equations, Maxwell's laws of electromagnetism, and armed with sophisticated methods of solving circuits will find nothing in this book that is at odds with his or her firmly based knowledge.

Hundreds of figures and photographs have been included, which serve to enhance understanding by appealing to both the verbal and visual senses. The photographs also highlight the actuality of the enormous machines and projects that are being undertaken today.

Another feature of this textbook is its self-paced, self-teaching approach. Simple one-question examples gradually become more comprehensive as they build on topics covered in previous chapters. The many examples raise students' confidence levels and prepare them for the challenging multiple-choice questions and problems at the end of each chapter.

A book that trains people for industry must be practical. Anyone who has built, assembled, maintained, or repaired electrical equipment is aware of the mechanical and thermal problems that continually crop up. Indeed, many of the problems that designers have had to solve do not relate to electricity but rather to mechanics and heat. The stiffness of shafts, the wear of bearings and brushes, inertia, noise levels, and the ever-present problem of temperature rise have always been closely associated with electrical technology. That is why the question of mechanics and heat occupies an important place in the text.

Another practical matter, which is discussed wherever appropriate, concerns standards and standards-setting bodies. The impact of standards on industry has been enormous from both safety and economic points of view. The fact that shaft diameters, mounting heights, frame sizes, and motor performance have been standardized has brought economic benefits to everyone.

This textbook should appeal to a variety of readers and cater to several levels of education. It is intended for vocational high schools, community colleges, technical institutes, and survey courses in universities for nonelectrical engineering students. It is also well adapted to industrial training programs, including those of electric utility companies. The casual industrial worker involved in the maintenance, upkeep, and repair of electrical apparatus will find the book helpful as a reference text. Technicians and engineers filling administrative, supervisory, or sales positions will also appreciate this reference.

This book can be used in several ways. It can be used in a two-semester course covering both circuits and machines. And, because circuit theory is often covered in a separate course, the book can be adopted as a machines textbook in a one-semester course. The circuits portion may then serve as a handy reference and refresher. Other instructors, such as those of specialized evening courses for adult education, may wish to stress chapters concerning the electric utility industry. At the university level, students with prior physics and mathematics backgrounds and the ability to learn by themselves can cover the entire book in one semester. Finally, some instructors may prefer to present only selected topics. This can be done readily with this textbook because of its cross-references, cited throughout.

I would like to acknowledge the assistance of the reviewers who undertook the task of going over my book in its preliminary stages. Their helpful advice and suggestions have made a significant contribution to its clarity, cohesion, and organizational structure.

The many photographs provided by various companies have also contributed to the book's content and level of interest, and I am grateful to their lenders. I also want to thank the staff at John Wiley for their professional and enthusiastic support.

Last but not least, I thank Lucie Veilleux for typing the manuscript and putting it into computerized form, and my son Karl for doing the drawings.

Contents

CHAPTER 1
REVIEW OF USEFUL FUNDAMENTALS   1

Introduction and Chapter Objectives   1

1.1   Units, the SI   2

ELEMENTARY MECHANICS

1.2   Force, the Newton (N)   4

1.3   Torque, the Newton Meter (N.m)   6

1.4   Work, the Joule (J)   7

1.5   Power, the Watt (W)   8

1.6   Prony Brake   10

1.7   Energy, the Joule (J)   11

1.8   Transformation of Energy   12

1.9   Efficiency of a Machine   13

1.10   Temperature Rise and Temperature Scales   14

1.11   Kinetic Energy   15

1.12   Typical Torque–Speed Curves   15

1.13   Power Output and Torque–Speed Curves   16

1.14   Load Torque   17

1.15   Acceleration and Deceleration of a Motor and its Load   18

SIMPLE MATHEMATICS

1.16   The Right Angle Triangle   19

1.17   Angles and Meaning of Sin θ, Cos θ, Tan θ 19

1.18   Sines and Cosines of Angles Greater than 90°   20

1.19   Meaning of Arcsin x, Arcos x, Arctan x 22

1.20   Cosine Law, Equilateral Triangles   22

1.21   Generating a Sine Wave   23

1.22   Case of a Rod that Turns at a Uniform Rate   26

1.23   Cycle   27

1.24   Frequency   27

1.25 Summary   27
Test Your Knowledge   28

CHAPTER 2
THE NATURE OF ELECTRICITY   33

Introduction and Chapter Objectives   33
2.1   Composition of Matter, Atoms and Molecules   33
2.2   Structure of the Atom, Nucleus, and Electrons   34
2.3   Free Electrons   34
2.4   Conductors and Insulators   35
2.5   Unit of Electric Charge, the Coulomb (C)   36
2.6   Sources of Electricity, Potential Difference (E)   36
2.7   Unit of Potential Difference, the Volt (V)   36
2.8   Electric Current (I)   36
2.9   Conventional Current Flow, the Ampere (A)   38
2.10 Measuring Voltage and Current; Voltmeter and Ammeter   39
2.11 Electric Power, the Watt (W)   39
2.12 Distinction Between Sources and Loads   40
2.13 Ohm's Law; the Ohm (Ω)   41
2.14 Application of Ohm's Law   41
2.15 Power Dissipated in a Resistance   42
2.16 Electric Energy, the Joule (J), the Kilowatthour (kW.h)   43
2.17 Storing Electric Energy   44
2.18 Summary   44
Test Your Knowledge   45
Questions and Problems   46

CHAPTER 3
SIMPLE DIRECT-CURRENT CIRCUITS   49

Introduction and Chapter Objectives   49
3.1   Series Circuits   49
3.2   Resistors in Series, Equivalent Resistance   50
3.3   Parallel Circuits   52
3.4   Resistors in Parallel, Equivalent Resistance   53

3.5   Series-Parallel Circuits   55
3.6   Power Flow in a Simple Transmission Line   57
3.7   Example of a Short-Circuit   58
3.8   Primary and Secondary Cells, Ampere Hour (A.h)   58
3.9   Cells in Series   60
3.10 Cells in Parallel   60
3.11 Cells Connected in Series-Parallel   60
3.12 Charging a Secondary Battery   61

SOLVING MORE COMPLEX CIRCUITS

3.13 Kirchhoff's Voltage and Current Laws   62
3.14 Voltage Rises and Voltage Drops   63
3.15 Applying Kirchhoff's Laws   63
3.16 Summary   65
Test Your Knowledge   66
Questions and Problems   67

CHAPTER 4
CONDUCTORS, RESISTORS
AND INSULATORS   69

Introduction and Chapter Objectives   69
4.1   Classification of Materials   69
4.2   Resistivity of Electrical Materials, the Ohm Meter (Ω.m) 70
4.3   Solid, Liquid, and Gaseous Insulators   71
4.4   Deterioration of Organic Insulators   71
4.5   Thermal Classification of Insulators   71
4.6   Breakdown of Solid and Gaseous Insulators   72
4.7   Round Conductors, Standard American Wire Gauge   74
4.8   Mils, Circular Mils   74
4.9   Stranded Cable   77
4.10 Square Wires and Other Conductor Shapes   77
4.11 Current Rating of Conductors; Ampacity   77
4.12 National Electrical Code and Electrical Installations   78
4.13 Calculation of Resistance   79
4.14 Resistance and Temperature   81

4.15 Classification of Resistors   81

4.16 Fuses   83

4.17 Contact Resistance   83

4.18 Resistance of the Ground   84

4.19 Summary   85

Test Your Knowledge   85

Questions and Problems   86

CHAPTER 5
MAGNETISM AND ELECTROMAGNETS   89

Introduction and Chapter Objectives   89

5.1   Magnetic Polarity, N and S Poles   89

5.2   Magnetic Attraction and Repulsion   89

5.3   Lines of Force; the Weber (Wb)   90

5.4   Properties of Lines of Force   91

5.5   Magnetic Domains; Induced and Remanent Magnetism   92

5.6   Permanent Magnets   93

5.7   Magnetic Field Created by a Current   93

5.8   Magnetic Field Produced by a Short Coil   94

5.9   Magnetomotive Force and Flux Density, the Tesla (T)   95

5.10 Magnetic Field of a Long Coil, Electromagnets   95

5.11 Typical Applications of Electromagnets   96

5.12 Properties of Magnetic Circuits   97

5.13 Magnetic Circuit With a Nonmagnetic Core   97

5.14 Magnetic Circuit With a Magnetic Core   98

5.15 Magnetic Circuit With an Air Gap   99

5.16 Reluctance of a Magnetic Circuit   100

5.17 Relative Permeability   100

5.18 Torque Developed by a Magnetic System   100

5.19 Reluctance Torque   102

5.20 Hysteresis Losses   103

5.21 Hysteresis Losses due to Rotation   104

5.22 Torque Produced by Hysteresis   104

5.23 Summary   105

Test Your Knowledge   105
Questions and Problems   106

CHAPTER 6
ELECTROMAGNETIC FORCES   109

Introduction and Chapter Objectives   109
6.1   Force on a Straight Conductor   109
6.2   Magnitude of the Force   111
6.3   Force Between Two Conductors   112
6.4   Forces Acting on a Coil   114
6.5   Blow-Out Coils   114
6.6   Torque Produced by a Rectangular Coil   115
6.7   Summary   116
Test Your Knowledge   117
Questions and Problems   117

CHAPTER 7
AC VOLTAGES AND CURRENTS   119

Introduction and Chapter Objectives   119
7.1   Positive and Negative Voltages   119
7.2   Graph of a Variable Voltage   120
7.3   Rate of Change of Voltage   120
7.4   Calculating the Rate of Change of Voltage   123
7.5   Positive and Negative Currents   123
7.6   Rate of Change of Current   124
7.7   Properties of a Sinusoidal Waveshape   124
7.8   Rate of Change of a Sinusoidal Voltage   126
7.9   Rate of Change of a Sinusoidal Current   126
7.10 Representing Sinusoidal Voltages and Currents by Phasors 127
7.11 Degree Scale Versus Time Scale   127
7.12 Relationship Between Two Sinusoidal Waveshapes   128
7.13 Typical Phasor Diagrams   130
7.14 Instantaneous Values   130

7.15	Summary	132
	Test Your Knowledge	133
	Questions and Problems	134

CHAPTER 8
ELECTROMAGNETIC INDUCTION

		135
	Introduction and Chapter Objectives	135
8.1	Voltage Induced in a Coil, Faraday's Law of Induction	135
8.2	Magnitude of the Induced Voltage	136
8.3	Induced Current	137
8.4	Direction of the Induced Current	138
8.5	Polarity of the Induced Voltage, Lenz's Law	139
8.6	Methods of Inducing a Voltage	139
8.7	Relationship Between an AC Flux and the Induced AC Voltage	142
8.8	Coil Revolving in a Stationary Field	142
8.9	Cutting Lines of Flux	144
8.10	Polarity of the Induced Voltage in a Straight Conductor	145
8.11	Eddy Currents	146
8.12	Eddy Currents in a Stationary Iron Core	148
8.13	Eddy Current Losses in a Revolving Core	149
8.14	Summary	149
	Test Your Knowledge	150
	Questions and Problems	151

CHAPTER 9
INDUCTORS AND INDUCTANCE

		153
	Introduction and Chapter Objectives	153
9.1	Mutual Induction	153
9.2	Mutual Inductance, the Henry (H)	154
9.3	Self-Induction and Self-Inductance	156
9.4	Energy Stored in a Coil	157
9.5	Equivalent Circuit of a Coil	159
9.6	Build-Up of Current in a Coil	159
9.7	Opening the Circuit of a Coil	161

9.8   Typical Inductance of Coils   162

9.9   Inductance of a Transmission Line   162

9.10 Inductor Carrying an AC Current   163

9.11 Summary   164

Test Your Knowledge   165

Questions and Problems   166

CHAPTER 10
CAPACITORS AND CAPACITANCE   169

Introduction and Chapter Objectives   169

10.1   Free Electrons in a Metal   169

10.2   Transfer of Charge and Potential Difference   169

10.3   Energy Stored in an Electric Field   171

10.4   Capacitance; the Farad (F)   171

10.5   Capacitors   172

10.6   Energy Stored in a Capacitor   174

10.7   Using a Source to Transfer a Charge   175

10.8   Capacitor as a Source or Load   176

10.9   Commercial Capacitors   176

10.10 Capacitors in Parallel and in Series   177

10.11 Charging a Capacitor; Time Constant   178

10.12 Discharging a Capacitor   179

10.13 Basic Capacitor Equation   180

10.14 Capacitance of Transmission Lines   181

10.15 Capacitor Connected to an AC Voltage   182

10.16 Summary   184

Test Your Knowledge   184

Questions and Problems   185

CHAPTER 11
DIRECT CURRENT GENERATORS   187

Introduction and Chapter Objectives   187

11.1   Generating a DC Voltage   187

11.2   Armature of a DC Generator   190

11.3   Field of a DC Generator   192

11.4   Assembly of a DC Generator   193

11.5   Schematic Diagram of a DC Generator   194

11.6   Current in the Armature; Armature Reaction   195

11.7   Separately Excited DC Generator; No-Load Saturation Curve   196

11.8   Value of the Induced Voltage   197

11.9   Shunt Generator at No-Load   198

11.10   Separately Excited Generator Under Load   199

11.11   Compound Generator   200

11.12   Rating and Voltage Regulation of a DC Generator   204

11.13   Actual Construction of a DC Armature   205

11.14   Commutation of Current   208

11.15   Armature Reaction and Commutation   209

11.16   Commutating Poles   209

11.17   Multipolar Machines   210

11.18   Summary   210

Test Your Knowledge   210

Questions and Problems   211

CHAPTER 12
DIRECT CURRENT MOTORS   213

Introduction and Chapter Objectives   213

12.1   Construction and Torque–Speed Characteristics of DC Motors   213

12.2   Motor Losses and Efficiency   216

12.3   Starting a DC Motor   218

12.4   Theory of the DC Motor   219

12.5   Torque of a DC Motor   221

12.6   Speed of a DC Motor   222

12.7   Acceleration of a Shunt Motor   223

12.8   Torque and Speed Characteristics of a Shunt Motor   225

12.9   Torque and Speed Characteristics of a Series Motor   227

12.10   Torque and Speed Characteristics of a Compound Motor   228

12.11   Speed Control of a DC Motor   228

12.12   Dynamic Braking and Plugging   230

12.13   Starters for DC Motors   231

12.14 Summary   232
Test Your Knowledge   232
Questions and Problems   233

CHAPTER 13
SIMPLE ALTERNATING CURRENT
CIRCUITS   235

Introduction and Chapter Objectives   235
13.1   Resistive Circuit   235
13.2   Power Dissipated in a Resistive Circuit   236
13.3   Effective Value of Voltage and Current   237
13.4   Inductive Circuit   239
13.5   Reactive Power in an Inductive Circuit, the Var   240
13.6   Inductive Reactance (X_L)   241
13.7   Capacitive Circuit, Capacitive Reactance (X_C)   243
13.8   Reactive Power in a Capacitive Circuit   244
13.9   Calculation of Capacitive Reactance   245
13.10 Active and Reactive Power   246
13.11 Phasor Diagrams and Effective Values   248
13.12 Recapitulation of R, L, C Circuits   248
13.13 Summary   248
Test Your Knowledge   249
Questions and Problems   250

CHAPTER 14
SOLVING SINGLE-PHASE AC CIRCUITS   251

Introduction and Chapter Objectives   251
14.1   Voltages and Currents in an RL Circuit   251
14.2   Addition of Phasors   253
14.3 X and Y Components of a Phasor   255
14.4   Constructing a Phasor from its X and Y Components   256
14.5   Calculating the Phasor Sum by Trigonometry   257
14.6   Impedance of a Circuit   258
14.7   Apparent Power of a Circuit, the Voltampere (VA)   259

14.8   Solution of a Parallel R, C Circuit   259

14.9   Solution of a Series R, L Circuit   260

14.10   Impedance of Two Elements in Series   261

14.11   Impedance of Two Elements in Parallel   262

14.12   Series Resonance; Power in AC Circuits   265

14.13   Measurement of Active and Reactive Power   266

14.14   Relationship between P, Q, and S 270

14.15   Power Factor   272

14.16   Summary   274

Test Your Knowledge   275

Questions and Problems   276

CHAPTER 15
THREE-PHASE CIRCUITS   279

Introduction and Chapter Objectives   279

15.1   Polyphase Systems   279

15.2   Generating a Three-Phase Voltage   280

15.3   Three-Phase Currents   283

15.4   Wye and Delta Connection   285

15.5   Currents and Voltages in a Wye Connection   286

15.6   Currents and Voltages in a Delta Connection   287

15.7   Power and Power Factor in a Three-Phase Line   287

15.8   Solving Three-Phase Circuits   288

15.9   Measuring Three-Phase Power   289

15.10   Phase Sequence   292

15.11   Summary   292

Test Your Knowledge   293

Questions and Problems   294

CHAPTER 16
TRANSFORMERS   295

Introduction and Chapter Objectives   295

16.1   The Transformer and Power Transmission   295

16.2   Simple Transformer   296

16.3   No-Load Operation   296
16.4   Transformer Under Load   299
16.5   Losses, Efficiency and Rating of a Transformer   300
16.6   Rated Voltage of Transformer   303
16.7   Polarity of a Transformer   303
16.8   Polarity Tests   304
16.9   Transformer Taps   305
16.10 Transformers in Parallel   307
16.11 Cooling Methods   307
16.12 Transformer Impedance   309

SPECIAL TRANSFORMERS

16.13 Distribution Transformers   311
16.14 Voltage Transformers   312
16.15 Current Transformers   313
16.16 Autotransformer   315
16.17 Connecting a Standard Transformer as an Autotransformer   316
16.18 Variable Autotransformer   318

TRANSFORMERS IN THREE-PHASE SYSTEMS

16.19 Delta-Delta Connection   319
16.20 Delta-Wye Connection   320
16.21 Wye-Delta Connection   321
16.22 Wye-Wye Connection   321
16.23 Open-Delta Connection   322
16.24 Three-Phase Transformers   322
16.25 Summary   324
Test Your Knowledge   325
Questions and Problems   327

CHAPTER 17
THREE-PHASE INDUCTION MOTORS   329

Introduction and Chapter Objectives   329
17.1   Forces Produced by a Moving Magnetic Field   329
17.2   Torque Produced by a Rotating Magnetic Field   331
17.3   Number of Poles and Construction of the Rotor   332

17.4   Slip and Synchronous Speed   333

17.5   Producing a Revolving Field   334

17.6   Effect of the Number of Poles   336

17.7   Construction of a Squirrel-Cage Induction Motor   337

17.8   Locked-Rotor Conditions   338

17.9   Acceleration of a Squirrel-Cage Motor   341

17.10 Squirrel-Cage Motor at No-Load   343

17.11 Squirrel-Cage Motor Under Load   344

17.12 Squirrel-Cage Motor Under Overload   345

17.13 Effect of Rotor Resistance   345

17.14 Wound-Rotor Induction Motor   346

17.15 Sector Motor   347

17.16 Linear Induction Motor   348

17.17 Summary   349

Test Your Knowledge   350

Questions and Problems   351

CHAPTER 18
SELECTION AND APPLICATION OF
THREE-PHASE INDUCTION MOTORS   353

Introduction and Chapter Objectives   353

18.1   Standardization and Classification of Induction Motors   353

18.2   Classification According to Environment and Cooling Methods   354

18.3   Classification According to Electrical and Mechanical
        Properties   356

18.4   Choice of Motor Speed   360

18.5   Plugging an Induction Motor   361

18.6   Effect of Inertia   363

18.7   Braking with Direct Current   363

18.8   Abnormal Conditions   364

18.9   Mechanical Overload   364

18.10 Line Voltage Changes   364

18.11 Single-Phasing   365

18.12 Frequency Variation   365

18.13 Typical Applications of Induction Motors   366

18.14 Summary   366
Test Your Knowledge   367
Questions and Problems   368

CHAPTER 19
SYNCHRONOUS MOTORS   371

Introduction and Chapter Objectives   371
19.1   Construction of Synchronous Motors   371
19.2   Acceleration of a Synchronous Motor   373
19.3   Effect of DC Excitation   376
19.4   Synchronous Motor Under Load   381
19.5   Relationship Between Torque and Torque Angle   382
19.6   Excitation of Synchronous Machines   383
19.7   Hunting of Synchronous Machines   385
19.8   Rating of Synchronous Motors   387
19.9   Starting Synchronous Motors   388
19.10 Application of Synchronous Motors   389
19.11 Summary   389
Test Your Knowledge   389
Questions and Problems   390

CHAPTER 20
SYNCHRONOUS GENERATORS   391

Introduction and Chapter Objectives   391
20.1   Stationary-Field AC Generator   391
20.2   Revolving-Field AC Generator   392
20.3   Number of Poles   394
20.4   Salient-Pole Rotors   396
20.5   Cylindrical Rotors   396
20.6   AC Generator Under Load   397
20.7   AC Generator on an Infinite Bus   397
20.8   Synchronous Machine Operating as a Generator   400
20.9   Active and Reactive Power Flow   403
20.10 Synchronization of an AC Generator   404

20.11 Equivalent Circuit of an AC Generator   406

20.12 Phasor Diagram of a Synchronous Generator   407

20.13 AC Generator Feeding an Isolated Load   410

20.14 Voltage Regulation of an Isolated Generator   412

20.15 Summary   414

Test Your Knowledge   414

Questions and Problems   416

CHAPTER 21
SINGLE-PHASE MOTORS   417

Introduction and Chapter Objectives   417

21.1   Single-Phase Induction Motor   417

21.2   Theory of the Revolving Field   418

21.3   Means of Starting Single-Phase Motors   419

21.4   Resistance Split-Phase Motor   423

21.5   Capacitor-Start Motor   423

21.6   Properties of Split-Phase Induction Motors   424

21.7   Capacitor-Run Motor   427

21.8   Shaded-Pole Motor   427

21.9   Series Motor   428

21.10 Choice of Single-Phase Motors   429

21.11 Synchros   430

21.12 Hysteresis Motors   431

21.13 Reluctance-Torque Motor   433

21.14 Stepper Motors   433

21.15 Precision Motor in a Watthourmeter   435

21.16 Summary   438

Test Your Knowledge   438

Questions and Problems   439

CHAPTER 22
INDUSTRIAL MOTOR CONTROL   441

Introduction and Chapter Objectives   441

22.1   Typical Control Devices   441

22.2   Normally Open and Normally Closed Contacts; Graphic Symbols   446
22.3   Control Diagrams   446
22.4   Automatic DC Starters   448
22.5   Starting Methods for Three-Phase Induction Motors   451
22.6   Manual Across-the-Line Starters   451
22.7   Magnetic Across-the-Line Starters   451
22.8   Inching and Jogging   454
22.9   Reduced-Voltage Starting   454
22.10 Primary Resistance Starting   454
22.11 Autotransformer Starting   456
22.12 Other Starting Methods   458
22.13 Cam Switches   459
22.14 Starting Three-Phase Synchronous Motors   461
22.15 Automatic Starter for a Synchronous Motor   464
22.16 Summary   466
Test Your Knowledge   466
Questions and Problems   467

CHAPTER 23
GENERATION, TRANSMISSION, AND DISTRIBUTION OF ELECTRICAL ENERGY   471

Introduction and Chapter Objectives   471
23.1   Basic Objectives of an Electric Utility System   472
23.2   Reactive Power and the Electric Utility System   474
23.3   System Demand and Power Control   474
23.4   Choice of Frequency   475
23.5   Choice of Voltage Level   476
23.6   System Stability   478

GENERATING STATIONS

23.7   Location of the Generating Station   479
23.8   Makeup of a Thermal Generating Station   480
23.9   Nuclear Generating Stations   484
23.10 Energy Released by Atomic Fission   484
23.11 Chain Reaction   484

23.12 Makeup of a Nuclear Generating Station   485
23.13 Hydropower Generating Stations   486
23.14 Makeup of a Hydropower Station   487

TRANSMISSION OF ELECTRICAL ENERGY

23.15 Components of a Transmission Line   492
23.16 Corona Effect and Bundled Conductors   494
23.17 Grounding   495
23.18 Electrical Properties of a Transmission Line   495
23.19 Voltage Regulation of a Transmission Line   496
23.20 Power-Handling Capacity   498

SUBSTATION EQUIPMENT

23.21 Circuit Breakers   502
23.22 Air-Break Switches   504
23.23 Disconnecting Switches   504
23.24 Grounding Switches   505
23.25 Lightning Arresters   505
23.26 Example of a Substation   507

LOW-VOLTAGE DISTRIBUTION IN BUILDINGS

23.27 Principal Components of an Electrical Installation   508
23.28 LV Distribution Systems   510
23.29 Grounding Electrical Installations   513
23.30 Electric Shock   513

THE COST OF ELECTRICITY

23.31 Tariff Based upon Energy   514
23.32 Tariff Based upon Demand   514
23.33 Demand Meter   514
23.34 Tariff Based upon Power Factor   516
23.35 Typical Rate Structures   517
23.36 Power Factor Correction   520
23.37 Summary   522
Test Your Knowledge   522
Questions and Problems   524

**APPENDIX A
CONVERSION CHARTS FOR UNITS**   527

APPENDIX B
PROPERTIES OF ROUND COPPER
CONDUCTORS   **533**

REFERENCES   **535**

ANSWERS TO PROBLEMS   **539**

INDEX   **543**

CHAPTER 1

Review of Useful Fundamentals

Introduction and Chapter objectives

The objective of this chapter is to review some of the mechanical terms that are encountered in electrical technology. Particular attention is devoted to the meaning of torque, and the conditions under which a machine will accelerate or slow down. A further objective is to review some of the geometric relations of right-angle and equilateral triangles. The study is then extended to cover trigonometric relations, and sine waves. These simple notions of geometry and trigonometry will be useful when we begin studying ac circuits and machines.

One would normally expect that a book on electricity would cover only electrical subjects. However, motors, generators, contactors, and other electrical devices produce thermal and mechanical effects, as well as electrical effects. Consequently, we cannot limit our study to electrical subjects alone, but must also have an understanding of elementary mechanics and heat.

Then there is the question of mathematics. Do we need mathematics to *understand* the properties of circuits and machines ? Actually, we don't, but mathematics are used as a tool. Mathematics enable us to predict the values of voltage, current and power in a circuit or machine by simply writing a few numbers and equations on a piece of paper. This is much faster than using instruments, machinery, wires, and power supplies to find the values by experiment. In this chapter we give a brief review of the simple mathematics we will be using later on.

With regard to units, most technical work is now done in SI units, but the foot-pound customary system is still in use. To accommodate the two schools of thought, we give both SI and customary-unit versions of some basic equations.

One final word before you undertake this chapter. You should read it rather quickly at first, to see what it is all about. Then, you can review this and that topic in greater depth as the need arises. For example, when you reach Chapter 12 on dc motors, you will want to review the parts dealing with torque, horsepower and speed. Later, when you begin Chapter 13 on ac circuits, you may want to take another look at trigonometry and elementary geometry.

1.1 Units, the SI

Lord Kelvin once said that when you can measure what you are speaking about, and express it in numbers, you know something about it. Units are the measuring sticks that enable us to express quantities such as mass, length, time, power, and voltage, in terms of numbers. It took hundreds of centuries to arrive at the exact measurement systems we use today. Thus, the yard was originally defined as the distance between the tip of King Edward's nose and the end of his outstretched hand. Today it can be expressed in terms of the distance that light travels in 1/327 857 019 of a second. This modern measurement of the yard is exact to better than one part in one billion!

The U.S. customary system of units comprises such familiar units as the foot, square yard, pound mass, pound force, second, volt, and kilowatt. Some of these units are related to each other by odd factors such as 12, 36, 32.17404, and these historically-developed multipliers are not very convenient to handle. For this and other reasons, a much simpler set of units was adopted by international agreement wherein the factors are all multiples of 10. This system of units, known as the *International System of Units* (abbreviated SI) is now recognized by all countries of the world. Indeed, most countries use it as their sole system of measurement.

The most commonly-used units of the SI are the meter, kilogram, second, and newton. (The newton corresponds to the pound force, which is the customary unit of force.) The SI comprises about 25 additional units which are used to measure quantities such as pressure, volume, magnetic flux, temperature, torque, and so forth. These units, and their relationship to the U.S. customary units, are listed in Table 1A. Study this table from time to time in order to become familiar with the names of these SI units. Once the names are firmly in your mind, it becomes much easier to work with the SI.

One final point: there is nothing mysterious or complicated about units. For any given quantity, such as mechanical pressure, the units of the various systems are similar, differing only in size. The various units of a given quantity are therefore related to each other by simple numbers, called conversion factors. Thus, 1 pound per square inch = 144 pounds per square foot = 6.894 757 kilopascals = 68.947 57 millibars. The pound per square inch is a U.S. customary unit, the kilopascal is an SI unit and the millibar is a special unit used by meteorologists. (The conversion factors relating the units of different systems are shown graphically in the Appendix.)

What unit should we use, for example, in measuring a length? For short lengths we may use a unit such as the inch. However, for large distances it is preferable to

**TABLE 1A   SOME COMMONLY-USED QUANTITIES AND UNITS
IN THE CUSTOMARY AND THE SI SYSTEMS.**

Quantity	Customary unit and/or symbol	SI unit and/or symbol	Relationship between SI and customary units
mass	pound mass (lbm)	kilogram (kg)	1 kg = 2.204 lbm
length	foot (ft)	meter (m)	1 m = 3.28 ft
time	second (s)	second (s)	-
area	square foot (ft^2)	(m^2)	1 m^2 = 35.31 ft^2
volume	cubic foot (ft^3)	(kg/m^3)	1 m^3 = 35.31 ft^3
flow	(ft^3/s)	(m^3/s)	1 m^3/s = 35.31 ft^3/s
speed	(ft/s)	(m/s)	1 m/s = 3.28 ft/s
speed of rotation	revolution per minute (r/min)	(r/min)	-
frequency	cycle per second	hertz (Hz)	1 cycle per second = 1 Hz
force	pound force (lbf)	newton (N)	1 lbf = 4.448 N
torque	foot-pound force (ft.lbf)	newton meter (N.m)	1 lbf.ft = 1.356 N.m
pressure	(lbf/ft^2)	pascal (Pa)	1 lbf/ft^2 = 47.88 Pa
electric energy	kilowatthour (kW.h)	joule (J)	1 kW.h = 3.6 × 10^6 J
mechanical energy	(ft.lbf)	joule (J)	1 ft.lbf = 1.356 J
thermal energy	British thermal unit (Btu)	joule (J)	1 Btu = 1055 J
electric power	watt (W)	watt (W)	-
mechanical power	horsepower (hp)	watt (W)	1 hp = 746 W
electric potential	volt (V)	volt (V)	-
electric current	ampere (A)	ampere (A)	-
electric charge	coulomb (C)	coulomb (C)	-
electric resistance	ohm (Ω)	ohm (Ω)	-
magnetomotive force	ampere-turn (At)	ampere (A)	1 At = 1 A
magnetic flux	line or maxwell	weber (Wb)	1 maxwell = 10^{-8} Wb
magnetic flux density	(line/in^2)	tesla (T)	1 line/in^2 = 15.5 × 10^{-6} T
inductance	henry (H)	henry (H)	-
capacitance	farad (F)	farad (F)	-
temperature	degree Fahrenheit (°F)	degree Celsius (°C)	°F = 1.8°C + 32

use a bigger unit, such as the mile. For example, the distance between New York and Montreal is better expressed as 350 miles than as 22 176 000 inches. But the width of a desk is more meaningful when stated as 32 inches than as 0.000 505 05 miles. In effect, we don't like to use very large or very small numbers when expressing the magnitude of something. That is why we use units that are multiples, (or submultiples) of a given unit. Thus, the yard is a *multiple* of the inch, equal to 36 inches. In the same way, the mil is a *submultiple* of the inch, equal to one thousandth of an inch.

In the SI, multiples and submultiples of units are related to each other by factors of ten. Thus, the centimeter is a submultiple of the meter, equal to 1/100 of a meter. On the other hand, the kilometer is a multiple of the meter, equal to 1000 meters.

Multiples and submultiples of 10 are conveniently expressed in exponent form. Thus, $100 = 10^2$; $1000 = 10^3$; and $1\ 000\ 000 = 10^6$. Similarly, $1/100 = 10^{-2}$; $1/1000 = 10^{-3}$; and $1/1\ 000\ 000 = 10^{-6}$.

Multiples and submultiples in the SI are designated by attaching a prefix to the "base" unit. The list of prefixes is given in Table 1B, together with the multiplier it represents. The most frequently encountered prefixes are *mega, kilo, milli* and *micro*. The others are rarely used except in scientific work. The only exception is the centimeter, but even here, many people prefer to express "short" lengths in millimeters rather than in centimeters. For example, in mechanical work, it is quite common to express a length of 9.76 meters as 9760 mm.

ELEMENTARY MECHANICS

1.2 Force

The most familiar force we know is the force of gravity. For example, whenever we lift a stone, we must exert a muscular effort to overcome the gravitational force that continually pulls the stone downwards. The magnitude of the force of gravity depends upon the mass of the body, and where it is located. For bodies that are relatively close to the surface of the earth (within 20 km, say) the force is given by the approximate equations:

SI UNITS		U.S. CUSTOMARY UNITS	
$F = 9.8\ m$	(1-1)	$F = m$	(1-2)

where   where
F = force of gravity, in newtons (N)   F = force of gravity in pounds force (lbf)
m = mass of body, in kilograms (kg)   m = mass of body in pounds (lb)

There are other kinds of forces, such as the force exerted by a stretched spring, or the force we exert when pushing a stalled car. All these forces are expressed in terms of the newton (symbol N), which is the SI unit of force.

TABLE 1B   MULTIPLES AND SUBMULTIPLES OF SI UNITS

Prefix	Multiplier	Symbol	Examples
exa	10^{18}	E	2 exameters $= 2 \times 10^{18}$ meters
			2 Em $= 2 \times 10^{18}$ m
peta	10^{15}	P	3 petajoules $= 3 \times 10^{15}$ joules
			3 PJ $= 3 \times 10^{15}$ J
tera	10^{12}	T	4 terawatts $= 4 \times 10^{12}$ watts
			4 TW $= 4 \times 10^{12}$ W
~~giga~~	10^{9} *n 1,000,000,000*	G	1 gigajoule $= 10^{9}$ joules
			1 GJ $= 10^{9}$ J
~~mega~~	10^{6} *n 1,000,000*	M	1 megapascal $= 10^{6}$ Pa
			1 MPa $= 10^{6}$ Pa
~~kilo~~	10^{3} or 1000	k	3 kilometers $= 3000$ meters
			3 km $= 3000$ m
~~hecto~~	10^{2} or 100	h	1 hectoliter $= 100$ liters
			1 hL $= 100$ L
~~deca~~	10	da	1 decatesla $= 10$ teslas
			1 daT $= 10$ T
~~deci~~	10^{-1} or 1/10 *.10*	d	1 cubic decimeter $= (\text{meter}/10)^3$
			$= \text{meter}^3/1000$
			1 dm^3 $= 10^{-3}$ m^3
~~centi~~	10^{-2} or 1/100 *.01*	c	4 centimeters $= (4/100)$ meter
			4 cm $= 0.04$ m
~~milli~~	10^{-3} or 1/1000 *.001*	m	1 millimeter $= 10^{-3}$ meter
			1 mm $= 0.001$ m
~~micro~~	10^{-6} *.000001*	μ	1 microfarad $= 10^{-6}$ farad
			1 F $= 10^{-6}$ F
~~nano~~	10^{-9} *.000000001*	n	1 nanosecond $= 10^{-9}$ second
			1 ns $= 10^{-9}$ s
~~pico~~	10^{-12}	p	1 picoampere $= 10^{-12}$ ampere
			1 pA $= 10^{-12}$ A
femto	10^{-15}	f	1 femtometer $= 10^{-15}$ meter
			1 fm $= 10^{-15}$ m
atto	10^{-18}	a	1 attojoule $= 10^{-18}$ joule
			1 aJ $= 10^{-18}$ J

1.3 Torque, the newton meter (N.m)

Whenever a force acts on a body so as to make it twist or rotate, (or tend to rotate) the body is said to be subjected to a torque. Torque is equal to the product of the force times the perpendicular distance between the axis of rotation and the direction of the force. For example, consider a cable wrapped around a winch having a radius r (Fig. 1-1). If the pull (force) on the cable is F, the torque exerted is given by:

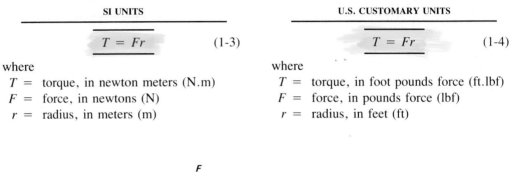

SI UNITS	U.S. CUSTOMARY UNITS
$T = Fr$ (1-3)	$T = Fr$ (1-4)

where
$T =$ torque, in newton meters (N.m)
$F =$ force, in newtons (N)
$r =$ radius, in meters (m)

where
$T =$ torque, in foot pounds force (ft.lbf)
$F =$ force, in pounds force (lbf)
$r =$ radius, in feet (ft)

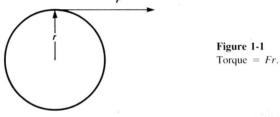

Figure 1-1
Torque $= Fr$.

Gasoline engines and electric motors develop a torque because whenever they are coupled to a mechanical load they exert a twisting action, causing the load to rotate.

Example 1-1:

A motor develops a starting torque of 150 ft.lbf. If the pulley on the shaft has a diameter of 1 ft, calculate the braking force needed to prevent the motor from turning.

Solution:

Because the radius is 0.5 ft, the required braking force is:
$$F = T/r = 150 \text{ ft.lbf}/0.5 \text{ ft} = 300 \text{ lbf}$$ Eq. 1-4

If the radius of the pulley were 2 ft, a braking force of 75 lbf would be enough to prevent the motor from starting.

1.4 Work, the joule (J)

Work is a very common word, but in mechanics it has a very precise meaning. Mechanical work is said to be done when a force F moves through a distance d, with the distance measured in the same direction as the force. The work is given by the equations:

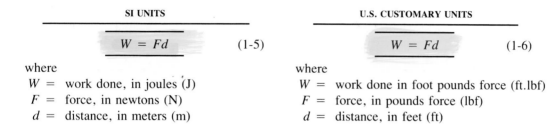

SI UNITS	U.S. CUSTOMARY UNITS
$W = Fd$ (1-5)	$W = Fd$ (1-6)

where
$W =$ work done, in joules (J)
$F =$ force, in newtons (N)
$d =$ distance, in meters (m)

where
$W =$ work done in foot pounds force (ft.lbf)
$F =$ force, in pounds force (lbf)
$d =$ distance, in feet (ft)

The SI unit of work is the joule (symbol J). It is equal to the work done when a force of 1 newton moves through a distance of 1 meter.

Example 1-2:

A 500 pound mass is lifted through a height of 60 feet (Fig. 1-2). Calculate the work done.

Solution:

In the customary system of units, the force of gravity, expressed in pounds force, is numerically equal to the mass, in pounds.
The force of gravity on the 500 lb mass is therefore equal to 500 pounds force. The work done is:

$$W = Fd = 500 \text{ lbf} \times 60 \text{ ft} = 3000 \text{ ft.lbf} \qquad \text{Eq. 1-2}$$

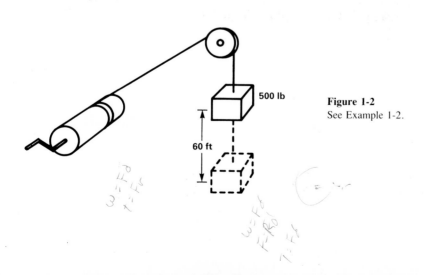

Figure 1-2
See Example 1-2.

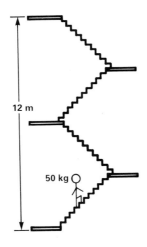

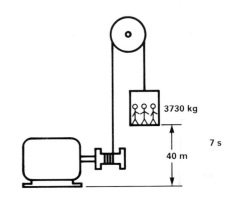

Figure 1-3
See Example 1-3.

Figure 1-4
See Example 1-4.

Example 1-3:

A person weighing 50 kilograms walks up a flight of stairs whose vertical height is 12 meters (Fig. 1-3). What is the work done by the person?

Solution:

1. The force of gravity acting downwards on the person is:
$$F = 9.8\ m = 9.8 \times 50 = 490 \text{ newtons} = 490 \text{ N} \qquad \text{Eq. 1-1}$$

2. The work done is:
$$W = Fd = 490 \times 12 = 5880 \text{ joules} = 5880 \text{ J} \qquad \text{Eq. 1-5}$$

1.5 Power, the watt (W)

Power is also a common word, but again, in mechanics, it has a very precise meaning. Power is the rate of doing work. It is given by the equations:

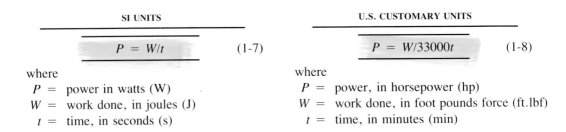

SI UNITS	U.S. CUSTOMARY UNITS
$P = W/t$ (1-7)	$P = W/33000t$ (1-8)

where
$P =$ power in watts (W)
$W =$ work done, in joules (J)
$t =$ time, in seconds (s)

where
$P =$ power, in horsepower (hp)
$W =$ work done, in foot pounds force (ft.lbf)
$t =$ time, in minutes (min)

The SI unit of power is the watt (symbol W). We often use the kilowatt (kW), equal to 1000 W, or the megawatt (mW), equal to one million watts. The power output of motors is usually expressed in horsepower (hp), a US customary unit that is equal to 746 W. In the SI, however, the mechanical power of a motor is expressed in watts, even if the motor happens to be a diesel engine. Furthermore, in the SI, the rate of heat loss of a radiator, or of any warm body, is also expressed in watts. As a result, whether power is electrical, mechanical or thermal, its value is always expressed in watts, and this is one of the important simplifications afforded by the SI.

Example 1-4:

An elevator motor lifts a mass of 3730 kg through a height of 40 m in 7 s (Fig. 1-4). Calculate the power developed by the motor, in kilowatts, and in horsepower.

Solution:

1. The lifting force is:
$F = 9.8 \, m = 9.8 \times 3730 = 36\,554$ N (N for newtons)   Eq. 1-1

2. The work done is:
$W = Fd = 36\,554 \times 40 = 1\,462\,160$ J (J for joules)   Eq. 1-5

3. The power is:
$P = W/t = 1\,462\,160/7 = 208\,880$ W $= 208.88$ kW   Eq. 1-7

4. Expressed in horsepower:
$P = 208.880/746 = 280$ hp.

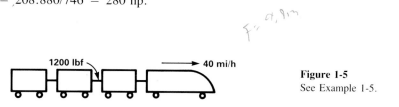

1200 lbf   40 mi/h

Figure 1-5
See Example 1-5.

Example 1-5:

An electric locomotive exerts a pull of 1200 lbf while moving along at a speed of 40 miles per hour (Fig. 1-5). Calculate the power developed by the driving motors. (note that 1 mile = 5280 ft)

Solution:

1. Work done in one hour is:
$W = Fd = 1200 \times 40 \times 5280 = 253.44 \times 10^6$ ft.lbf

2. The power is:
$P = W/33\,000 \, t = 253.44 \times 10^6/33\,000 \times 60 = 128$ hp.

1.6 Prony brake

The mechanical power of a motor depends upon the torque it develops, and its speed of rotation. The power is given by the equations:

SI UNITS	U.S. CUSTOMARY UNITS
$P = nT/9.55$ (1-9)	$P = nT/5252$ (1-l0)

where
P = mechanical power, in watts (W)
n = speed, in revolutions per minute (r/min)
T = torque, in newton meters (N.m)
9.55 = constant, to take care of units

where
P = mechanical power, in horsepower (hp)
n = speed, in revolutions per min (r/min)
T = torque, in foot pounds force (ft.lbf)
5252 = constant, to take care of units

We can both load up and measure the mechanical power of a motor by means of a prony brake. In its simplest form, it consists of a flat pulley mounted on the shaft of the motor, and a braking device. The device possesses brake shoes that rub on the surface of the pulley. The resultant braking force is transmitted to a radius arm having a length r. (Fig. 1-6). The braking action can be increased or decreased by tightening or slackening a thumb-screw V. A spring balance measures the force F

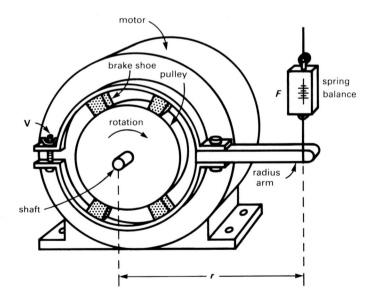

Figure 1-6
A motor can be loaded up and its power measured by means of a prony brake.

exerted at the distance r from the center of the motor shaft. Consequently, the product Fr gives the torque developed by the motor while it is running. A tachometer measures the speed n, and so the power output can immediately be calculated, using Eq. 1-9 or 1-l0.

The mechanical power of the motor is entirely converted into heat, and so the pulley heats up very quickly unless it is cooled by some means.

Example 1-6:

During a prony brake test on a 5 hp motor, the following readings were taken: $F = 6$ lbf, $r = 14$ inches, $n = 1720$ r/min. Calculate the torque and power developed by the motor under these conditions.

Solution:

$$T = Fr = \ 6 \text{ lbf} \times 14/12 \text{ ft} = 7 \text{ ft.lbf} \qquad\qquad \text{Eq. 1-4}$$

$$P = nT/5252 = 1720 \times 7/5252 = 2.29 \text{ hp} \qquad\qquad \text{Eq. 1-10}$$

The motor is not fully loaded because it could actually develop a rated power output of 5 hp. To increase the power output, screw V must be tightened so that the brake shoes press more firmly against the pulley.

Example 1-7:

A small synchronous motor running at 1800 r/min has a nominal power output of 120 W. If the prony brake arm has a length of 10 centimeters, calculate the braking force F needed so that the motor develops its rated power output.

Solution:

1. $P = nT/9.55$; therefore $120 = 1800 \ T/9.55$
   and so $T = 0.637$ N.m

2. $T = Fr$; therefore $0.637 = F \times 10/100$
   and so $F = 6.37$ N.

The required scale reading is 6.37 N or about 1.43 lbf. (Note that from Appendix A or Table 1A, 1 lbf = 4.448 N)

1.7 Energy, the joule (J)

A body is said to possess energy whenever it is capable of doing work. Indeed, all work is produced as a result of using up energy. Energy can exist in one of the forms listed below:

1. Mechanical energy (energy in a waterfall, a coiled spring or a moving car);
2. Thermal energy (heat released by a stove, by friction, by the sun);
3. Chemical energy (energy contained in dynamite, coal, or an electric storage battery);
4. Electrical energy (energy produced by a generator, or manifested by lightning);
5. Atomic energy (energy released when the nucleus of an atom is split).

Because energy can be made to do work, energy is essentially work that is stored up. Consequently, energy, like work, is expressed in joules. We therefore come to the remarkable result that all forms of energy - mechanical, electrical, chemical, atomic and thermal - can be expressed in the same SI unit, the joule (J).

1.8 Transformation of energy

Energy can neither be created or destroyed, but it can be converted from one form to another by means of appropriate devices called *machines*. In Fig. 1-7 for example, the chemical energy contained in coal may be transformed into thermal energy by burning the coal in a steam boiler. The thermal energy contained in the steam can be transformed into mechanical energy by using a turbine. Finally, mechanical energy can be transformed into electrical energy by means of a generator. In this example, the boiler, turbine and generator are the machines that do the energy transformation.

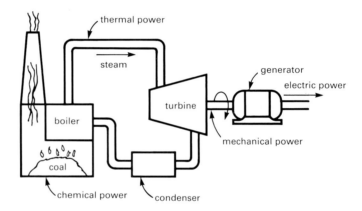

Figure 1-7
Energy is converted from one form to another by means of machines.

Unfortunately, whenever energy is transformed, the output is always less than the input because all machines have losses. These losses appear in the form of heat, causing the temperature of the machine to rise. In the case of an electric motor (Fig. 1-8), the mechanical power output (expressed in watts) is less than the electrical power input (also expressed in watts). The difference between the two increases as the motor is loaded up, which means that the losses become greater with increasing load. It follows that a motor is hotter when running at full load than at no-load. Furthermore, if the motor is loaded beyond its rated capacity, the losses may become so great that the motor overheats and the windings may burn out. This fundamental fact applies to all electric machines, no matter how they are constructed, or how big they are.

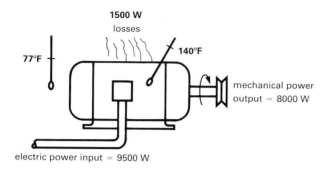

Figure 1-8
The losses in a motor cause its temperature to rise above the ambient temperature.

1.9 Efficiency of a machine

The efficiency of a machine is the ratio of the power output it delivers, to the power input it receives. Because the power output is always less than the input, the efficiency must be less than one. The efficiency is usually given in percent and its value is calculated by the equation:

$$eff = 100\, P_o\, /\, P_i \qquad (1\text{-}11)$$

where
eff = efficiency, in percent (%)
P_o = power output of machine, in watts (W)
P_i = power input of machine, in watts (W)

For a given machine, say a 10 hp electric motor, the efficiency varies with the load. It starts from zero at no-load, increases rapidly as the load increases and then stays almost flat over a considerable load range. If the machine is loaded much above its normal rating, the efficiency again begins to fall, but only slightly. This typical variation of efficiency with load is shown by the graph of Fig. 1-9.

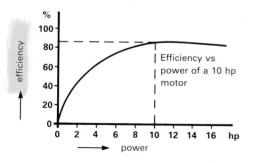

Figure 1-9
Typical efficiency versus power output of a 10 hp motor.

The efficiency of electric motors improves with size. Thus, a 1/4 hp motor has a full-load efficiency of about 60 percent whereas a 10 000 hp. motor has an efficiency of about 98%. Small may be beautiful, but it is always accompanied by low efficiency. For example, the motor in an electric razor has an efficiency of about 10%, and there is no way it can be significantly improved.

Example 1-7:

A 150 hp motor has a full-load efficiency of 91%. Calculate the electric power it consumes, and the losses at full load.

Solution:

1. The mechanical power output is:
   P_o = 150 hp = 150 × 746 = 111 900 W

2. The electric power input is:
   P_i = P_o /*eff* = 111 900/0.91 = 122 967 W

3. The losses are:
   P_i − P_o = (122 967 − 111 900) = 11 067 W = 11.0 kW

1.10 Temperature rise and temperature scales

When a body produces heat, its temperature rises above that of its surroundings. The difference between the temperature of the body and the ambient temperature is called temperature rise. Thus, if the temperature of a motor winding is 87°C, when the ambient temperature is 20°C, the temperature rise of the winding is (87-20) = 67°C. In order to ensure that electrical equipment will last a long time, upper limits have been set on temperature rise. For example, the temperature rise of transformer windings is usually limited to 55°C, when the coils are immersed in oil. On the other hand, the permissible temperature rise may be as high as 105°C for specially-insulated dry-type transformer windings.

The SI unit of temperature (and of temperature rise) is the degree Celsius (°C). The relationship between the SI and the customary unit of temperature is given by:

$$F = 1.8 C + 32 \qquad (1\text{-}12)$$

where
   F = temperature, in degrees Fahrenheit (°F)
   C = temperature, in degrees Celsius (°C)

Example 1-8:

A thermometer placed on a motor frame (Fig. 1-8) registers a temperature of 140°F. The ambient temperature is 77°F. Express these temperatures in °C, and calculate the temperature rise.

Solution:

1. The temperature of the motor frame is:
$$F = 1.8\ C + 32$$
$$140 = 1.8\ C + 32$$
$$C = (140 - 32)/1.8 = 108/1.8 = 60°$$

2. The ambient temperature is:
$$F = 1.8\ C + 32$$
$$77 = 1.8\ C + 32$$
$$C = (77 - 32)/1.8 = 45/1.8 = 25°$$

3. The temperature rise of the frame is:
$$60 - 25 = 35°\ \text{C or } 140 - 77 = 63\ °\text{F}$$

1.11 Kinetic energy

A fast-moving train, a falling stone, or a revolving flywheel all possess a form of stored mechanical energy called kinetic energy. The amount of energy depends upon the mass and speed of the body: the heavier it is and the faster it goes, the greater is its kinetic energy.

The kinetic energy in a moving body does not come out of the blue, but is due to the mechanical work that was previously done upon it. In other words, to bring any body up to speed, we must do mechanical work. If there is no friction or any other braking force, the amount of kinetic energy is exactly equal to the mechanical work that was previously supplied. For a given body, the kinetic energy increases as the square of the speed. Thus, if the speed is tripled, the stored kinetic energy increases nine times.

What are some of the consequences of kinetic energy? We have all observed that when we cut the power to an electric motor, the rotor and its connected load continue to turn for a certain time. If the friction is small and the rotating parts are heavy, it may take several minutes before the system comes to a halt. In effect, the system continues to run as long as it has kinetic energy. As the speed falls, the kinetic energy decreases progressively, until it finally becomes zero when the motor stops. The loss in kinetic energy, from start to finish, is exactly equal to the heat produced by friction. We can bring the system to a faster stop by increasing the friction, or braking torque. Nevertheless, the drop in kinetic energy is still the same, and consequently the total amount of heat produced remains unchanged.

Owing to the phenomenon of kinetic energy, it is impossible to bring a motor up to speed in zero time. Similarly, we cannot cause it to stop in zero time. The reason is that the accelerating torque and the braking torque would have to be infinite in each case.

1.12 Typical torque-speed curves

The torque developed by machines such as electric motors, gasoline engines, and pneumatic motors usually varies with the speed. Figure 1-10 shows the typical

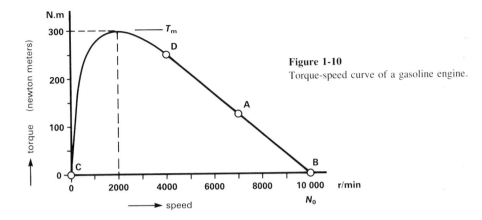

Figure 1-10
Torque-speed curve of a gasoline engine.

torque-speed curve of a gasoline engine when operating at a fixed setting of the throttle. Note that when the speed is zero, the torque is also zero, and consequently a gasoline engine must be equipped with a starter to get it going. The torque then increases with the speed, reaching a maximum of 300 N.m at 2000 r/min. Beyond this peak, the torque decreases gradually and finally becomes zero at 10 000 r/min.

The engine develops its rated power output (in horsepower) at operating point A. When it runs at no load, the operating point is B. Under normal conditions, the engine runs between no-load and full-load, which is to say somewhere along line AB. The engine can, however, deliver more than its rated power for brief periods. Such an overload situation corresponds to point D.

The torque-speed curve of an electric motor is quite similar. For example, Fig. 1-11 shows the characteristic of a three-phase induction motor. However, unlike the gasoline engine, this motor develops a substantial torque at zero speed (point C). Consequently, it is a self-starting motor.

One interesting feature is that the operating characteristic between no-load and full-load (line AB) is essentially a straight line in both cases. This linear torque-speed characteristic holds true for all electric motors.

Portion CDA of the curve represents the starting characteristic of the motor (Fig. 1-11), or of the gasoline engine (Fig. 1-10). This characteristic is important because it tells us the maximum load that can be brought up to speed. However, the starting characteristic is only important during the short time the motor is accelerating.

1.13 Power output and torque-speed curves

According to Eqs. 1-9 and 1-10, the power output of a motor is proportional to the product of its torque T and speed n. This simple fact enables us to visualize the power that a motor develops, by looking at its torque-speed curve. Referring to Fig. 1-12, suppose we select any operating point (1). The corresponding torque is T_1 and the corresponding speed is n_1. The power which the motor develops at this

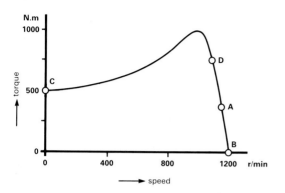

Figure 1-11
Torque-speed curve of an induction motor.

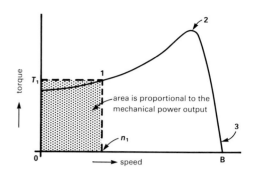
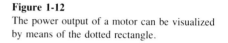

Figure 1-12
The power output of a motor can be visualized by means of the dotted rectangle.

point is proportional to the product T_1 times n_1. But this product is equal to the area of the dotted rectangle in Fig 1-12. Consequently, the area of the rectangle is a measure of the power that the motor develops.

As we move along the curve, the successive "power" rectangles change in size. This gives us a visual means of evaluating the mechanical power the motor develops. Thus, as we move along the torque-speed curve of Fig 1-12, we can actually "see" that the power at low speeds (1) is small, that it reaches a maximum at operating point (2) and that it again falls off at point (3). Finally, the output power vanishes (the rectangle has zero area) when the motor reaches no-load point B.

1.14 Load torque

Mechanical loads, just like motors, often possess distinctive torque-speed curves. Note for example, the T-n curve of a fan, and of a reciprocating pump (Figs. 1-13, 1-14).

However, some loads such as punch-presses, lathes, vibrators, and saws have

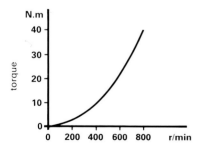

Figure 1-13
Torque-speed curve of a fan.

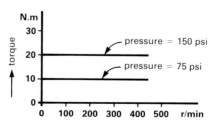

Figure 1-14
Torque-speed curve of a reciprocating pump.

highly unusual torque-speed characteristics which cannot be described by a single curve. The reason is that such loads cover a whole spectrum of torques and speeds. Nevertheless, at any instant, the mechanical power absorbed by a load is given by Eqs. 1-9 or 1-10, where T and n are the *instantaneous* torque and speed.

1.15 Acceleration and deceleration of a motor and its load

A motor and its load together constitute a system which obeys some simple, but important mechanical laws. These laws may be stated as follows:

1. The speed of the motor is always equal to that of the load. This is so because the motor shaft is assumed to be directly coupled to the load shaft.

2. If the motor torque exceeds the load torque, the speed will increase. The reason is that the power developed by the motor then exceeds the power demanded by the load. The excess mechanical power causes the system to accelerate, which increases its kinetic energy.

3. If the motor torque is less than the load torque, the speed will decrease. The reason is that the power developed by the motor is now less than that demanded by the load. The system decelerates and in so doing, it releases kinetic energy. The power absorbed by the load is then equal to the power delivered by the motor, *plus* the power released by the falling kinetic energy of the system.

4. If the motor torque is equal to the load torque, the speed remains constant. We repeat: When the torque developed by the motor is equal to the torque exerted by the load, the speed remains constant. The system does not come to a halt, as we might be inclined to believe.

With reference to items 2 and 3 above, how quickly does the speed change when the torque developed by the motor is not equal to the load torque? It depends upon two things:

    a) the difference between the torques; the bigger the difference, the faster the speed will change.

    b) the so-called inertia of the revolving parts; the bigger the *inertia*, the slower the speed will change.

A large rotor or a heavy, massive load possesses a lot of inertia, and such revolving masses do not quickly change their speed. For example, it may take several minutes to bring a large flywheel up to speed, even if the bearings are frictionless.

SIMPLE MATHEMATICS

1.16 The right angle triangle

Consider the right angle triangle of Fig. 1-15 having sides AB and BC, and hypotenuse AC. Let their respective lengths be a, b and c, and let the angle between side AB and the hypotenuse be θ. From the Pythagorean proposition, we have the relationship:

$$c^2 = a^2 + b^2 \tag{1-13}$$

from which we deduce the following equations:

$$c = \sqrt{a^2 + b^2} \qquad a = \sqrt{c^2 - b^2} \qquad b = \sqrt{c^2 - a^2} \tag{1-14}$$

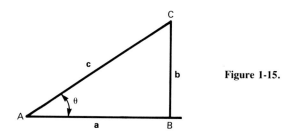

Figure 1-15.

1.17 Angles, and meaning of sin θ, cos θ, tan θ

In Fig. 1-15, and for a given angle θ, the ratio b/c is constant, independent of the size of the triangle. Thus, for an angle of 30°, we find that b/c is *always* equal to 0.5, and for an angle of 45°, b/c is *always* equal to 0.707. In other words, for any given angle, there is a corresponding ratio b/c. This means that the ratio b/c can be used to designate the magnitude of an angle. Because of its importance, this ratio has been given a special name called the *sine* of the angle θ, or simply sin θ. We can therefore write sin θ = b/c. The following table gives some typical values of sin θ for angles between zero and 90°.

θ	0	15	30	45	60	75	90
sin θ	0	0.259	0.500	0.707	0.866	0.966	1.00

The ratio b/c is not the only ratio that is constant for a given angle θ. We also find that the ratio a/c is constant, and so, too, is the ratio b/a. Thus, for an angle of 30°, the ratio a/c is *always* 0.866 and the ratio b/a is *always* 0.588. Again, because of its importance, the ratio a/c is called the cosine of the angle θ, or simply cos θ. Similarly, the ratio b/a is called the tangent of angle θ, or tan θ.

Recognizing that ABC is a right angle triangle, we can state the following simple rules for the values of the ratios we have just discussed.

$$\sin \theta \ = \ b/c \ = \ \frac{\text{side opposite to } \theta}{\text{hypotenuse}}$$

$$\cos \theta \ = \ a/c \ = \ \frac{\text{side adjacent to } \theta}{\text{hypotenuse}}$$

$$\tan \theta \ = \ b/a \ = \ \frac{\text{side opposite to } \theta}{\text{side adjacent to } \theta}$$

1.18 Sines and cosines of angles greater than 90°

It is impossible to construct a right angle triangle in which one of the angles is greater than 90°. Nevertheless, we can determine the sine of an angle greater than 90°, by using two crossed lines, or axes, together with a set of rules.

In Figure 1-16 X'OX, and Y'OY are two axes that cross at point 0, and carry a scale of some kind. Values measured along OX and OY are positive, while those measured along OX' and OY' are negative. Point O is called the *origin*.

Suppose a line having a length c extends outwards from point O, at any desired angle θ, (Fig. 1-17). The value of c is always considered to be positive. From the extremity of c, a dotted line b is drawn perpendicular to the horizontal axis X'OX. The magnitude and sign of b is found by measuring its length along the vertical axis YOY'. Thus, the sign of b is positive in Fig. 1-17. It is also positive in Fig. 1-18. However, b is negative in Figs. 1-19 and 1-20.

A final rule states that angles are positive (+) when measured *counterclockwise* starting from line OX. However, if an angle is negative, it is measured clockwise starting from line OX.

Referring to Fig. 1-17, $\sin \theta_1$ = b/c and $\cos \theta_1$ = a/c. Both quantities are positive, because a and b are positive.

In Fig. 1-19, $\sin \theta_3$ (= b/c) and $\cos \theta_3$ (= a/c) are both negative because a and b are negative.

In Fig. 1-20, $\sin \theta_4$ is negative because b is negative. But $\cos \theta_4$ is positive because a is positive.

Using these rules we can draw up a table showing the values of sin θ for any angle between zero and 360 degrees. Typical values are given in Table 1C.

The sine, cosine and tangent of an angle can be found very quickly by using a hand calculator. Thus, you can readily verify that sin 27.5° = 0.4617 ; cos 243° = −0.454 and tan 350° = tan −10° = −0.1763. We assume the reader has a hand calculator available to make such trigonometric calculations.

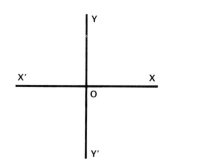

Figure 1-16
X and Y axes.

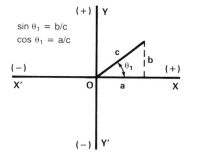

Figure 1-17
Sine and cosine of an angle
between zero and 90°.

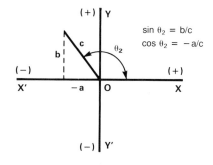
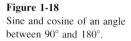

Figure 1-18
Sine and cosine of an angle
between 90° and 180°.

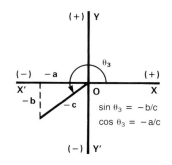

Figure 1-19
Sine and cosine of an angle
between 180° and 270°.

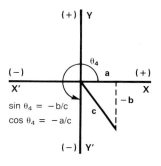

Figure 1-20
Sine and cosine of an angle
between 270° and 360°.

TABLE 1C   Values of sin θ

angle	sin θ	angle	sin θ	angle	sin θ	angle	sin θ
15	0.259	105	0.966	195	− 0.259	285	− 0.966
30	0.500	120	0.866	210	− 0.500	300	− 0.866
45	0.707	135	0.707	225	− 0.707	315	− 0.707
60	0.866	150	0.500	240	− 0.866	330	− 0.500
75	0.966	165	0.259	255	− 0.966	345	− 0.259
90	1.000	180	0	270	− 1.000	360(0)	0

1.19 Meaning of arcsin x, arcos x, arctan x

We have seen that the sine of any angle θ can be found either graphically, or by means of a hand calculator. Conversely, we can find the magnitude of an angle θ when its sine value is given. The notation used is θ = arcsin x which reads "θ is the angle whose sine is x". Sometimes the notation θ = sin^{-1} x is used instead of θ = arcsin x, but the meaning is the same.

The notation is extended to cosine and tangent functions. Thus θ = arcos y has the same meaning as θ = cos^{-1} y, namely, "θ is the angle whose cosine is y". Using a hand calculator, we can readily find that arcsin 0.5 = 30°, arcos (−0.8) = 143.13°, and arctan 0.6 = 30.96°.

1.20 Cosine law, equilateral triangles

In solving electrical circuits, we sometimes encounter triangles that are not right-angled (Fig. 1-21). In such cases, the relationship between the three sides is given by the equation:

$$c^2 = a^2 + b^2 - 2ab \cos \theta \qquad (1\text{-}15)$$

In words, the square of the side opposite angle θ is equal to the sum of the squares of the sides adjacent to θ minus twice the product of those sides multiplied by cos θ.

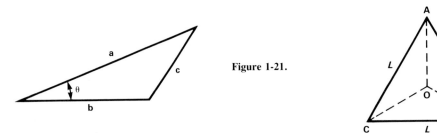

Figure 1-21.

Figure 1-22.

Example 1-9:

Given that a = 52, b = 36 and θ = 22° in Fig. 1-21, calculate the magnitude of c.

Solution:

From Eq. 1-15 we have

$$c^2 = a^2 + b^2 - 2ab\cos\phi$$
$$= 52^2 + 36^2 - 2 \times 52 \times 36 \cos 22$$
$$= 2704 + 1296 - 3644 \times 0.92718 = 528.6$$
$$c = \sqrt{528.6} = 23$$

Consider the equilateral triangle of Fig. 1-22, whose sides have a length L. The bisectors of the 60-degree angles A, B, C meet at a common point O. Using Eq. 1-15, we can prove that the lengths OA, OB and OC are each equal to $L/\sqrt{3} = L/1.732$. This geometrical relationship will be particularly useful when we study three-phase circuits.

1.21 Generating a sine wave

Consider a slender rod of length c that is free to turn around point 0 (Fig. 1-23). Suppose that a bright light on the left projects a shadow of the rod on a flat screen PQ. For a position such as 1, the height of the shadow is b_1. For another position such as 2, the height of the projection (shadow) is b_2. To distinguish projections that are above line AOB from those that are below, they are given positive and negative signs, respectively. Thus, in Fig. 1-23, b_1 is (+), and b_3 is (−).

The position of the rod is given by the angle the rod makes with respect to axis AOB. The standard practice is to measure the angle θ between line OB and the rod. Thus, for positions 1 and 3, the angles are respectively θ_1 and θ_3 degrees. The length of any projection b is given by

$$b = c \sin \theta, \qquad (1-16)$$

because clearly the ratio $b/c = \sin \theta$

Suppose the rod has a length of 100 cm (about 40"). Using the values of sin θ in Table 1C, we can tabulate the values of projection b for different angles.

θ	0	15	30	45	60	90	120	135	180	195	210	270	300	360
b	0	+25.9	+50	+70.7	+86.6	+100	+86.6	+70.7	0	−25.9	−50	−100	−86.6	0

If we plot these values on a graph, as shown in Fig. 1-24, we begin to see the outline of a wave-like shape.

We can obtain many more points on the graph, by taking additional values of θ, such as 6°, 12°, 18°, and so forth. When such 6° intervals are used, we obtain the

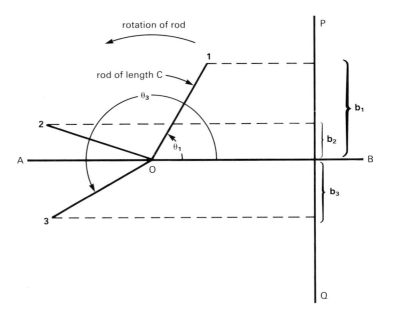

Figure 1-23
The projections b vary with the position of rod c.

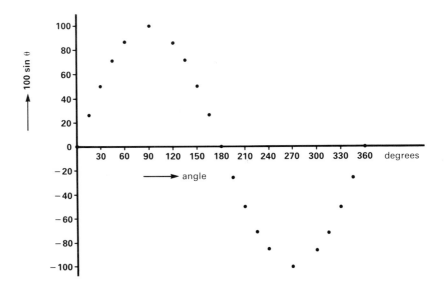

Figure 1-24
The successive dots show the magnitude of the projections at 15° intervals.

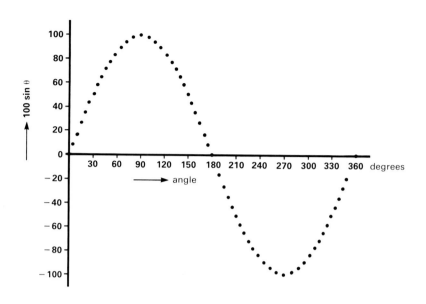

Figure 1-25
Same as Figure 1-24, but with intervals of 6 degrees.

graph of Fig. 1-25. Finally, if we plotted the points at 1° intervals, we would get a nearly-smooth curve, enabling us to read the value of b for any angle.

Such a smooth curve is called a *sine wave*. A sine wave is composed of an infinite series of points which enable us to determine the value of the projection on the screen for any angle.

It is obvious that once the rod has turned through 360°, the values of the projections repeat, and this is true for any number of complete turns. The waveshape thus repeats itself at 360° intervals (Fig. 1-26). Note that angles in excess of 360° simply

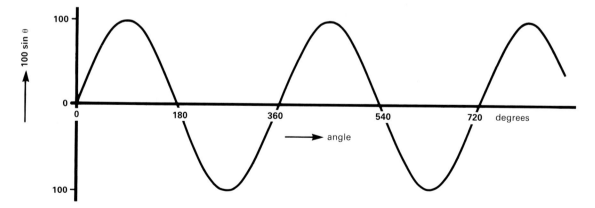

Figure 1-26
Graph of a sine wave having a peak amplitude of 100.

mean that the rod has made more than one turn. For example, 7350° corresponds to 7350/360 = 20.4166 turns, which corresponds to (20 turns + 0.4166 turns), which corresponds to 0.4166 turns, which corresponds to 0.4166 x 360 = 150°. As far as projection b is concerned, an angle of 7350° is the same as an angle of 150°. In effect, sin 7350° = sin 150 = 0.5.

1.22 Case of a rod that turns at a uniform rate

Figure 1-23 gives the projection of the rod for any angle θ. Suppose that the rod rotates at a constant speed of one revolution per minute. Each angle now corresponds to a specific time interval. Thus, 360° corresponds to 1 min, 180° to 1/2 min (30 s), and so forth. We can therefore show the value of the projection as a function of time (Fig. 1-27).

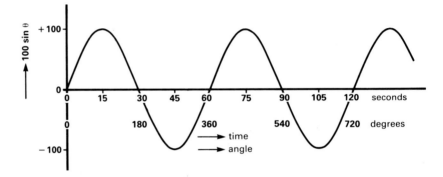

Figure 1-27
Projections expressed both in terms of time and the angle of rotation.

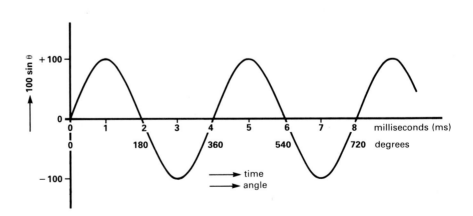

Figure 1-28
The rate of rotation of the rod is 15 000 times faster than in Figure 1-27, but one revolution is still completed in 360°.

If the rod turns very quickly, say 250 revolutions per second, the waveshape of Fig. 1-27 will be the same, but the time scale will be different. Thus, 360° now corresponds to 1/250 s or 4 milliseconds (4 ms), and 90° corresponds to 1 ms. (Fig. 1-28).

However, no matter what the speed of rotation happens to be, the projection of the rod can always be expressed as a function of the angle of rotation. Thus, one revolution always corresponds to 360°, irrespective of the speed of rotation.

We will find that the properties of the "rod" and the sine wave are very useful in describing alternating voltages and currents. The "rod" is then called a *phasor*, a concept that will be more fully discussed in Chapter 7.

1.23 Cycle

Returning to Fig. 1-23, whenever the rod makes a complete turn it is said to have completed one *cycle*. The start of the cycle can correspond to any angle such as θ_1.

The concept of a cycle can be related to the sine wave of Fig. 1-26. Thus, a cycle is any portion of the wave that has a "length" of 360°. The interval from 0° to 360° is therefore one cycle. Similarly, the interval from 30° to 390°, or from 282.9° to 642.9°, is also one cycle.

1.24 Frequency

Suppose the rod of Fig. 1-23 rotates at a constant speed of 250 turns per second. This means that it completes 250 cycles per second, and the corresponding sine wave is shown in Fig. 1-28. The *frequency* of a sine wave is equal to the number of cycles per second. The unit of frequency is the *hertz* (Hz) and it is equal to one cycle per second. The frequency in Fig. 1-28 is therefore 250 Hz. The concept of frequency and cycle will often be used in the study of ac circuits and machines.

1.25 Summary

We have reviewed the meanings of work, force, torque, power, and energy because these mechanical properties are always related to electrical machines and equipment.

The SI system of units is rapidly gaining ground in favor of the customary system of units, especially in technical work. The newton is a particularly important unit, because it is the SI equivalent of the pound force.

SI units are ranked in multiples and submultiples of 10, and carry corresponding prefixes. Conversions from one unit to another can readily be made by using the conversion charts in Appendix A.

This chapter has pointed out the potential usefulness of simple mathematics. Mathematics enable us to predict the properties of many devices without having to carry out physical experiments. Fortunately, the same rules of algebra and trigonometry will be used over and over again, as we proceed in our study of circuits, machines, and devices.

TEST YOUR KNOWLEDGE

1-1   The foot is an SI unit.

☐ true   ☑ false

1-2   The prefix kilo stands for

a. 100   b. 36   c. 1000   d. 1/1000

1-3   The prefix mega stands for

a. 144   b. 10^6 c. 10 000   d. 0.001

1-4   A length of 2 m is equal to

a. 2 miles   b. 200 cm   c. 20 mm   d. 6 ft

1-5   The symbol r/min stands for

a. revolution per minute   b. return per minute
c. minimum radius   d. radian per minute

1-6   The SI units of mass and force are respectively

a. gram and kilogram   b. kilogram and pascal
c. kilogram and joule   d. kilogram and newton

1-7   When we tighten a bolt with a wrench, the bolt is subjected to

a. a force   b. a pressure   c. a torque   d. power

1-8   When we push hard against a brick wall, we do a lot of work

☐ true   ☑ false

1-9   The SI unit of mechanical power is expressed in

a. newton meters   b. watts   c. horsepower   d. joules

1-10   One horsepower is equivalent to

a. 100 W   b. 400 N.m   c. 746 W   d. 476 W

1-11   A prony brake is used to

a. prevent a machine from turning   b. measure the power of a spring
c. measure the power of a motor   d. produce heat

1-12   A machine is a device that can

  a. convert one form of energy into   b. do useful work
  another
  c. rotate when power is applied   d. run by itself

1-13   A machine that has an efficiency of 60 percent has a loss of

  a. 30 percent   b. 0.6   c. 40 percent   d. torque

1-14   A car runs at a speed of 30 km/h. If its speed increases to 90 km/h, the
  kinetic energy increases

  a. 3 times   b. 27 times   c. 9 times   d. not at all

1-15   Referring to Fig. 1-13, the torque at 600 r/min is about

  a. 20 N.m   b. 10 N.m   c. 75 psi   d. 20 psi

1-16   The pump in Fig. 1-14 is operating at 400 r/min and against a pressure of
  150 psi. The torque is

  a. 20 N.m   b. 10 N.m   c. 75 psi   d. 20 psi

1-17   An electric motor drives a load at a speed of 1600 r/min. The motor exerts a
  torque of 60 N.m, and the load torque is 70 N.m. Under these conditions,
  the speed of the motor (and load)

  a. is steady at 1600 r/min   b. is increasing
  c. is decreasing   d. is accelerating

1-18   A right-angle triangle has a hypotenuse of 30 cm and a side of 18 cm. The
  length of the third side is

  a. 48 cm   b. 12 cm   c. 24 cm   d. 18 cm.

1-19   The sine of an angle has a value of 0.966. The angle is

  a. 90°   b. 45°   c. 75°   d. 15°

1-20   The sine of an angle of 105° is

  a. 0.966   b. −0.966   c. 0.259   d. 0.500

1-21   A motor develops 500 hp. The equivalent power in SI units is

  a. 0.6702 W   b. 373 kW   c. 373 kJ   d. 373 kg

1-22   A speed of 1800 r/min is equal to

  a. 109 000 r/h   b. 30 r/h   c. 30 r/s   d. 1800 kHz

1-23   An automobile has a mass of 1200 kg. It exerts a downward force on the road equal to

      a. 1200 kg   b. 11.76 kN   c. 2600 lbf

1-24   A wrench 10 cm long exerts a torque of 12 N.m. The pull exerted on the wrench is

      a. 120 kg   b. 12.24 kg   c. 120 N

1-25   In Fig. 1-2 express the mass and height in SI units

      a. 1102 kg ; 18.29 m   b. 226.7 kg ; 18.29 m
      c. 196.8 m ; 1102 kg.

1-26   An electric motor draws 20 kW from a line for a period of 8 h. The energy absorbed by the motor is

      a. 2.5 kW/h   b. 9.6 MJ   c. 576 MJ

1-27   The motor shown in Fig. 1-6 is turning at 1700 r/min, and the scale reading is 25 N. If the radius arm r = 30 cm, the mechanical power of the motor is

      a. 1335 W   b. 18 hp   c. 12.75 kW

1-28   In Fig. 1-8, the efficiency of the motor is

      a. 84.2%   b. 0.1875   c. 57.14 W/°F

1-29   A room temperature of 77°F corresponds to

      a. 25°C   b. 60.55°C   c. 170.6°C

1-30   In Fig. 1-10, point D corresponds to 4000 r/min and 250 N.m. Calculate the mechanical power of the engine.

      a. 104.7 hp   b. 190.4 hp   c. 104.7 kW

1-31   Point A in Fig. 1-10 corresponds to 7000 r/min and 150 N.m. The power developed by the engine at point A is greater than that at point D.

      ☐ true   ☐ false

1-32   In Fig. 1-12 as we move along the torque-speed curve from point 1 to point 3, the mechanical power increases progressively.

      ☐ true   ☐ false

1-33   In Fig. 1-14, it takes more power to drive the pump at 400 r/min and a pressure of 75 psi then at 300 r/min and 150 psi.

      ☐ true   ☐ false

1-34 In Fig. 1-15, if a = 12 cm, b = 26 cm, the value of c is

    a. 170 cm   b. 0.2863 m   c. 23.065 cm

1-35 In Fig. 1-22, if L = 433 mm, the length OC is

    a. 2.5 cm   b. 216.5 mm   c. 0.25 m

1-36 In Fig. 1-23, the rod has a length of 200 cm. If θ_1 = 75° the projection b_1 has a length of

    a. 1932 mm   b. 15 km   c. 2.66 cm

1-37 In Fig. 1-27, an interval of 1 degree corresponds to a time of

    a. 6 s   b. 166.6 ms   c. 60 μs

1-38 In Fig. 1-28, an interval of 1 degree corresponds to a time of

    a. 111.1 μs   b. 90 μs   c. 11.11 μs

1-39 In Fig. 1-23, the length of the rod is 200 cm, and projection b_3 is -100 cm. The value of angle θ_3 is

    a. 150°   b. 330°   c. 210°

1-40 In Fig. 1-21, a = 240 mm, b = 160 mm, θ = 30°. The value of side c is

    a. 346.4 mm   b. 80 cm   c. 129.18 mm

CHAPTER **2**

The Nature of Electricity

Introduction and Chapter Objectives

In describing circuits and machines, we will continually be using terms such as voltage, current, resistance, electric power, and electric energy. This chapter explains the meaning of these terms and how they are related. The relationship can be expressed by a few simple equations, the most important of which is Ohm's law. A further objective of this chapter is to describe the basic properties of matter. This leads us to an understanding of conductors and insulators, as well as the properties of sources and loads.

2.1 Composition of matter, atoms and molecules

Let us take a block of aluminum and break it in two. Then let us select one of the pieces and again break it in two. If we were able to continue this process millions and millions of times, we would eventually reach a point at which it would be impossible to subdivide the tiny speck of aluminum any further, without altering its fundamental properties. In other words, if we were to subdivide it one more time, the resulting fragments would no longer be aluminum. This smallest possible bit of aluminum is called an *atom* of aluminum.

All substances such as aluminum, carbon, copper, hydrogen, and oxygen, are composed of atoms, each having a particular structure which is characteristic of the substance. For most substances, however, the smallest possible bit is a *molecule*. A molecule is composed of two, three, or more atoms that act together as a group. Thus, a molecule of water is composed of two atoms of hydrogen and one atom of oxygen. A molecule of salt is composed of one atom of sodium and one atom of chlorine. Finally, a molecule of rubber consists of at least 5000 atoms of carbon and 8000 atoms of hydrogen strung out in a continuous chain.

2.2 Structure of the atom, nucleus, and electrons

If we could look inside an atom, we would see that it is composed of a small central core surrounded by particles rotating around the core, like planets around the sun. The core is called the *nucleus* of the atom, and the rotating particles are called *electrons*. Each electron carries a single negative charge of electricity, and this is the smallest electric charge that is known to exist. The nucleus carries a positive electric charge, equal to the sum of the negative charges of the orbiting electrons. It follows that the atom is electrically neutral because the negative charges exactly balance the positive charges. Experiments have shown that electric charges having the same sign repel each other, and opposite signs attract. Consequently, a force of attraction exists between each electron and the central nucleus. However, the force is greatest on those electrons that are closest to the nucleus.

The electrons fly around the nucleus in one or more distinct orbits. The first orbit is said to be full when it contains 2 electrons. The second and third orbits are full when they contain respectively 8 and 18 electrons. An atom of aluminum possesses three distinct orbits containing 2, 8, and 3 electrons (Fig. 2-1). It follows that its nucleus has a charge of +13. On the other hand, an atom of copper contains 29 electrons, one of which occupies the fourth orbit (Fig. 2-2).

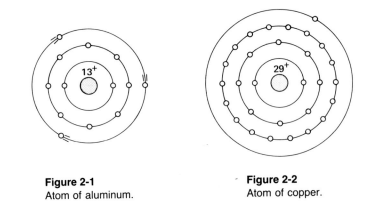

Figure 2-1
Atom of aluminum.

Figure 2-2
Atom of copper.

The atoms in a solid block of material are fixed in place, but they vibrate in their position of rest. Furthermore, compared with the dimensions of the electrons and the nucleus, the distance between individual atoms (or molecules) is enormous. As a result, even a dense solid such as aluminum is largely composed of empty space.

2.3 Free electrons

We saw that an atom of aluminum has only 3 electrons in its outer orbit, compared with the "normal" number of 18. Because this outer orbit is far from being complete, the electrons it contains are only weakly bound to the nucleus. The result is that in a solid piece of aluminum, these relatively free electrons continually jump from one atom to the next. They zigzag in every direction at an average speed of

1000 km/s (Fig. 2-3). However, despite their high velocity, these free electrons never leave the piece of metal, and so the sum of the positive charges on the stationary nuclei are still balanced by the sum of the negative charges on all the electrons. As a result, the piece of aluminum remains electrically neutral. Figure 2-3 is obviously not drawn to scale because it is known that one cubic millimeter of aluminum contains nearly 10^{20} (100 000 000 000 000 000 000) free electrons. This does not include the electrons in the first and second orbits that are permanently bound to an individual nucleus.

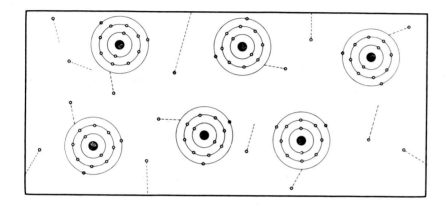

Figure 2-3
Free electrons moving inside a piece of aluminum.

If we forcibly remove some of the free electrons from a piece of aluminum, the positive charge on the nuclei is no longer balanced by the negative charge on the electrons, and so the net charge of the piece of aluminum is positive. Similarly, if we add electrons to a block of aluminum that is initially neutral, the block will possess a net negative charge.

Thus, any body is positive or negative depending upon whether it has a deficiency or a surplus of electrons.

2.4 Conductors and insulators

Materials such as copper, aluminum, and iron contain billions upon billions of free electrons because the electrons in the outer orbits of individual atoms are only weakly bound to the nucleus. On the other hand, materials such as rubber, air, and oil have a strong hold on all their electrons – even those in the outer orbits of individual atoms. Such materials possess very few free electrons.

Materials having many free electrons are said to be good conductors of electricity.

Materials having very few free electrons are poor conductors of electricity. They are called insulators.

The number of free electrons per cubic centimeter is precisely the factor that determines whether a material is a good conductor of electricity or an insulator. Electrons move freely in a conductor and only with great difficulty in an insulator. Thus, if an insulator is placed between two conductors, it will prevent any movement of free electrons between them. Metals, in general, are excellent conductors.

2.5 Unit of electric charge, the coulomb (C)

The electric charge on an electron is so small that it is preferable, in day-to-day calculations, to use a larger value. The practical unit of electric charge is called the coulomb (symbol C). It is equivalent to the charge of
6 241 450 380 000 000 000 or $6.241\ 450\ 38 \times 10^{18}$ electrons.

2.6 Sources of electricity, potential difference (E)

How can we produce electricity? The secret is to establish a difference in the relative number of electrons between two points. Devices that are able to create a surplus of electrons at one point with respect to another point are called *generators* (or *sources*) of electricity.

This unequal distribution of electrons between two points can be achieved in several ways. It can be done chemically, as in a dry cell; mechanically, as in a rotating generator; thermally, as in a thermocouple; or optically, as in a photoelectric cell. In each case the point (or terminal) having a lack of electrons possesses a positive charge and a so-called positive polarity (+). Conversely, the point (or terminal) having a surplus of electrons possesses a negative charge, and consequently, a negative polarity (−).

Whenever one terminal of a device possesses a shortage of electrons and the other terminal a surplus, *a difference of electric potential* is said to exist between the terminals. It is represented by the symbol E. This difference of potential, or *electromotive force* (emf), may be considered to be a kind of electric pressure. By analogy, it may be compared to the difference of hydraulic pressure created between the inlet and outlet of a water pump. A generator can therefore be considered to be a sort of electron pump.

2.7 Unit of potential difference, the volt (V)

The SI unit of electric potential difference (and of electromotive force) is the volt (symbol V). The kilovolt (kV) is a multiple of the volt, equal to 1000 V. Similarly, the millivolt (mV) is a submultiple of the volt, equal to 1/1000 V. The difference of potential, or voltage* between two points is measured with a voltmeter (Fig. 2-4).

2.8 Electric current (I)

Consider the storage cell in Fig. 2-5. The central terminal has a deficiency of electrons and it is therefore positive. On the other hand, the outer terminal has a surplus of electrons and so its polarity is negative. Because of this shortage and surplus of electrons, a voltage appears between the terminals. The voltage ranges between 1.5 V and 2 V, depending on the type of cell.

---

*   Voltage is a colloquial term for difference of potential.

Figure 2-4a
Panelboard ac voltmeter used in a
utility company substation.
(*Courtesy General Electric*)

Figure 2-4b
Portable dc voltmeter having three
scales: 150 V, 300 V and 750 V.
(*Courtesy Weston Instruments*)

Let us now connect a conductor of electricity across the terminals, as shown in
Fig. 2-6. The surplus electrons at the negative terminal will repel the free electrons
inside the conductor, urging them toward the positive terminal. At the same time,
the positive terminal attracts these free electrons. The result is that the free electrons
inside the conductor will begin to move from the negative to the positive terminal.
They continue to zigzag at high speed, bumping from one atom to the next, but to
this random motion is now added a gradual, mass drift from the negative to the positive terminal. This mass drift of electrons is called an *electric current*. It is represented by the symbol *I*.

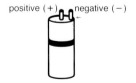

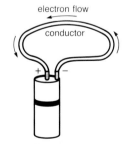

Figure 2-5
A storage cell produces a
difference of potential between its
terminals.

Figure 2-6
Electrons flow from the (−) to the
(+) terminal of a conductor.

Referring to Fig. 2-6, the free electrons come out of the negative terminal of the cell and re-enter by the positive terminal. What happens to the electrons that enter the positive terminal? They flow through the body of the cell and eventually re-emerge at the negative terminal. Current flow is therefore a circular flow of electrons whereby they continuously move through the conductor, then through the source, and back again. Inside the conductor, electrons move from the negative to the positive terminal. However, inside the source, they move from the positive to the negative terminal.

The path followed by the electron flow is called an *electric circuit*. Thus, the combination of the cell and conductor shown in Fig. 2-6 constitutes an electric circuit. Most circuits are more complicated than this one because they offer several paths through which current may flow.

2.9 Conventional current flow, the ampere (A)

Before electrons were known to exist, scientists assumed that current in a conductor flowed from the positive to the negative terminal. Today we know that this *conventional* direction of current flow is opposite to the actual electron flow. Nevertheless, in this book we have adopted the conventional direction of current flow because it is almost universally recognized in electrical technology. Thus, from now on, we consider that current in a conductor always flows from the positive to the negative terminal (Fig. 2-7). In addition, current inside a source always flows from the (−) to the (+) terminal.

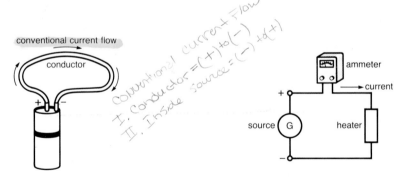

Figure 2-7
Conventional current flow in a conductor.

Figure 2-8
Current is measured with an ammeter.

Current flow in a conductor may be compared to the flow of water in a pipe, which can be measured in gallons per second. Similarly, current flow can be measured in coulombs per second. By definition, when one coulomb of electricity moves past a point each second, the flow is equal to one ampere. The SI unit of current is the ampere (symbol A) and it is equivalent to the flow of 6.24×10^{18} electrons per second.

2.10 Measuring voltage and current; voltmeter and ammeter

Consider Fig. 2-8 in which an electric heater is connected to a source G. We can measure the current flow by inserting an ammeter directly into the line, so that all the electrons pass through the instrument. A pointer indicates the current flow in amperes. The ammeter is designed so that it does not appreciably change the current flow that existed before the ammeter was connected into the line.

To measure the difference of potential between the terminals of the heater, we connect a voltmeter across the terminals, as shown in Fig. 2-9. A pointer indicates the difference of potential in volts. A very tiny current flows through the voltmeter, but this current is negligible compared with the current flowing through the heater. Consequently, the voltmeter reading is very close to the difference of potential that existed between the terminals of the heater before the voltmeter was connected into the circuit.

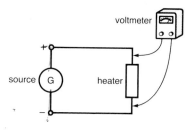

Figure 2-9
The difference of potential is
measured with a voltmeter.

Figure 2-10
This multi-range volt-ohm-milliameter has a digital readout
and an accuracy of 0.1 %. (*Courtesy Weston Instruments*)

It is important to remember that the circuit must be de-energized, and the line has to be "cut" in order to connect the current-measuring ammeter. The voltmeter, on the other hand, is simply connected across two terminals, without otherwise disturbing the circuit. **Caution**: an ammeter must never be connected, like a voltmeter, across the terminals of a source.

Multi-scale voltmeters and ammeters are available that can measure from microvolts to kilovolts, and from microamperes to amperes. Some instruments have a digital readout (Fig. 2-10); others use a pointer that moves across a calibrated scale (Fig. 2-4).

2.11 Electric power, the watt (W)

Consider Fig. 2-11 in which a generator G is connected to a heater. As current flows through the heater, the electrons continually bump against the stationary atoms, and these repeated impacts produce heat. Because heat is a form of energy that the heater cannot produce by itself, it follows that the energy must come from

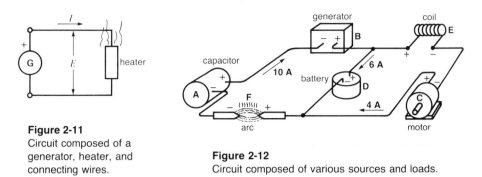

Figure 2-11
Circuit composed of a
generator, heater, and
connecting wires.

Figure 2-12
Circuit composed of various sources and loads.

the generator. Scientists have proved that the power released in the form of heat is exactly equal to the product of the voltage E across the heater, times the current I that flows through it. The electric power delivered by the generator is also equal to the difference of potential (or electromotive force) across its terminals times the current it delivers.

We can generalize by saying that the electric power associated with a device is always given by the voltage E across it, times the current I that flows through it. We can therefore write:

$$P = EI \qquad (2\text{-}1)$$

where

P = power, in watts (W)

E = difference of potential across the device, in volts (V)

I = current flowing in the device, in amperes (A)

Equation (2-1) can be applied to any device, whether it is a heater, a dc motor, a transformer, a transistor, a coil or an electroplating bath. Furthermore, the instantaneous electric power a device receives (or delivers) is *always* equal to the voltage E times the current I that exists at that particular instant.

2.12 Distinction between sources and loads

The flow of electric power enables us to identify a source and a load. A device that receives electric power is called a load; a device that delivers electric power is a source. But how can we tell whether a device is a source or a load? We can tell one from the other by observing two things: (1) the polarity of the terminals and (2) the direction of conventional current flow through the device.

The rule is very simple:

  a)   If current flows out of the positive terminal, the device is a source
  b)   If current flows into the positive terminal, the device is a load.

Figure 2-12 shows an unusual electric circuit that contains devices we haven't studied yet. However, the direction of the currents, and the polarities of the terminals are given. Using the simple rule above, we immediately see that devices A, B, C are sources, and devices D, E, F are loads. The sources deliver electric power, and the loads absorb it.

2.13 Ohm's law; the ohm (Ω)

Consider a conductor of electricity, such as the heating element of an electric stove (Fig. 2-11). If we apply a voltage across its terminals, a certain current will flow. If we double the voltage, we discover that the current also doubles. Again, if we triple the voltage the current triples, and so forth. The ratio of the voltage divided by the current is therefore a constant for this particular heater. We can therefore write:

$$\frac{E_{heater}}{I_{heater}} = \text{constant (for this particular heater)}$$

The constant is a property of the heater which we call its *resistance*. The symbol for resistance is R. A heater which allows a large current to flow through it when a moderate voltage is applied across its terminals is said to have a low resistance. On the other hand, a heater (or other conductor) that allows only a small current to flow through it is said to have a high resistance.

This relationship between the voltage, current, and resistance of a conductor is known as Ohm's law. Ohm's law states:

$$R = \frac{E}{I} \qquad (2\text{-}2)$$

where

$E =$ voltage applied across the conductor, in volts (V)

$I =$ resulting current that flows through the conductor, in amperes (A)

$R =$ resistance of the conductor, in ohms (Ω)

The SI unit of resistance is the ohm (symbol Ω). By definition, the electric resistance between two terminals of a conductor is equal to one ohm when a constant difference of potential of one volt applied across the terminals produces a current of one ampere.

2.14 Application of Ohm's law

Now that we know Ohm's law relating E, I, R, it is easy to find any one of these three values when we know the two others. Thus, from Eq. 2-2 we obtain two other equations

$$I = \frac{E}{R} \qquad \text{or} \qquad \text{amperes} = \frac{\text{volts}}{\text{ohms}} \qquad (2\text{-}3)$$

$$\text{and} \quad \boxed{E = IR} \quad \text{or} \quad \text{volts} = \text{amperes} \times \text{ohms} \qquad (2\text{-}4)$$

Example 2-1:

An electric iron is connected to a 120 V source, and the resulting current is 10 A. Calculate the resistance of the heater element.

Solution:

$$R = \frac{E}{I} = \frac{120 \text{ V}}{10 \text{ A}} = 12 \text{ ohms} = 12 \text{ }\Omega \qquad \text{Eq. 2-2}$$

Example 2-2:

A long copper wire having a resistance of 5 Ω carries a current of 20 A. Calculate the difference of potential across the ends of the wire.

Solution:

$$E = IR = 20 \text{ A} \times 5\Omega = 100 \text{ V} \qquad \text{Eq. 2-4}$$

Example 2-3:

In Example 2-1, calculate the current flowing in the electric iron if the line voltage drops to 108 V.

Solution:

The resistance of the electric iron is a property of the device which, like its weight or color, is fixed. Because $R = 12 \text{ }\Omega$, we can write:

$$I = E/R = 108 \text{ V} / 12 \text{ }\Omega = 9 \text{ A} \qquad \text{Eq. 2-3}$$

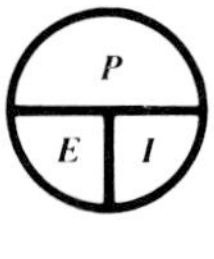

Figure 2-13

Readers who are using Ohm's law for the first time may find it useful to refer to Fig. 2-13 to solve problems. The symbols E, I, and R are arranged so that when one of them is covered, its relationship to the remaining two symbols becomes evident. For example, when E is covered, the symbols IR remain, indicating that $E = IR$. Similarly, when I is covered, the symbols E/R appear, indicating that $I = E/R$. The equation $P = EI$ can be represented and used in the same way (Fig. 2-14).

Figure 2-14

2.15 Power dissipated in a resistance

We can calculate the power dissipated in a resistance by applying Eq. 2-1, which states that $P = EI$. However, it is often useful to combine Eq. 2-1 and Ohm's law as follows:

$$P = EI = (IR)\ I = I^2R$$

whence
$$\boxed{P = I^2R} \qquad (2\text{-}5)$$

used for power losses in transmission lines

Alternatively, we can write $P = EI = E(E/R) = E^2/R$

whence $$P = E^2R \tag{2-6}$$

Equation 2-5 tells us that the power dissipated in a conductor is proportional to the square of the current it carries. This equation is particularly useful in calculating the power losses in transmission lines, and the windings of motors and generators.

Equation 2-6 indicates that the power dissipated in a heater, for example, is proportional to the square of the voltage applied across its terminals. Thus, if the line voltage in a home decreases by 10 percent, the power output of all heating elements (hot water heaters, electric ranges, and so forth) will decrease 20 percent.

Equations 2-5 and 2-6 do not add anything new because both are derived from $P = EI$ and $E = IR$. Nevertheless, we will often apply Eq. 2-5.

Example 2-4:

A transmission line that is 15 km long has a resistance of 3 Ω. Calculate the power loss in the line when it carries a current of 60 A.

Solution:

$$P = I^2R = (60)^2 \times 3 = 3600 \times 3 = 10\ 800\ \text{W} = 10.8\ \text{kW} \qquad \text{Eq. 2-5}$$

2.16 Electric energy, the joule (J), the kilowatthour (kW.h)

In Chapter 1 we learned that mechanical power is the rate at which mechanical energy is used up. A similar definition applies to electric power. Electric power P is the rate at which electrical energy W is used up; it is electrical energy divided by time. Consequently, electrical energy is electric power multiplied by time. We can therefore rewrite Eq. 1-7 in the form:

$$W = Pt \tag{2-7}$$

where

$W = $ electrical energy, in joules (J)

$P = $ electric power, in watts (W)

$t = $ time during which the power is consumed (or produced) in seconds (s).

The SI unit of electrical energy is the joule; it is equal to one watt-second. However, if power is expressed in kilowatts and time in hours, the electric energy can also be stated in kilowatthours (kW.h). The kilowatthour* is not an SI unit, but it is a unit used by most electric utility companies to measure the energy consumed in business, factories and homes.

For example, a 100 W lamp that burns for 20 seconds consumes energy equal to 100 watts × 20 seconds = 200 watt-seconds, which is equal to 2000 joules. In

---

*   1 kW.h = 1000 W × 3600 s = 3 600 000 W.s = 3 600 000 J = 3.6 megajoules = 3.6 MJ

the same way, a generator that develops 150 kW during a 20-hour period delivers a quantity of energy equal to 150 kW × 20 h = 3000 kW.h.

2.17 Storing electric energy

Enormous quantities of chemical energy are stored in underground deposits of oil and coal. When these fuels are burned, they release thermal energy which, in turn, can be converted into electrical energy.

We can also store large quantities of mechanical energy by erecting power dams. The potential energy stored in the water behind the dam is later converted into electrical energy as the water rushes through the generator turbines.

The nuclear energy in a uranium mine is another form of stored energy.

Unfortunately, no practical method exists for storing large quantities of electrical energy. Batteries, for example, do not store electrical energy. They store chemical energy, which is released in the form of electrical energy as the chemical elements are transformed. Only two devices are able to store electricity in its natural state; they are coils and capacitors. However, to store even moderate amounts of electrical energy, the size and cost of these devices make them economically unfeasible. For example, a coil 2 meters long, 1 meter in diameter, and weighing 2 tons can keep a 100 W lamp burning for only about 2 minutes.

Because it is virtually impossible to store large quantities of electricity, utility companies must produce it at the same rate it is being consumed. If the rate of consumption is only slightly different from the rate of generation, the electrical system reacts dramatically. Overvoltages, excessive currents, and speed changes are produced which immediately trip circuit breakers, bringing the electrical system to a halt. If we could one day store electrical energy as easily as chemical energy is stored in a liter of oil, the generation, transmission, and distribution of electricity would be profoundly changed.

2.18 Summary

This chapter contains many topics that will be used throughout the book. We learned that materials are made up of atoms and molecules. An atom is composed of a positively charged nucleus that is surrounded by negatively charged electrons. In its normal state, the atom is electrically neutral because the (+) charge on the nucleus is exactly balanced by the sum of the (−) charges of the electrons. The unit of electric charge is the coulomb, equal to the charge of $6.24 × 10^{18}$ electrons.

All materials contain free electrons. Conductors of electricity contain enormous quantities of free electrons; insulators contain very few.

A body or terminal that has a surplus of electrons is said to have a negative polarity. Conversely, if it has a deficiency of electrons it has a positive polarity.

A potential difference E exists between two terminals (or points) when one of the terminals has a surplus of electrons with respect to the other. The unit of potential difference is the volt (V).

If a difference of potential exists between two terminals, an electric current will flow when they are connected by a conductor. The current consists of free electrons that flow in a closed loop. The electrons flow from the (−) to the (+) terminal of the conductor.

The direction of conventional current flow is opposite to the actual electron flow. As a result, the conventional current flows from the (+) to the (−) terminal of a conductor. The unit of electric current is the ampere (A). It is equal to one coulomb per second.

All electric circuits are composed of sources and loads. Sources deliver electric power, and loads absorb it. In a load, current flows *into* the (+) terminal. In a source, current flows *out of* the (+) terminal. These two simple rules will be helpful later on, as we learn more about circuits and machines.

The electric power absorbed by the load, or delivered by a source, is equal to the voltage across its terminals, times the current that flows through it. Thus, $P = EI$, where P is the power in watts.

Ohm's law states that the ratio of the voltage across a conductor to the current that flows through it is equal to the resistance of the conductor. Thus, $R = E/I$.

The unit of electrical energy is the joule, equal to 1 watt-second. The kilowatthour (kW.h) is another unit of electrical energy commonly used by electric utilities. It is equal to 3600 J. Finally, we learned that electrical energy cannot be economically stored in large quantities. This means that electrical energy has to be generated at the same rate as it is being consumed.

TEST YOUR KNOWLEDGE

2-1   An atom of iron has 26 electrons. The charge on the nucleus is

a. zero   b. − 26   c. + 26   d. exactly 13

2-2   Electrons repel each other.

☐ true   ☐ false

2-3   A material having 1000 free electrons per cubic millimeter is a good insulator.

☐ true   ☐ false

2-4   The unit of potential difference is

a. the voltage   b. the ampere(A)
c. the volt (E)   d. the volt (V)

2-5   The unit of electric current is

a. the watt   b. the electron
c. the ampere   d. the flow

2-6   The voltage across the terminals of a device is 60 V, and the current through it is 5 A. Current is flowing into the (+) terminal. The device is

a. a load which consumes 12 W   b. a source that is delivering 300 W
c. a load which consumes 300 W

2-7   A conductor becomes hotter when the current through it increases

☐ true   ☐ false

2-8   A resistance dissipates 12 kW when 240 V is applied to its terminals. The current in the resistance is

a. 2.88 mA   b. 50 A   c. 0.02 A   d. 4.8 Ω

2-9   A current of 20 mA is equal to

a. 20 000 A   b. 0.2 A   c. 0.02 A

2-10   A potential difference of 2 450 000 V is equal to

a. 2.45 MV   b. 2.45 mV   c. 245 kV

2-11   The SI unit of electrical energy is

a. the watt   b. the kilowatt   c. the joule   d. the kilowatthour

2-12   The polarity of a terminal is positive if it has

a. a surplus of electrons   b. an absence of electrons
c. a deficiency of electrons compared
with its neutral state

QUESTIONS AND PROBLEMS

2-13   An atom of iron has 26 electrons. How many electrons are there in the outer orbit? Draw a diagram of the atom, similar to that of Fig. 2-1.

2-14   A molecule of salt is composed of one atom of sodium and one atom of chlorine. The nucleus of the sodium atom has a charge of +11, while that of chlorine is +17. a) How many electrons are there in a molecule of salt? b) How many electrons are there in the two inner orbits next to the nuclei?

2-15   A circuit is composed of a 12 V battery, connected to a car heater. Draw a diagram showing how a voltmeter and ammeter should be connected into the circuit.

2-16   An electric heater draws a current of 8 A when it is connected to a 120 V source. Calculate (a) the power dissipated and (b) the resistance of the heater.

2-17   In problem 2-16, if the voltage drops to 60 V, calculate the new values of power and resistance.

2-18   A soldering iron is connected to a 120 V outlet. Knowing that its resistance is 50 Ω, calculate the current it draws from the line.

2-19   An average lightning stroke produces a current of 16 kA at a voltage of 200 MV. If the duration of the stroke is 30 μs calculate a) the power in megawatts, b) the average resistance of the arcing path, and c) the energy released by the stroke, in kilowatthours.

2-20   The magnetic coils of a 500 MW generator have a resistance of 62.6 mΩ. Calculate the voltage needed to cause a current of 4060 A to flow.

CHAPTER 3

Simple Direct Current Circuits

Introduction and Chapter Objectives

In this chapter we apply Ohm's law in a systematic way to predict the voltages and currents in sources and loads. The study will cover direct-current circuits that are connected in series, in parallel, and in series-parallel. Battery circuits are also covered, together with some of the important properties of primary and secondary cells. A further objective is to introduce the transmission line as a means of carrying power over a distance.

The concept of a circuit diagram is also developed, which enables us to represent motors, generators, lamps, heaters, and their interconnections in a simple way.

A final objective is to introduce Kirchhoff's current and voltage laws. They enable us to solve any dc circuit, no matter how complex. This portion of the chapter is not essential to an understanding of succeeding chapters; consequently, it may be put aside until needed later.

3.1 Series circuits

Electric devices are said to be connected in series when the terminal of one device is connected to the terminal of the next, like a chain. The devices may be either sources or loads. Figure 3-1 shows a series circuit composed of a motor, a lamp and a heater. If this circuit is connected to a generator G (Fig. 3-2) a current will flow. Such series circuits possess three main properties:

1. The current in each device is the same.

    In Fig. 3-2, the currents in the three devices and the generator are identical. The reason is that the electrons cannot escape from the conductors in which they flow. Thus, the four ammeters give the same reading:

    $$I_{motor} = I_{lamp} = I_{heater} = I_{generator}$$

    $$I_M = I_L = I_H = I_G$$

2. The sum of the voltages across the individual loads is equal to the voltage across the source.

    In Fig. 3-2, voltmeters are connected across each device. If we take careful measurements, we discover that:

    $$E_{\text{motor}} + E_{\text{lamp}} + E_{\text{heater}} = E_{\text{generator}}$$

    $$E_{\text{M}} + E_{\text{L}} + E_{\text{H}} = E_{\text{G}}$$

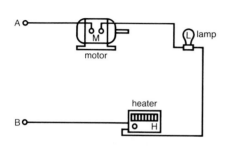

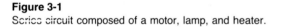

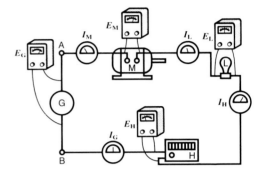

Figure 3-1
Series circuit composed of a motor, lamp, and heater.

Figure 3-2
Series circuit connected to a generator G.

3. The sum of the powers absorbed by the individual loads is equal to the power supplied by the source.

    This is true because of the law of conservation of energy.

    Using Fig. 3-2, we can therefore state:

    $$P_{\text{motor}} + P_{\text{lamp}} + P_{\text{heater}} = P_{\text{generator}}$$

    However, we recall that the power of each device (source or load) is equal to the product of the current it carries times the voltage across it. Consequently, we can write:

    $$E_{\text{M}}I_{\text{M}} + E_{\text{L}}I_{\text{L}} + E_{\text{H}}I_{\text{H}} = E_{\text{G}}I_{\text{G}}$$

    These three rules apply to any dc series circuit, no matter what the elements are.

    In drawing electric circuits, it is useful to simplify the pictures by using a wiring diagram symbol for each device. Thus, the "real" circuit of Fig. 3-2 can be represented by the much simpler schematic diagram of Fig. 3-3.

3.2 Resistors in series, equivalent resistance

Heaters, lamps, toasters, and other devices that do not generate an emf all possess a certain resistance, which is expressed in ohms. From a circuit standpoint, these devices are considered to be *resistors* because they oppose the flow of current to a

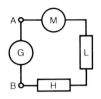

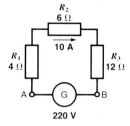

Figure 3-3
This circuit diagram is a simpler representation of the circuit in Fig. 3-2.

Figure 3-4
Series circuit composed of three resistors connected to a source G.

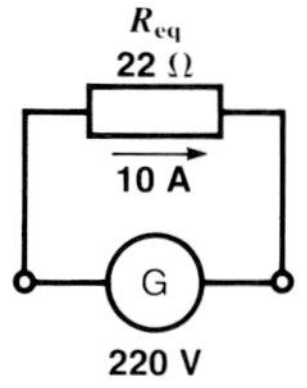

Figure 3-5
This circuit is equivalent to the series circuit of Fig. 3-4.

greater or lesser degree. Even a good conductor of electricity, such as a length of copper wire, may be considered to be a resistor because it offers *some* opposition to current flow.

Consider three resistors of 4 Ω, 6 Ω, and 12 Ω connected in series across a 220 V source (Fig. 3-4). The total resistance R_{eq} of the group is equal to the sum of their individual resistances:

$$R_{eq} = 4 + 6 + 12 = 22 \ \Omega$$

Therefore, a single equivalent resistor R_{eq} of 22 Ω could replace the three individual resistors. The reason is that if this resistor were connected across the 220 V source, it would draw the same current as the three resistors do (Fig. 3-5). Thus, the current flowing in the individual resistors is:

$$I = \frac{E}{R_{eq}} = \frac{220 \ V}{22 \ \Omega} = 10 \ A$$

The 22 Ω resistor is called an equivalent resistance R_{eq} because it represents the overall effect of the three resistors in the circuit.

In general, a group of resistors R_1, R_2, R_3, ..., R_n connected in series can be replaced by a single equivalent resistance R_{eq}. The value of the equivalent resistance is given by:

$$R_{eq} = R_1 + R_2 + R_3 + \ ... \ + R_n \qquad (3\text{-}1)$$

What is the difference of potential across the individual resistors in Fig. 3-4? It can be found by applying Ohm's law ($E = IR$) to each element. Knowing that the current is 10A, and referring to Fig. 3-4, we find:

for R_1, $E_1 = IR_1 = 10 \times 4 = 40 \ V$

for R_2, $E_2 = IR_2 = 10 \times 6 = 60 \ V$

for R_3, $E_3 = IR_3 = 10 \times 12 = 120 \ V$

If voltmeters were connected across each resistor, they would give readings as shown in Fig. 3-6.

Note that $E_1 + E_2 + E_3 = 220$ V. Thus, the sum of the voltages across the resistors is equal to the voltage of the source, as it should be.

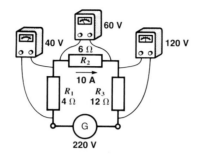

Figure 3-6
Voltmeters are used to measure the voltage across each resistor.

Let us now consider the power aspects of the circuit. Using $P = I^2R$, we can calculate the power dissipated in each resistor. Thus:

in R_1, $P_1 = I^2R_1 = 10^2 \times 4 = 400$ W

in R_2, $P_2 = I^2R_2 = 10^2 \times 6 = 600$ W

in R_3, $P_3 = I^2R_3 = 10^2 \times 12 = 1200$ W

Total power dissipated in the three resistors is:

$P_{\text{load}} = 400 + 600 + 1200 = 2200$ W.

On the other hand, the power delivered by the source is:

$P_{\text{source}} = EI = 220 \times 10 = 2200$ W

Thus, the total power dissipated by the resistors is equal to the power supplied by the source.

3.3 Parallel circuits

Electric devices are said to be connected in parallel when their terminals are directly connected to two common terminals. Figure 3-7 shows the schematic diagram of a parallel circuit composed of a motor M, lamp L and heater H. These devices are effectively connected to the two common terminals A and B. The conductors leading from terminals A and B to terminals 1, 2, 3 and 4, 5, 6 are assumed to have negligible resistance. It follows that terminals 1, 2, 3 are *electrically* at the same potential as terminal A, while terminals 4, 5, 6 are *electrically* at the same potential as terminal B. Thus, from an electrical standpoint, the four terminals 1, 2, 3, A form one common terminal. The same is true of terminals 4, 5, 6, B.

Let us connect this parallel circuit to a source G. Voltages and currents will immediately appear in each device, as indicated in Fig. 3-8.

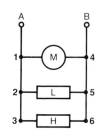

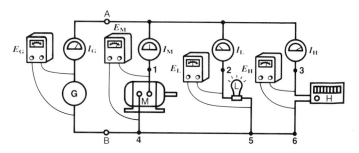

Figure 3-7
Parallel circuit composed of a
motor, lamp, and heater.

Figure 3-8
Parallel circuit connected to a source G.

Parallel circuits, like series circuits, possess three main properties:

1. The difference of potential (or voltage) across each device is the same.
    Thus, in Fig. 3-8, we have:

    $$E_{\text{motor}} = E_{\text{lamp}} = E_{\text{heater}} = E_{\text{generator}}$$

    $$E_{\text{M}} = E_{\text{L}} = E_{\text{H}} = E_{\text{G}}$$

2. The sum of the currents flowing in the individual loads is equal to the current
    supplied by the source.
    Thus, in Fig. 3-8 we have:

    $$I_{\text{motor}} + I_{\text{lamp}} + I_{\text{heater}} = I_{\text{generator}}$$

    $$I_{\text{M}} + I_{\text{L}} + I_{\text{H}} = I_{\text{G}}$$

    This result is to be expected because the current drawn by each load must inevitably come from the source.

3. The sum of the powers consumed by the individual loads is equal to the power
    supplied by the source.
    As in the case of series circuits, we obtain:

    $$P_{\text{motor}} + P_{\text{lamp}} + P_{\text{heater}} = P_{\text{generator}}$$

    $$E_{\text{M}}I_{\text{M}} + E_{\text{L}}I_{\text{L}} + E_{\text{H}}I_{\text{H}} = E_{\text{G}}I_{\text{G}}$$

These three rules apply to any parallel circuit, no matter what the elements may be.

3.4 Resistors in parallel, equivalent resistance

Consider a 6 Ω and a 3 Ω resistor connected in parallel across the terminals of a 30
V battery (Fig. 3-9). The current in the 6 Ω resistor is

$$I_1 = \frac{E}{R_1} = \frac{30}{6} = 5\ \text{A} \qquad\qquad \text{Eq. 2-3}$$

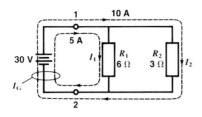

Figure 3-9
Resistors connected in parallel
across a battery.

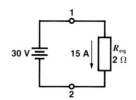

Figure 3-10
This circuit is equivalent to the
parallel circuit of Fig. 3-9.

Similarly, the current in the 3 Ω resistor is

$$I_2 = E/R_2 = 30/3 = 10 \text{ A}$$

The total current supplied by the battery is therefore

$$I_G = I_1 + I_2 = 5 + 10 = 15 \text{ A}$$

If the two resistors were hidden under a box so that only the wires leading into the box were accessible, a technician would measure a current of 15 A and a voltage of 30 V. From these measurements, he would be led to believe that the box contains a resistor of $R = E/I = 30/15 = 2 \Omega$.

We conclude that a single resistor of 2 Ω is equivalent to a 6 Ω and a 3 Ω resistor connected in parallel (Fig. 3-10).

We can prove that whenever two resistors R_1 and R_2 are connected in parallel, they can be replaced by an equivalent resistance R_{eq} given by:

$$\frac{1}{R_{eq}} = \frac{1}{R_1} + \frac{1}{R_2} \tag{3-2}$$

Example 3-1:

Two resistors of 6 Ω and 3 Ω are connected in parallel (Fig. 3-9). Calculate the value of the equivalent resistance.

Solution:

$$\frac{1}{R_{eq}} = \frac{1}{R_1} + \frac{1}{R_2} = \frac{1}{6} + \frac{1}{3} = 0.1666 + 0.3333 = 0.4999$$

therefore $R_{eq} = 1/0.4999 = 2 \Omega$

Note that the equivalent resistance (2 Ω) is lower than the lowest of the two resistances (3 Ω). The reason is that when we connect a second resistor in parallel with the 3 Ω resistor, we offer an additional path for current flow. Consequently, the resistance "seen" by the source is less than 3 Ω.

We can generalize Eq. 3-2 by stating that if a group of resistors $R_1, R_2, R_3, \ldots,$ R_n are connected in parallel, the value of the equivalent resistance is given by:

$$\frac{1}{R_{eq}} = \frac{1}{R_1} + \frac{1}{R_2} + \frac{1}{R_3} + \cdots + \frac{1}{R_n} \qquad (3\text{-}3)$$

Example 3-2:

Calculate the resulting resistance when three resistors of 7 Ω, 30 Ω, and 42 Ω are connected in parallel.

Solution:

$$\frac{1}{R_{eq}} = \frac{1}{7} + \frac{1}{30} + \frac{1}{42} = 0.1429 + 0.0333 + 0.0238 = 0.2$$

$$R_{eq} = \frac{1}{0.2} = 5 \; \Omega$$

3.5 Series-parallel circuits

To solve a series-parallel circuit, we begin by looking for resistors that are directly in series or directly in parallel. We then replace these resistors by their equivalent resistance. The resulting circuit will therefore contain fewer resistors than before. If this circuit still contains resistors that are directly in series or directly in parallel, they, too, are replaced by a new equivalent resistance. Proceeding in this way, from circuit to circuit, we eventually end up with a circuit containing only one resistor. We can easily calculate the voltage and current in this resistor. We then work backward, moving from one circuit to the next higher up, until we have found the current and voltage in every resistor. The method is best explained by means of a numerical example.

Example 3-3:

In the circuit of Fig. 3-11, calculate the current, voltage, and power for each resistor.

Solution:

We immediately recognize that the 30 Ω and 6 Ω resistors are directly in parallel. Their equivalent resistance is:

$$\frac{1}{R_{eq}} = \frac{1}{6} + \frac{1}{30} \qquad \text{Eq. 3-2}$$

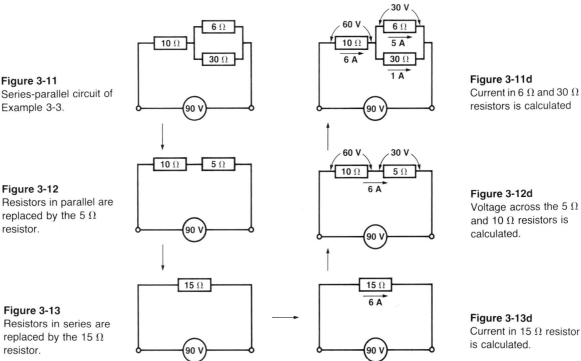

Figure 3-11
Series-parallel circuit of
Example 3-3.

Figure 3-11d
Current in 6 Ω and 30 Ω
resistors is calculated

Figure 3-12
Resistors in parallel are
replaced by the 5 Ω
resistor.

Figure 3-12d
Voltage across the 5 Ω
and 10 Ω resistors is
calculated.

Figure 3-13
Resistors in series are
replaced by the 15 Ω
resistor.

Figure 3-13d
Current in 15 Ω resistor
is calculated.

thus, $R_{eq} = 5\ \Omega$

Using the 5 Ω resistor, the circuit simplifies to that of Fig. 3-12. However, we now see that the 10 Ω and 5 Ω resistors are in series, and so we can replace them by an equivalent resistance

$$R_{eq} = R_1 + R_2 = 10 + 5 = 15\ \Omega$$

Thus, the series-parallel circuit of Fig. 3-11 has been reduced to a single resistance of 15 Ω (Fig. 3-13).

We now work backward, using the same diagrams but inserting the voltages and currents as we go along. (To prevent confusion, we have duplicated Figs. 3-13 to 3-11 and labeled them 3-13d to 3-11d.)

In Fig. 3-13d, the current flowing in the 15 Ω resistor is:

$$I = E/R = 90/15 = 6\ \text{A} \hspace{4em} \text{Eq. 2-3}$$

Clearly, the current in Fig. 3-12d is also 6 A. The voltage across the 10 Ω resistor is:

$$E_1 = IR = 6 \times 10 = 60\ \text{V}$$

The voltage across the 5 Ω resistor is

$$E_2 = 6 \times 5 = 30\ \text{V}$$

Moving backward to Fig. 3-11d, the voltage and current in the 10 Ω resistor are obviously the same as in Fig. 3-12d, namely 60 V and 6 A,respectively.

Next, we remember that the 5 Ω resistor replaces the 6 Ω and 30 Ω resistors in parallel. Because the voltage is the same in a parallel circuit, the voltage is 30 V across both the 6 Ω and 30 Ω resistors (Fig. 3-11d). The respective currents are

$I_1 = E/R = 30/6 = 5$ A

$I_2 = 30/30 = 1$ A

Knowing the voltages and currents (Fig. 3-11d), we can calculate the power dissipated in each resistor:

in the 10 Ω resistor: $P = EI = 60$ V $\times$ 6 A $= 360$ W   Eq. 2-1
in the 6 Ω resistor : $P = EI = 30$ V $\times$ 5 A $= 150$ W
in the 30 Ω resistor: $P = EI = 30$ V $\times$ 1 A $= 30$ W
Total power dissipated $= 360 + 150 + 30 = 540$ W

As a check on our calculations, we calculate the power supplied by the source:

$P = EI = 90 \times 6 = 540$ W

Thus, the generator power is indeed equal to the total power dissipated by the three resistors.

3.6 Power flow in a simple transmission line

The purpose of a transmission line is to carry electric power from a source to a distant load. Figure 3-14 shows a two-conductor line connecting a generator G to a motor M. The generator is driven by a gasoline engine and it generates a terminal voltage of 100 V. The potential difference across the motor terminals is assumed to be 90 V, and the line current is 50 A. Note that current flows out of the (+) terminal of the generator and into the (+) terminal of the motor. Therefore, the generator is definitely a source and the motor is definitely a load.

The transmission line voltage at the motor end is less than at the generator end, and the so-called *voltage drop* is $100 - 90 = 10$ V. The voltage drop (also called *IR* drop) occurs because the two line conductors have a certain amount of resistance. We can calculate the value of this resistance by reasoning as follows. The voltage drop for both conductors is 10 V. Therefore, the voltage drop for one conductor is 5 V. Knowing that the line current is 50 A, the resistance for one conductor is $R = E/I = 5/50 = 0.1$ Ω.

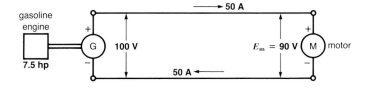

Figure 3-14
Motor connected to a source by means of a transmission line.

The voltage between the line conductors decreases uniformly from 100 V at the generator end to 90 V at the motor end. Thus, the difference of potential at the center of the line is 95 V.

Let us now calculate the power of each element in the circuit:

1. The generator supplies power PG = EI = 100 × 50 = 5000 W.

2. The motor receives power P_M = EI = 90 × 50 = 4500 W.

3. The difference between P_M and P_G (500 W) is due to line losses. The losses are dissipated as heat in the transmission line. (We also could have calculated the line losses from P = I^2R. Thus, the loss in one conductor is P = I^2R = 50^2 × 0.1 = 250 W, or 500 W for both conductors.)

3.7 Example of a short-circuit

Suppose that the lines in Fig. 3-14 accidentally touch each other inside the terminal box of the motor. The current in the line is now limited only by the resistance of the two conductors, namely 0.2 Ω. Consequently, the current will suddenly rise to I = E/R = 100/0.2 = 500 A. The generator output will jump to P = EI = 100 × 500 = 50 000 W, which is far above normal. This power is entirely dissipated as heat in the conductors, and they will become intensely hot in a matter of seconds (Fig. 3-15). Such a short-circuit is dangerous because it may cause a fire or burn out the generator.

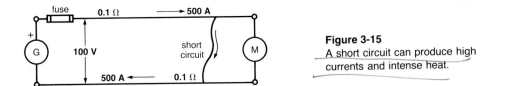

Figure 3-15
A short circuit can produce high currents and intense heat.

To prevent such disasters we place a *fuse* in series with the line. When the current reaches an abnormally high value, the fuse melts, and the current is interrupted automatically before any damage is done. A fuse is therefore a very important safety device.

3.8 Primary and secondary cells, ampere hour (A.h)

Electric cells are sources of electricity that transform chemical energy into electrical energy. When several cells are connected together to produce either a higher voltage, a greater current, or more power, the combination is called an *electric battery*.

There are about a dozen basic types of cells of commercial importance. They produce dc voltages that range from about 1.5 V to 2.5 V on open-circuit, that is, with no load connected to the terminals. The energy a cell can deliver depends upon its size and chemical composition. For example, a standard carbon-zinc flashlight cell delivers about 150 kJ per kilogram; a car battery delivers about 80 kJ per kilogram.

The capacity of a cell or battery is usually expressed in terms of the number of ampere hours (A.h) it can deliver. Thus, if a 12 V car battery has a capacity of 40 A.h, it can supply 1 A for 40 h, or 2 A for 20 h, and so forth. Assuming an average terminal voltage of 12 V, this represents a quantity of energy equal to 40 A.h × 12 V = 480 W.h = 480 × 3600 W.s = 1 728 000 J = 1728 kJ.

When a cell delivers electrical energy, it undergoes a progressive chemical change. In the so-called *primary* cell, the chemical change is permanent. As a result, when the chemical material in such a cell is completely transformed, the cell becomes "dead" and has to be thrown away.

In the case of a *secondary* cell, the original chemical composition can be partly restored by circulating current through the cell in a direction opposite to that of the discharge current. Secondary cells can be recharged several hundred times before they, too, have to be discarded.

When a secondary cell such as a lead-acid cell is being recharged, it releases hydrogen, a flammable gas. Battery rooms must therefore be well ventilated to prevent possible explosions. Particular care must also be taken to keep away from the extremely corrosive and blinding sulfuric acid.

All batteries possess a certain amount of internal resistance, usually a few milliohms. As a result, the terminal voltage under load is less than at no-load. However, for normal battery loading, the drop in voltage is seldom more than 10 percent of the open-circuit voltage.

Example 3-4:

A dry cell produces an emf of 1.5 V on open-circuit. It has an internal resistance of 0.2 Ω. Calculate the current and the terminal voltage when the cell is connected to a resistor of 1 Ω:(Fig. 3-16).

Solution:

The cell can be represented by a source of 1.5 V in series with a resistance of 0.2 Ω (Fig. 3-17). Note that A and B are the actual terminals of the cell. Point X is not accessible but exists only in the circuit diagram.

1. The total resistance of the circuit is:
    $R = 1.0 + 0.2 = 1.2\ \Omega$ Eq. 3-1

2. The current is $I = E/R = 1.5/1.2 = 1.25$ A

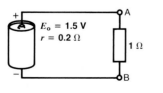

Figure 3-16
See Example 3-4

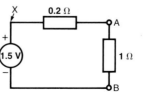

Figure 3-17
Equivalent circuit of Fig. 3-16.

3. The difference of potential across the load (and across the cell terminals) is:
    $$E = IR = 1.25 \times 1.0 = 1.25 \text{ V}$$

4. The internal voltage drop of the battery is $(1.5 - 1.25) = 0.25$ V

3.9 Cells in series

Cells are connected in series to create a source of higher emf. The emf of the battery is equal to the sum of the emfs of the individual cells. Furthermore, the internal resistance of the battery is equal to the sum of the internal resistances of the individual cells.

Connections are made from the positive terminal of one cell to the negative terminal of the next until only two terminals are left — one positive, the other negative. Figure 3-18 shows three cells connected in series.

The cells should be of the same type and have the same ampere hour capacity. If the cells are not identical, some will become discharged earlier than others, and so the battery is rendered useless before all the cells have released their available energy. This also applies to cells connected in parallel.

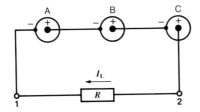

Figure 3-18
Cells connected in series.

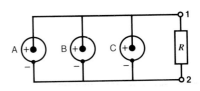

Figure 3-19
Cells connected in parallel.

3.10 Cells in parallel

Cells have to be connected in parallel whenever the load current is greater than the rated current of a single cell. Cells are also connected in parallel whenever we have to store a large quantity of energy. In a parallel connection, all the positive terminals are connected together to make a single positive terminal. Similarly, all the negative terminals are connected to make a single negative terminal (Fig. 3-19). If, by accident, we should connect one cell in reverse, all the cells will become discharged in a few seconds by a heavy circulating current. The reason is that a reversed cell produces the same effect as a short-circuit across the cells. If the cells happen to be primary cells, they will all be destroyed in less than a minute.

3.11 Cells connected in series-parallel

Cells are connected in series-parallel, first, to produce the desired emf and, second, to store the required amount of energy. Figure 3-20 shows two groups of four cells connected in series. The groups themselves are connected in parallel.

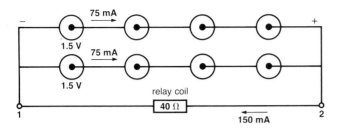

Figure 3-20
Cells connected in series-parallel. See Example 3-6.

Example 3-6:

The coil of a relay is rated at 6 V and has a resistance of 40 Ω. The coil has to operate off a battery for at least 400 h. Dry cells rated at 1.5 V and having a capacity of 40 A.h are available. How many cells are needed, and how should they be connected?

Solution:

Four cells have to be connected in series to produce 6 V (4 × 1.5 V = 6 V). The load current drawn by the relay coil is

$$I = E/R = 6/40 = 0.15 \text{ A} = 150 \text{ mA}$$

During the 400 h operating period, the battery must deliver at least 400 h × 0.15 A = 60 A.h. Because one group of four cells in series can deliver only 40 A.h, we must use two such groups in parallel, thus yielding 80 A.h. This capacity is more than we need, but it is better to be on the safe side. The current flowing through the battery and load is shown in Fig. 3-20.

3.12 Charging a secondary battery

When a secondary battery supplies power to a load, current flows out of the positive terminal (Fig. 3-21). To recharge the battery, we have to force current *into* the positive terminal, as shown in Fig. 3-22. This is done by connecting the (+) terminal of a battery charger to the (+) terminal of the battery. The charger voltage

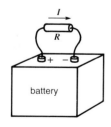

Figure 3-21
Current flow when battery
is discharging.

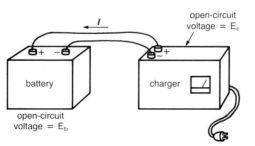

Figure 3-22
Current flow when battery is charging up.

E_c is slightly greater than battery voltage E_b. As a result, current flows into the (+) terminal of the battery, causing it to charge up.

According to our basic definition for a source and load, the battery is actually a load in Fig. 3-22. It receives energy from the charger and this energy restores the battery to its former chemical state.

We now see that a source, such as a battery, can also behave as a load, depending upon the direction of current flow. We shall encounter this surprising dual behavior in other circuits and machines.

SOLVING MORE COMPLEX CIRCUITS

3.13 Kirchhoff's voltage and current laws

Most circuits can be solved using the methods covered in the previous sections. Other circuits, however, are more complex, and a systematic method must be used to solve them. This method is based upon two laws, known as Kirchhoff's voltage law (KVL) and Kirchhoff's current law (KCL).

Kirchhoff's voltage law states that the algebraic sum of the voltages around any closed loop is zero. In essence, this means that the sum of all the voltage rises is equal to the sum of all the voltage drops as we move around a circuit loop. This is much the same as a mountain climber who starts from a given point, takes a random walk over hills and valleys, and then comes back to the starting point. The sum of all the uphill vertical distances he covers is obviously equal to all the downhill vertical distances.

Kirchhoff's current law states that the algebraic sum of the currents that arrive at a point is zero. This means that the sum of the currents that arrive at a terminal is equal to the sum of the currents that leave it.

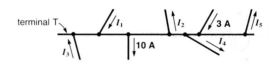

Figure 3-23
Circuit illustrating Kirchhoff's current law.

To understand how KCL is applied, consider seven currents that flow into and out of a common terminal T, as shown in Fig. 3-23. The sum of the currents that arrive at the terminal is $I_1 + I_3 + 3$. The sum of the currents that leave is $I_2 + I_5 + 10 + I_4$. Using KCL the equation is:

$$I_1 + I_3 + 3 = I_2 + I_5 + 10 + I_4$$

3.14 Voltage rises and voltage drops

To understand how KVL is applied, we first define what is meant by a voltage rise and a voltage drop. Consider a voltage E having the polarities given in Fig. 3-24. Imagine that we move from terminal A to terminal B. Because this is a move from $(+)$ to $(-)$, it is considered to be a voltage drop. If we move from B to A, this is a move from $(-)$ to $(+)$, which is considered to be a voltage rise. The magnitude of the voltage rise (or drop) is equal to E.

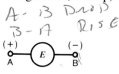

Figure 3-24
Defining voltage rise and voltage drop.

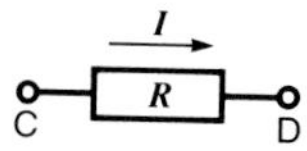

Figure 3-25a
Current flowing in a resistor produces a voltage across its terminals.

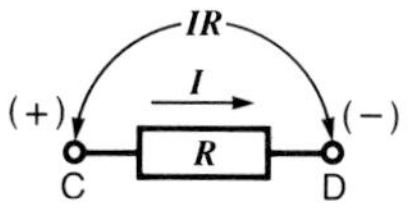

Figure 3-25b
Currents produce a voltage rise or voltage drop.

Voltage rises and voltage drops are also produced when currents flow through a resistance.

Thus, suppose a current I is flowing through a resistor R in the direction shown in Fig. 3-25a. Clearly, the voltage across the resistor is equal to IR. Because current always flows from $(+)$ to $(-)$ in a resistor, we indicate these polarities, as shown in Fig. 3-25b. Now if we move from terminal C to terminal D, we go from $(+)$ to $(-)$ which gives us a *voltage drop*. If we move from D to C, we obtain a *voltage rise*. The magnitude of the voltage rise (or drop) is equal to IR.

3.15 Applying Kirchhoff's laws

Let us apply these principles to solve the circuit of Fig. 3-26. It is composed of two resistors, and of three sources — 10 V, 20 V, and 30 V — having the polarities shown. It is not obvious in which direction the currents will flow, so let us *assume* that I_1 flows from B to C, and I_2 flows from B to D (Fig. 3-27). This enables us to

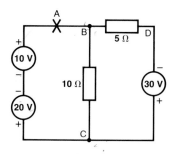

Figure 3-26
Solving a more complex circuit.

put $(+)$ and $(-)$ signs on the 10 Ω and 5 Ω resistors. Furthermore, the voltage across each resistor is equal to $10I_1$ and $5I_2$, as shown.

We are now ready to apply Kirchhoff's voltage law because every element in the circuit has polarity marks, and the magnitude of the voltage across each element is indicated.

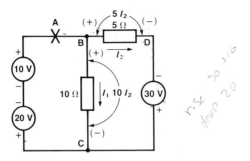

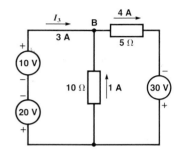

Figure 3-27
Applying KVL to the circuit of Fig. 3-26.

Figure 3-28
Currents in the circuit of Fig. 3-26.

Think of yourself as being an electrical-mountain climber starting at any point, such as A. You are allowed to move along any loop you please, but you must come back to your starting point. As you cross the electrical hills and valleys, you record all the voltage rises and voltage drops. Then you equate their sum.

Referring to Fig. 3-27, and starting from point A, let us move clockwise around the outer loop. It is composed of the three sources and the 5 Ω resistor. As we move from B to D, across the 5 Ω resistor, we experience a voltage drop equal to $5I_2$. Then, as we cross the 30 V source, we have a voltage rise. Next, as we pass C and cross the 20 V source it gives us a voltage drop. Finally, we cross the 10 V source, which gives us a voltage rise and brings us back to the starting point. We list the voltage rises and voltage drops, and find their sum, as follows:

$$\text{voltage rises} = 30 + 10 = 40$$
$$\text{voltage drops} = 5I_2 + 20$$

Equating these sums, we obtain:

$$20 + 5I_2 = 40$$

and so $5I_2 = 20$ therefore $I_2 = 4$ A

To find I_1 we again start at point A, but travel over the left-hand loop. It consists of the 10 V and 20 V sources and the 10 Ω resistor. This time, to show our versatility, let us move counterclockwise (ccw) around the loop. We again record the voltage rises and voltage drops, and find their sum:

$$\text{voltage rises} = 20 + 10I_1$$
$$\text{voltage drops} = 10$$

Equating the sums, we find

$$20 + 10I_1 = 10$$
$$10I_1 = -10$$
$$I_1 = -1A$$

Current I_1 is equal to 1 A, but it has a negative sign. This means that the current actually flows in the opposite direction to what we had assumed. The actual current flows are shown in Fig. 3-28.

The remaining current I_3 can be found by using Kirchhoff's current law. The sum of the currents that arrive at point B is $(I_3 + 1)$. The sum of the currents that leave it is obviously 4 A. Consequently,

$$I_3 + 1 = 4$$

and so

$$I_3 = 3 A$$

3.16 Summary

In this chapter, we have learned that most circuits are composed of elements that are connected either in series, in parallel, or in series-parallel. In a series circuit, the current is the same in each element. In a parallel circuit, the voltage is the same across each element. In every circuit, no matter how complex, the sum of the powers consumed by the loads is equal to the sum of the powers supplied by the sources.

We have also become familiar with circuit diagrams. They are models that enable us to visualize the voltages and currents in a real circuit.

We have learned how to connect cells to make an electric battery having a desired voltage and ampere hour capacity. The cells in a battery must have identical ratings.

The study of an elementary transmission line enabled us to understand the meaning of voltage drop and line losses, and the importance of fuse protection.

Finally, we found that any dc circuit can be solved using Kirchhoff's voltage and current laws. Fortunately, most circuits are so simple that the ordinary series-parallel circuit rules can be used to solve them.

TEST YOUR KNOWLEDGE

3-1   A series circuit is composed of several loads and a single source. The sum of the currents in the loads is equal to the current supplied by the source.

☐ true   ☐ false

3-2   Four *identical* resistors are connected in series across a 120 V source. The current in one of the resistor is 8 A. The power dissipated by one resistor is

a. 240 W   b. 960 W   c. 240 J   d. 15 W

3-3   A current of 360 A feeds 20 identical resistors connected in parallel. The current in each resistor is

a. 360 A   b. 18 A   c. 7200 A

3-4   Three resistors of 7 Ω, 42 Ω and 30 Ω are connected in parallel. The equivalent resistance is

a. 79 Ω   b. 0.2 Ω   c. 5Ω

3-5   Thirty resistors of 60Ω each are connected in parallel. The equivalent resistance is

a. 2 Ω   b. 0.5 Ω   c. 60 Ω

3-6   A primary cell can be recharged by using a battery charger

☐ true   ☐ false

3-7   When recharging a secondary battery, the positive battery terminal is connected to the negative terminal of the battery charger.

☐ true   ☐ false

3-8   In Fig. 3-20 there is a voltage drop in going from terminal 1 to terminal 2.

☐ true   ☐ false

3-9   Three batteries have voltages of 6 V, 12 V, and 20 V, respectively. If they are connected in series in every possible way, the highest and lowest voltages are

a. 38 V and 6 V   b. 20 V and 6 V   c. 38 V and 2 V

3-10   When identical cells are connected in series, the ampere hour capacity of the battery is equal to the ampere hour capacity of one cell.

☐ true   ☐ false

3-11   When identical cells are connected in parallel, the ampere hour capacity of the battery is equal to the sum of the ampere hour capacities of the cells.

☑ true   ☐ false

QUESTIONS AND PROBLEMS

3-12   In Fig. 3-18 each cell produces 2 V, and load resistance R is 4 Ω. Calculate (a) the load current I_L and (b) the power supplied by each cell.

3-13   In Fig. 3-19, each cell produces 2 V, and load resistance R is 2 Ω. Calculate (a) the load current I_L and (b) the power supplied by each cell.

3-14   In Fig. 3-18 each cell produces an open-circuit voltage of 2 V and has an internal resistance of 0.1 Ω. If the capacity of each cell is 60 A.h and R is 1.7 Ω calculate: (a) the voltage across the resistor, (b) the voltage across the terminals of each cell, and (c) the time to discharge the battery.

3-15   In Fig. 3-19, each cell produces an open-circuit voltage of 1.5 V and has an internal resistance of 0.3 Ω. If the capacity of each cell is 2 A.h and R is 100 Ω, calculate (a) the voltage across the resistor and (b) the time to discharge the battery.

3-16   Primary cells are available having a rating of 1.5 V, 20 A.h. Using these cells, we wish to make a 30 V battery having a rating of at least 110 A.h. How many cells are required, and how must they be connected?

3-17   A home is equipped with nine 100-watt lamps. Calculate the total current if the line voltage is 120 V.

3-18   A 25-ohm resistor is connected in series with a relay coil having a resistance of 80 Ω. If 42 V is applied to the circuit, calculate (a) the current in the coil and (b) the power dissipated in the resistor.

3-19   Two resistors of 20 Ω and 30 Ω are connected in series across a 1.5 kV source. Calculate (a) the voltage across the 30-ohm resistor and (b) the power supplied by the source.

3-20   Two resistors of 60 Ω and 40 Ω are connected in series. The voltage across the 40 Ω resistor is 50 V. Calculate (a) the voltage of the source and (b) the power dissipated in the 60 Ω resistor.

3-21   Fill in the empty spaces in the following table. Each case involves only one resistor.

Case Number	1	2	3	4	5
P	100 W		200 W	100 kW	2 W
E		10 V		1 kV	
R		1000 Ω	800 Ω		10^6 Ω
I	10 A				

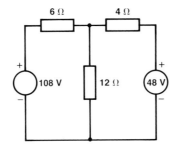

Figure 3-29
See Problem 3-22.

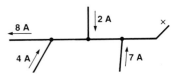

Figure 3-30
See Problem 3-23.

3-22   In Fig. 3-29, calculate the current in the 12 Ω resistor, using KVL and KCL.

3-23   In Fig. 3-30, determine the magnitude and direction of the current that flows in conductor x.

Conductors,
Resistors,
and Insulators

Introduction and Chapter Objectives

In this chapter we will study the properties of electrical materials. We will cover the subjects of dielectric strength, wire size, and the change of resistance with temperature. We will also discover that the life expectancy of electrical equipment is directly related to temperature and the quality of the insulation. Finally, we will discuss commercial resistors, fuses, and the resistance of ground electrodes.

4.1 Classification of materials

There is hardly any material that is not used either directly or indirectly in electrical equipment or as part of an electrical system. Copper, iron, paper, cement, gold, cotton, and numerous plastics are all used in the construction of motors, heaters, transmission lines, lamps, relays, and switches. All these materials can be grouped into two categories — conductors and insulators. The conductors, in turn, can be divided into three other categories: good conductors, not-so-good conductors (called *resistors*) and semiconductors. Semiconductors are special non-linear materials that behave as either conductors or insulators, depending upon voltage, current, and temperature.

There is no clear dividing line between conductors and resistors. For example, carbon is considered to be a conductor when used as a brush in a dc motor, but is classified as a resistor when used as a current-limiting element. Table 4A lists some of the more common electrical materials and how they are usually classified.

70

TABLE 4A   CLASSIFICATION OF MATERIALS

Conductors	Resistors	Insulators
silver	iron	air
copper	carbon	rubber
aluminum	tungsten	porcelain
brass	nichrome	paper
carbon	molybdenum	plastics
mercury	salt water	mineral oil

4.2 Resistivity of electrical materials, the ohm meter (Ω.m)

Conductors, resistors, and insulators may be distinguished from each other by the relative ease with which they conduct an electric current. They also may be classified according to their relative opposition, or resistivity, to the flow of electric current.

Figure 4-1 gives an idea of the relative resistance of three identical samples of copper, nichrome and rubber. Note that nichrome (an alloy composed of 80% nickel and 20% chromium) is about 60 times more resistive than copper is. Rubber is many million times more resistive than copper is.

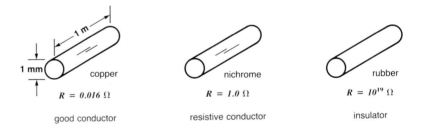

copper	nichrome	rubber
$R = 0.016\ \Omega$	$R = 1.0\ \Omega$	$R = 10^{19}\ \Omega$
good conductor	resistive conductor	insulator

Figure 4-1
Relative resistance of three materials having identical dimensions.

The SI unit of resistivity is the ohm meter (symbol Ω.m). It is numerically equal to the resistance of a sample of material that is one meter long and whose cross section is one square meter. For example, the resistivity of copper is 15.88×10^{-9} Ω.m at 0°C, and that of rubber is about 10^{13} Ω.m.

4.3 Solid, liquid, and gaseous insulators

Solid, liquid, and gaseous insulators are used in electrical equipment. For example, solid insulators such as rubber, paper, and synthetic plastics are used to cover electrical wires and cables. Treated paper and fiber are used to prevent electrical coils from coming in contact with the iron cores of rotating machines. Some insulators are composed of several materials, such as asbestos, glass, and plastics, so they can withstand very high temperatures while retaining great strength.

In large transformers, liquid insulators such as mineral oil serve to insulate the high-voltage windings from each other and from ground. The oil also carries the heat away from the windings and prevents air from coming in contact with them. This prevents oxidation and deterioration of the insulation, and increases the useful life of the transformer.

Under normal conditions, one of the best insulators is the air that surrounds us. Its thermal properties are better than those of porcelain, it can act as a cooling agent, and, best of all, it costs nothing. However, at extremely high temperatures (between 5000°C and 50 000°C) the molecules of air begin to break down, releasing free electrons. Air then becomes a relatively good conductor, with a resistivity approaching that of salt water. This molecular breakdown, called *ionization*, occurs whenever an electric arc passes through air.

Sulfur hexafluoride (SF_6) is another important insulating gas, whose insulating properties are about ten times better than air. It is used in high-voltage circuit breakers, transformers, and short cable runs when space reduction is particularly important.

Hydrogen is a gaseous insulator used in very large turboalternators. It prevents air from coming in contact with the windings. It also acts as an excellent cooling agent to carry away the heat from the innermost parts of the machine.

4.4 Deterioration of organic insulators

Time and temperature are the agents that cause the gradual deterioration and ultimate failure of insulators. The rate of deterioration is further accelerated if the insulating material is in a polluted or humid atmosphere, or if it is subjected to continuous vibration.

Apart from accidental electrical and mechanical failures, the life expectancy of electrical apparatus is limited by the temperature of its insulation: the higher the temperature, the shorter its life. Tests made on many insulating materials have shown that the useful life of electrical apparatus diminishes by approximately half for every 10°C increase above its rated operating temperature. This means that if a motor has a normal life expectancy of eight years at a rated temperature of 100°C, it will have a life expectancy of four years at a temperature of 110°C, of two years at 120°C, and of only one year at 130°C.

4.5 Thermal classification of insulators

A user expects newly purchased electrical equipment to last a reasonable length of time before it has to be repaired. In general, it is felt that a useful life of about ten years is adequate. Based on these considerations and knowing that temperature is the main factor that determines equipment life, standards-setting organizations have established five temperature grades for insulation and insulation systems. They correspond respectively to a hot-spot temperature of 105°C, 130°C, 155°C, 180°C, and 220°C.* The hot-spot temperature is the temperature of the hottest spot inside a winding.

In addition to the hot-spot temperature limits, the standards specify that the ambient temperature shall not exceed 40°C.

What is the relative advantage of one insulation class over another? First, a higher temperature class enables us to get more output from a machine of a given size. For example, a motor rated at 150 hp with a 105°C insulation system can yield an output of almost 200 hp if rewound with a 155°C insulation system. However, the motor runs hotter and the high-temperature insulation is more costly.

Second, a higher temperature class enables us to build machines that can operate for years in very high ambient temperatures (above 40°C) without breaking down.

4.6 Breakdown of solid and gaseous insulators

Suppose an insulator (solid, liquid, or gas) is placed between two metal plates spaced 1 mm apart. Let us apply a voltage across the plates, using an external source (Fig. 4-2). A current will flow in the circuit, but it will be exceedingly small because the insulator possesses very few free electrons. However, if we gradually raise the voltage, we eventually reach a critical point at which the insulation suddenly breaks down. In effect, the insulator suddenly becomes a conductor, and a large current will flow in the circuit.

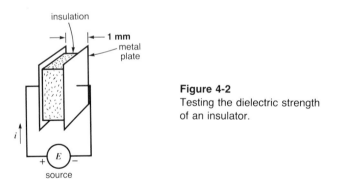

Figure 4-2
Testing the dielectric strength
of an insulator.

The breakdown voltage required to produce this insulation failure depends upon the nature of the insulator and its thickness. The ratio of breakdown voltage to insulator thickness is called *dielectric strength*. It is generally expressed in kilovolts per millimeter (kV/mm). Table 4B gives the dielectric strength of several insulators.

When a solid insulator breaks down, its chemical composition is permanently damaged, and so it must be replaced. On the other hand, liquid and gaseous insulators are inherently self-healing and can usually withstand many arcing discharges.

---

*   These insulation grades were formerly designated class A (105°C), class B (130°C), class F (155°C), class H (180°C), and class C (220°C and up).

TABLE 4B   PROPERTIES OF INSULATING MATERIALS

Insulator	Dielectric strength	Maximum Operating temperature
	kV/mm	°C
dry air	3	2000
SF$_6$	30	—
glass	100	600
mineral oil	10	110
paper (treated)	14	120
rubber	12 to 20	65
Mylar^R	400	150

Example 4-1:

A two-conductor 120 V cable used in house wiring has a 0.03 inch thermoplastic covering over each conductor (Fig. 4-3). If the dielectric strength is 25 kV/mm, calculate the voltage that could be applied between the conductors before breakdown occurs. (Note: 1 inch = 25.4 mm, by definition)

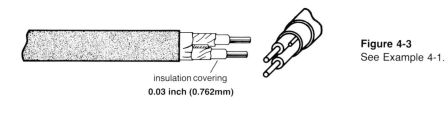

insulation covering
0.03 inch (0.762mm)

Figure 4-3
See Example 4-1.

Solution:

1. The total distance between the conductors is 0.03 × 2 = 0.06 inch

2. The distance in millimeters is 0.06 × 25.4 = 1.52 mm

3. The approximate breakdown voltage is 1.5 mm × 25 kV/mm = 38 kV

The breakdown voltage is much higher than that needed to withstand 120 V. A thinner insulation covering could be used, but it would not have the mechanical strength and resistance to abrasion that the thicker covering has. Thus, in low-voltage installations, the thickness of the insulation is dictated more by mechanical robustness than by voltage breakdown. In high-voltage equipment, the reverse is true.

4.7 Round conductors, Standard American Wire Gauge

In the United States and Canada, the size of round conductors is standardized according to what is known as the Standard American Wire Gauge (abbreviation AWG). This standard is also called the Brown & Sharp Gauge (B & S).

According to this standard, each wire bears a gauge number that corresponds to a definite diameter. The diameter of the wire diminishes as the gauge number increases: for example, a No. 6 gauge wire is smaller than a No. 4 gauge wire. Table 4C gives the diameter and cross section of round wires corresponding to the respective gauge numbers. It also gives the weight and resistance of round copper wires. You will find it useful to memorize the following rules which apply to the AWG system:

1. A conductor which has twice the cross section of another has a gauge number that is three numbers smaller. Example: the cross section of a No. 15 wire is double that of a No. 18 wire.
2. A conductor which has ten times the cross section of another has a gauge number that is ten numbers smaller. Example: a No. 4 gauge wire has the same cross section as ten No. 14 gauge wires.

Figure 4-4 shows how to measure the conductor size using a slotted wire gauge. After stripping all insulation (including any thin varnish covering), the bare wire is inserted into the slot that gives the best fit. The slot number corresponds to the wire size.

4.8 Mils, circular mils

The diameter of a conductor is sometimes expressed in mils instead of millimeters. The mil is a unit of length equal to one thousandth of an inch, or 0.0254 mm.

Similarly, the cross section of round wires is sometimes expressed in circular mils instead of in square millimeters. The circular mil (cmil or CM) is equal to the area of a circle having a diameter of 1 mil (Fig. 4-5). A circular mil is equal to 0.000 506 707 mm^2.

The cross section of a round wire expressed in circular mils is equal to the square of its diameter expressed in mils.

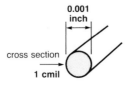

0.001 inch

cross section

1 cmil

Figure 4-5
The area enclosed by the circle is 1 circular mil.

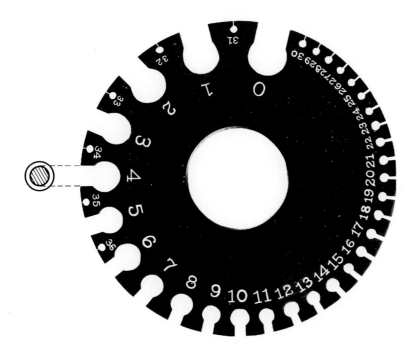

Figure 4-4
Method of measuring the size of a conductor using an AWG wire gauge. A
No. 4 bare conductor is about to be slipped into the No. 4 slot. The wire
gauge and its slots are shown full size.

Example 4-2:

Calculate the cross section of a round wire having a diameter of 0.102 inch.

Solution:

1. The diameter is equal to 102 thousandths of an inch, or 102 mils

2. The cross section is therefore:
    A = 102 mils × 102 mils = 10 404 circular mils

3. Expressed in square millimeters, the area is:
    A = 10 404 × 0.000 506 707 = 5.27 mm^2

   Note that conductors bigger than 0000 (or 4/0) do not bear a gauge number; they
are usually identified by their cross section in thousands of circular mils (MCM).
Thus, a 250 MCM conductor has a cross section of 250 000 cmil, or 126.6 mm^2.

TABLE 4C   PROPERTIES OF ROUND COPPER CONDUCTORS

Gauge number AWG/ B & S	Diameter of bare conductor		Cross section		Resistance mΩ/m or Ω/km		Weight g/m or kg/km	Typical diameter of insulated magnet wire used in relays magnets motors transformers etc.
	mm	mils	mm²	cmils	25°C	105°C		
250MCM	12.7	500	126.6	250 000	0.138	0.181	1126	
4/0	11.7	460	107.4	212 000	0.164	0.214	953	
2/0	9.27	365	67.4	133 000	0.261	0.341	600	
1/0	8.26	325	53.5	105 600	0.328	0.429	475	
1	7.35	289	42.4	87 700	0.415	0.542	377	
2	6.54	258	33.6	66 400	0.522	0.683	300	
3	5.83	229	26.6	52 600	0.659	0.862	237	
4	5.18	204	21.1	41 600	0.833	1.09	187	
5	4.62	182	16.8	33 120	1.05	1.37	149	
6	4.11	162	13.30	26 240	1.32	1.73	118	
7	3.66	144	10.5	20 740	1.67	2.19	93.4	
8	3.25	128	8.30	16 380	2.12	2.90	73.8	
								mm
9	2.89	114	6.59	13 000	2.67	3.48	58.6	3.00
10	2.59	102	5.27	10 400	3.35	4.36	46.9	2.68
11	2.30	90.7	4.17	8 230	4.23	5.54	37.1	2.39
12	2.05	80.8	3.31	6 530	5.31	6.95	29.5	2.14
13	1.83	72.0	2.63	5 180	6.69	8.76	25.4	1.91
14	1.63	64.1	2.08	4 110	8.43	11.0	18.5	1.71
15	1.45	57.1	1.65	3 260	10.6	13.9	14.7	1.53
16	1.29	50.8	1.31	2 580	13.4	17.6	11.6	1.37
17	1.15	45.3	1.04	2 060	16.9	22.1	9.24	1.22
18	1.02	40.3	0.821	1 620	21.4	27.9	7.31	1.10
19	0.91	35.9	0.654	1 290	26.9	35.1	5.80	0.98
20	0.81	32.0	0.517	1 020	33.8	44.3	4.61	0.88
21	0.72	28.5	0.411	812	42.6	55.8	3.66	0.79
22	0.64	25.3	0.324	640	54.1	70.9	2.89	0.70
23	0.57	22.6	0.259	511	67.9	88.9	2.31	0.63
24	0.51	20.1	0.205	404	86.0	112	1.81	0.57
25	0.45	17.9	0.162	320	108	142	1.44	0.51
26	0.40	15.9	0.128	253	137	179	1.14	0.46
27	0.36	14.2	0.102	202	172	225	0.908	0.41
28	0.32	12.6	0.080	159	218	286	0.716	0.37
29	0.29	11.3	0.065	128	272	354	0.576	0.33
30	0.25	10.0	0.0507	100	348	456	0.451	0.29
31	0.23	8.9	0.0401	79.2	440	574	0.357	0.27
32	0.20	8.0	0.0324	64.0	541	709	0.289	0.24
33	0.18	7.1	0.0255	50.4	689	902	0.228	0.21
34	0.16	6.3	0.0201	39.7	873	1140	0.179	0.19
35	0.14	5.6	0.0159	31.4	1110	1450	0.141	0.17
36	0.13	5.0	0.0127	25.0	1390	1810	0.113	0.15
37	0.11	4.5	0.0103	20.3	1710	2230	0.091	0.14
38	0.10	4.0	0.0081	16.0	2170	2840	0.072	0.12
39	0.09	3.5	0.0062	12.3	2820	3690	0.055	0.11
40	0.08	3.1	0.0049	9.6	3610	4720	0.043	0.1

4.9 Stranded cable

A stranded wire or cable is composed of six or more conductors, so that it can provide flexibility. Its cross section is equal to the sum of the cross sections of the strands. It does not include the area of the empty spaces between the strands. Thus, a No. 10 stranded wire possesses the same net cross section (and same resistance) as a No. 10 solid wire.

4.10 Square wires and other conductor shapes

Square wires have gauge numbers that correspond to those of round wires. According to the AWG, if a square wire has the same width as the diameter of a round wire, the two bear the same gauge number (Fig. 4-6). It follows that the cross section of a square wire is about 1.25 times greater than that of a round wire having the same gauge number.

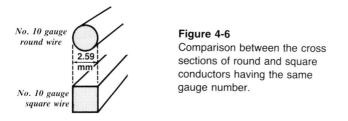

Figure 4-6
Comparison between the cross sections of round and square conductors having the same gauge number.

Rectangular conductors whose dimensions can be made to order are also available. They are used in the windings of large transformers and rotating machines. Some have a hollow center so that water can be circulated inside the conductor to keep it cool.

Very large bare conductors called busbars are used to carry the heavy currents in factories and substations. Some are rectangular with such typical dimensions as 8 mm × 150 mm, and others are pipe-shaped.

4.11 Current rating of conductors; ampacity

Even the best conductors have some resistance, consequently, they all heat up when they carry an electric current. The heat produced in insulated conductors is transmitted through the insulating layers and finally dissipated to the surrounding air. The greater the current in the wire, the higher the temperature will be. However, to ensure a reasonable life, the temperature of the insulation must not be too high. We therefore arrive at the following important conclusion:

The maximum current a conductor can carry depends upon the highest temperature its insulation can withstand for the desired lifetime of the conductor.

Although bare conductors present no insulation problem, the current and temperature must still be kept within reasonable limits because high temperatures produce excessive oxidation and flaking of the conductor. They also create a potential fire hazard.

4.12 National Electrical Code and electrical installations*

The life expectancy of the electrical wiring in factories, buildings, and homes must be particularly long because we cannot afford to replace the conductors every ten years. For this reason, the National Electrical Code specifies rather low maximum temperatures for wire and cable used in electrical installations. Depending on the type of insulation, the National Electrical Code typically recognizes maximum temperatures of 60°C, 75°C, and 90°C.

A wire of a given size rated at 90°C can carry a higher current than one having a lower temperature rating. For example, a No. 6 wire whose insulation is rated at 60°C can carry a current of 80 A in free air. The same wire with insulation rated at 90°C can carry a current of 100 A. Both conductors have the same life expectancy under the specified temperature conditions. Table 4D, taken in part from the National Electrical Code, gives an idea of the *ampacities* (ampere capacities) of various conductor sizes.

**TABLE 4D   ALLOWABLE AMPACITIES FOR INSULATED
CONDUCTORS RATED 0-2000 V
SINGLE CONDUCTORS IN FREE AIR,
BASED ON AN AMBIENT TEMPERATURE OF 30°C**

Gauge Number	Temperature rating of insulation covering	
(AWG)	60°C	90°C
12	25 amperes	40 (25) amperes
10	40	55 (40)
8	55	70
6	80	100
4	105	135
2	140	180
1/0	195	245
3/0	260	330
250 MCM	340	425

Note: The above values are derived from the National Electrical Code. The Canadian Electrical Code gives identical values except for those shown in parenthesis.

---

*   The general remarks made in this section also apply to the Canadian Electrical Code.

When several insulated conductors are placed in the same conduit, the heat dissipated by each raises the temperature of the others. The ampacity of each conductor must then be reduced so as not to exceed the maximum permissible temperature. For example, the National Electrical Code specifies that when three No. 6 conductors rated at 60°C are run in a conduit, the ampacity per conductor is only 55 A compared with 80 A when they are suspended separately in free air.

4.13 Calculation of resistance

For a given temperature, the resistance of a conductor depends upon its length, cross section, and resistivity. The relationship is given by the equations:

SI UNITS	U.S. CUSTOMARY UNITS

$$R = \rho l/A \qquad (4\text{-}1)$$

$$R = \rho l/A \qquad (4\text{-}2)$$

where

R = resistance of the conductor, in ohms (Ω)

ρ = resistivity of the conductor, in nanohm meters (nΩ.m)

l = length of the conductor, in meters (m)

A = cross section of the conductor in square meters (m^2)

where

R = resistance of conductor, in ohms (Ω)

ρ = resistivity of conductor, in ohm circular mil per foot (Ω.cmil/ft)

l = length of conductor, in feet (ft)

A = cross section of conductor in circular mils (cmil)

We recall that the SI unit of resistivity is the ohm meter. However, owing to the low resistivity of metallic conductors, we prefer to use a submultiple, the nanohm meter (nΩ.m), equal to 10^{-9} Ω.m. The resistivity of some common metals is given in Table 4E.

Example 4-3:

Calculate the resistance of a copper conductor having a length of 2 km and a cross section of 22 mm^2. Assume the resistivity is 18 nΩ.m.

Solution:

Knowing that:

l = 2 km = 2000 m

A = 22 mm^2 = 22 $\times$ 10^{-6}m^2

ρ = 18 nΩ.m = 18 $\times$ 10^{-9} Ω.m

TABLE 4E   RESISTIVITY AND TEMPERATURE COEFFICIENT OF SOME CONDUCTORS

Conductor	Resistivity at 0°C		Temperature coefficient	Melting point
	n Ω.m.	Ωcmil/ft	per °C	°C
silver	15.0	9.02	0.004 11	960
copper	15.9	9.56	0.004 27	1083
aluminum	26.0	15.6	0.004 39	660
tungsten	49.6	29.8	0.005 5	3410
manganin	482	290	0.000 015	1020
nichrome	1080	649	0.000 11	1400
carbon	8000 to 30 000	—	− 0.000 3	3600

We find:

$$R = \rho \; \frac{l}{A}$$

$$= \frac{18}{10^9} \times \frac{2000}{22 \times 10^{-6}} = 1.64 \; \Omega$$

Example 4-4:

Calculate the resistance of 82 ft of No. 20 AWG nichrome wire at a temperature of 0°C.

Solution:

According to Table 4C, the cross section of No. 20 gauge round wire is 1020 cmils. Furthermore, the resistivity of nichrome is 649 Ω.cmil/ft (Table 4E). The resistance of the conductor is therefore:

$$R = \rho \; \frac{l}{A} = 649 \times \frac{82}{1020} = 52.2 \; \Omega \qquad \text{Eq. 4-2}$$

4.14 Resistance and temperature

The resistance of most metallic conductors increases with temperature. The increase depends upon the so-called *temperature coefficient* of the conductor. The resistance can be calculated from the equation:

$$R_t = R_o (1 + \alpha t) \qquad (4\text{-}3)$$

where

R_t = resistance of the conductor at t °C, in ohms

R_o = resistance of the conductor at 0 °C, in ohms

α = temperature coefficient of the conductor, at 0 °C.

t = temperature of the conductor, in °C

Example 4-5:

An aluminum transmission line has a resistance of 80 Ω at a winter temperature of 0°C. Calculate its resistance in the summer when the conductor temperature is 36°C.

Solution:

From Table 4E, the temperature coefficient for aluminum is 0.00439. The resistance of the line at 36°C is:

$$
\begin{aligned}
R_{36°} &= R_o (1 + \alpha t) \qquad\qquad\text{Eq. 4-3}\\
&= 80 (1 + 0.00439 \times 36)\\
&= 80 (1 + 0.158)\\
&= 92.6 \ \Omega
\end{aligned}
$$

The resistance in summer is therefore almost 15 percent higher than in winter.

The resistance of some alloys, such as manganin, changes very little with temperature; consequently, these materials are used in precision voltmeters and ammeters and to make resistance standards.

Other alloys, such as nichrome, possess both high resistivity and a low temperature coefficient. They are used in heater elements and commercial resistors.

4.15 Classification of resistors

Resistors can be grouped into three classes depending upon whether they operate at low, medium, or high temperature. Low temperatures are considered to be those below 155°C, medium temperatures lie between 275°C and 415°C, and high temperatures are those above 600°C.

Low-temperature resistors are mainly used in electronic circuits. They are usually enclosed in boxes, and so they are rated to operate in ambient temperatures as high as 70°C. The allowable temperature rise of these resistors is therefore $(155 - 70) = 85$°C.

These resistors are small because they typically dissipate between 1/4 W and

10 W. They are made by depositing a thin layer of carbon on a ceramic tube or by winding nichrome wire on a porcelain support.

Some carry four colored bands which permit us to identify the ohmic value and precision of the resistor. Thus, red and blue correspond respectively to the numbers 2 and 6. The complete color code is given in Table 4F.

TABLE 4F   COLOR CODING OF RESISTORS

Color	Numerical value or multiplier	Coding
black	0	first band = first number
brown	1	second band = second number
red	2	third band = number of zeros
orange	3	fourth band = precision
yellow	4	
green	5	Example : 420 000 $\Omega \pm 10\%$
blue	6	
violet	7	4 2 0000 $\pm 10\%$
grey	8	
white	9	
gold	$\pm$ 5 %	
silver	$\pm$ 10 %	yellow red yellow silver

There are many types of resistors in the low-temperature category. Their properties and performances are determined by standards organizations such as EIA (Electronic Industries Association), and IEC (International Electrotechnical Commission), or by military standards (MIL specs).

Medium temperature resistors (275°C to 415°C) are those most often found in industry. They include fixed and variable resistors (called rheostats), that are designed to operate in an ambient temperature of 40°C. The power rating varies from 10 W to many kilowatts. These resistors are used to start motors, to dim lights in theaters, to vary the current in magnetic coils, and in numerous other applications, including space heating in factories and homes.

High-temperature resistors (600°C and up) are used in electric stoves and furnaces, in infrared lamps, and in incandescent light bulbs. Typical temperatures are: incandescent lamp, 2600°C ; industrial electric furnace, 1100°C ; red-hot stove element, 950°C.

4.16 Fuses

The melting point of a conductor is put to practical use in the construction of fuses. These devices usually consist of a fuse link enclosed in a fiber tube. The fuse link is indented at one, two, or three places along its length to create short, narrow bridges of relatively high resistance (Fig. 4-7). When the current exceeds the rated value, the bridges melt, thereby interrupting the circuit. Plug fuses found in homes are designed along the same principles as industrial fuses (Fig. 4-8).

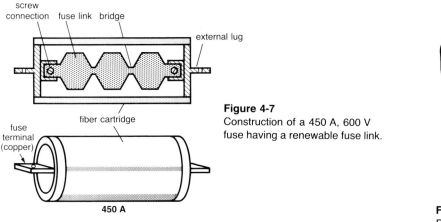

Figure 4-7
Construction of a 450 A, 600 V fuse having a renewable fuse link.

Figure 4-8
Plug fuse rated at 15 A, 250 V.

When a short-circuit occurs, the current becomes very high and the tremendous heat causes the fusible element to literally explode. The fiber tube must withstand the high internal pressure, and special precautions must be taken to prevent the arc from being sustained by the vaporized metal. To meet these requirements, the fuse is made longer as the operating voltage increases. Furthermore, the amount of fusible metal is kept to a strict minimum. Zinc, which melts at 420°C, is most often used.

High rupturing capacity (HRC) fuses use a thin copper or silver wire as the fusible element.

4.17 Contact resistance

When two current-carrying conductors meet at a terminal, they may cause appreciable contact resistance and substantial I^2R losses. The heat will gradually carbonize the surrounding insulation and oxidize the metallic joint. This, in turn, raises the contact resistance until a catastrophic failure occurs. Note that a low contact resistance is no guarantee that a connection will not overheat.

Example 4-6:

The terminals of a circuit-breaker are bolted to a bus-bar and the contact resistance is 0.0003 Ω. If the current is 5000 A, calculate the heat dissipated.

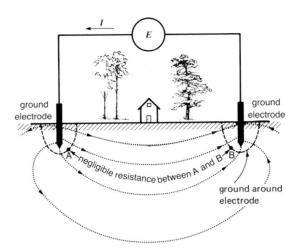

Figure 4-9
Current flow between two ground electrodes.

Solution:

The heat released is:

$$P = I^2R$$

$$= 0.0003 \times (5000)^2 = 7500 \text{ W}$$

Eq. 2-5

4.18 Resistance of the ground

The resistance of the ground is very important in electrical installations. First, for safety reasons, most electrical installations are connected to ground (earthed). Second, the ground is often used as a return conductor on some transmission lines. Unfortunately, the ground also offers a path for leakage currents, which can corrode buried metallic pipes and structures.

The resistivity of the ground ranges between 5 Ω.m and 5000 Ω.m depending upon its composition (clay, sand, granite, etc.) and the degree of moistness. Despite its high resistivity, the ground is an excellent conductor because of the enormous cross section it offers to current flow. For example, if we apply a voltage E between two electrodes driven into the ground, the resulting current flows through the entire volume of the earth, following a path similar to that shown in Fig. 4-9. As a result, the resistance between the electrodes remains quite low, even though they may be many kilometers apart. The resistance of a single electrode is typically less than 25 Ω ohms. Experience has shown that this electrode resistance is mainly concentrated in a 10-meter radius around each electrode. Beyond this circle, the resistance is negligible. Consequently, the distance between electrodes does not change the resistance between them unless the electrodes are very close.

We can reduce the resistance to a fraction of an ohm by driving the electrodes deeper into the ground or by impregnating the surrounding soil with chemicals, such as copper sulfate. When a very low ground electrode resistance is required, a steel grid, covering several square meters, can be buried in the ground.

4.19 Summary

This chapter has taught us that the resistivity of electrical materials varies over a tremendous range. Thus, whereas copper has a resistivity of 15.88×10^{-9} Ω.m at 0°C, good insulators have resistivities between 10^6 Ω.m and 10^{15} Ω.m. The resistivity of soil, sand, and stone ranges between 5 Ω.m and 5000 Ω.m, which is almost in the insulator class. Nevertheless, the earth is an excellent conductor because it offers a huge cross section to current flow.

We also discovered that the resistance of metallic conductors increases with temperature. Later on, we will find that this property enables us to calculate the temperature inside a coil.

The temperature and quality of insulation are crucially important in electric machines and devices because they determine how long the machines will last. That is why insulation systems are standardized into five temperature classes. As a rough rule of thumb, whenever these standards are exceeded, each increment of 10°C reduces the life of the equipment by half.

We learned that commercial resistors are also governed by temperature standards. In succeeding chapters we shall see many applications of these resistors to control the voltage and current of electrical equipment.

TEST YOUR KNOWLEDGE

4-1   The following metals are most commonly used as electrical conductors

a. silver and copper   b. copper and aluminum   c. iron and copper

4-2   The life expectancy of an insulator is 16 years at a temperature of 105°C. If the temperature rises to 135°C, the life expectancy will be

a. 9.3 years   b. 1.5 years   c. 4 years

4-3   A piece of writing paper that is 10 cm wide, 15 cm long and 0.05 mm thick has a dielectric strength of 8 kV/mm. If placed between two copper plates and subjected to an increasing voltage, it will break down at

a. 160 kV   b. 400 V   c. 120 kV

4-4   A round aluminum conductor has a diameter of about 0.032 inch. The corresponding AWG number is

a. No. 40   b. No. 28   c. No. 20

4-5   A No 12 copper conductor has a length of 3400 m. Its weight is:

a. 115.25 kg   b. 100.3 kg   c. 0.115 kg

4-6   A No 15 copper wire has a length of 48 ft. Using Table 4C its resistance at 25°C is

a. 155.1 mΩ   b. 155.1 Ω   c. 508.8 mΩ

4-7   A round conductor has a diameter of 0.025 inch. Its cross section is

a. 635 mil^2 b. 0.00625 cmil   c. 625 cmil

4-8   The allowable ampacity of No. 12 wire used in house wiring is limited in order to

a. prevent fires   b. prevent excessive voltage drop
c. ensure a long life for the wiring system

4-9   A resistor is color coded brown-black-red-gold. Its value and precision are

a. 102 Ω ±5 %   b. 1000 Ω ±5 %   c. 100 Ω ±5 %

QUESTIONS AND PROBLEMS

4-10   Name three insulators that can withstand temperatures above 220°C.

4-11   A transformer is insulated with class H insulation. What is the maximum allowable hot-spot temperature?

4-12   An electric motor is built using a 105°C insulation system. The motor runs in a very hot location, where the ambient temperature is far above 40°C. As a result, the motor gets too hot and breaks down every 5 months. If the motor is rewound using a 155°C insulation system, how long would you expect it to last, assuming no mechanical failure?

4-13   Calculate the approximate breakdown voltage of

a. a sheet of glass ⅛ inch thick
b. an air gap of 1 inch

4-14   A stranded aluminum cable has a cross section of 500 MCM. Express this in square millimeters.

4-15   Why do we sometimes prefer to use a No 10 stranded wire instead of a No 10 solid wire?

4-16   Without referring to the wire table, what size conductor is eight times bigger than No. 10 gauge wire?

4-17   Using Table 4C, calculate the thickness of the insulation covering a No. 15 magnet wire.

4-18   A 600 A, 250 MCM extra-flexible welding cable is composed of 6300 copper wires. What is the gauge number of the strands?

4-19   A coil made of No. 22 wire has a resistance of 400 Ω at 25°C. Using Table 4C, calculate (a) the length of the wire and (b) the weight of the coil.

4-20   Using Table 4C, and applying the rules of the AWG system, calculate the cross section and diameter of a No. 45 round wire.

4-21   A coil made of copper wire has a resistance of 400 ohms at 27°C. Calculate its resistance at (a) 0°C and (b) 105°C.

4-22   A 2-conductor feeder made of No. 4 copper wire has a length of 800 m. Calculate (a) the total line resistance at 25°C and (b) the voltage drop and line losses for a line current of 120 A.

4-23   Two small resistors have the following color code:

(a) red - red - red - silver
(b) brown - red - blue - gold

Determine their value and precision.

4-24   A rectangular aluminum busbar is ¼ inch thick, 6 inches wide, and 200 feet long. Using Table 4E, calculate its resistance at 80°C.

CHAPTER 5

Magnetism and Electromagnets

Introduction and Chapter Objectives

Most electrical devices and machines make use of magnetism to generate voltages, to develop forces, and to produce torque. In this chapter we will cover the basic principles of magnetism and magnetic materials. This includes the study of permanent magnets and magnetic domains, and how a coil produces a magnetic field. Saturation curves, reluctance, magnetic permeability, and the origin of hysteresis losses are also explained. Finally, we will offer some practical applications of electromagnets.

5.1 Magnetic polarity, N and S poles

If we tie a string around the center of a permanent magnet and suspend it from the ceiling, one end of the magnet will always point toward the geographic North Pole, the other to the geographic South Pole. The two ends of the bar magnet are therefore different. The end that points toward the north is called a *magnetic north pole* (N). Similarly, the end that points toward the south is called a *magnetic south pole* (S).

5.2 Magnetic attraction and repulsion

Suppose we have two bar magnets A and B whose magnetic north and south poles have been identified in the manner we have just explained. If we bring the N pole of magnet A near the S pole of magnet B, we discover that the poles are attracted to each other. On the other hand, if we bring the S pole of magnet A near the S pole of magnet B, the poles repel each other. By doing more experiments with magnets of different types and sizes, we always observe the same behavior. We can therefore state a basic law of magnetism: **poles of opposite polarity attract each other, and poles of the same polarity repel each other**.

The law of repulsion and attraction can explain why one end of a bar magnet always points toward the geographic North Pole. The reason is that a magnetic S pole exists near the geographic North Pole. Similarly, a magnetic N pole exists near the geographic South Pole. These magnetic poles are created by enormous electric currents that circulate within the earth's core.

Navigators formerly applied this magnetic effect by using a compass to circumnavigate the earth. A compass is a sensitive instrument that consists of a very small bar magnet pivoted on its center. One end of the compass needle always points north, and this permitted the sailors to find their way.

5.3 Lines of force; the weber (Wb)

If a compass is placed near a bar magnet, the N pole of the compass will be repelled by the N pole and attracted by the S pole of the magnet. Similarly, the S pole of the compass needle is attracted by the N pole and repelled by the S pole of the magnet. The effect of these forces causes the compass needle to point in a definite direction which depends upon where the compass is placed.

Suppose we have dozens of tiny compasses to experiment with. Starting from a point A on the bar magnet, let us string out a series of compasses so that the needles follow each other, head to tail. We discover that the needles trace a curved path that extends from A to B (Fig. 5-1). Similarly, if we start at another point such as X, we obtain a new path which ends up at point Y.

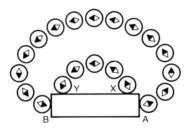

Figure 5-1
Compass needles indicate the shape of the flux line.

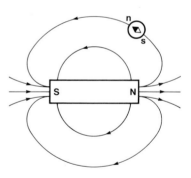

Figure 5-2
Magnetic field around a bar magnet.

Figure 5-2 shows some of the many paths that can be traced out in this way. They are called *flux lines*, or magnetic lines of force. The entire space surrounding the bar magnet is filled with these lines. The sum total of the lines is called a *magnetic field*.

Although magnetic lines of force do not actually exist, they are very useful in representing a magnetic field. In the same way that the weather forecaster on television uses arrows and curves to illustrate the direction and strength of the wind, we use lines to indicate the direction and strength of a magnetic field.

In some measurement systems, the line of force (or maxwell) is the unit of magnetic flux. However, the SI unit of magnetic flux is the weber symbol (Wb). One weber is equal to 10^8 lines; consequently, 1 line $= 10^{-8}$ Wb $= 0.01$ microwebers (μWb).

5.4 Properties of lines of force

To simplify the study of magnetism, lines of force are assumed to have a direction. The direction is that indicated by the N pole of a compass needle. As shown in Fig. 5-3, the flux lines around a bar magnet are therefore directed from the N pole to the S pole. Consequently, a flux line always comes out of a N pole, and enters by a S pole. Each flux line is assumed to pass right through the bar magnet, forming an individual, closed loop.

Michael Faraday, a 19th century British physicist, also found it useful to assume that flux lines have the following properties:

1. They behave as if they were tightly stretched elastic bands that mutually repel each other.

2. They never cross each other.

3. They always tend to follow the shortest (or easiest) path.

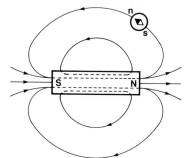

Figure 5-3
Flux lines come out of the N pole and enter by the S pole.

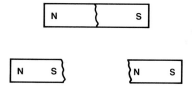

Figure 5-4
When a bar magnet is cut in two, N and S poles are created in each part.

The flux lines are not affected in any way if we bring a piece of material, such as aluminum, glass, copper, or cement, near the magnet. The flux lines pass through such materials as if they weren't there. The only substances that affect the lines are iron, nickel, cobalt, and their alloys. Such substances are called *magnetic materials*, and we shall learn more about them later.

If we cut a bar magnet in two and separate the two parts, we discover that this does not create a single N pole and a single S pole, as we might expect. Instead, each part again has a north and a south pole as shown in Fig. 5-4. The two poles continue to exist on each part, even if the parts are moved away from each other. This leads us to another important law of magnetism: for every N pole there is always a corresponding S pole.

5.5 Magnetic domains; induced and remanent magnetism

If we could look inside a block of iron, we would discover that it is composed of millions of so-called *magnetic domains*, packed side by side. Each domain is a tiny permanent magnet having one N and one S pole. The domains have irregular shapes and vary in size. However, many are big enough to be seen under an ordinary microscope. Under normal conditions, the N - S poles of the magnetic domains are oriented in every which way. The result is that the magnetic fields cancel each other, and so the block of iron has no external N or S pole.

However, if we bring the N pole of a permanent magnet near a block of iron, thousands of magnetic domains will swing around as if they were compass needles. The S-poles of these magnetic domains will point toward the N pole of the magnet. This orderly line-up of magnetic domains causes the block of iron to develop an external N and S pole, as shown in Fig. 5-5. Thus, by bringing a permanent magnet near a piece of iron, the iron itself becomes a magnet. In effect, the N pole of the permanent magnet induces a S pole in the iron. Because unlike poles attract, the block of iron will be drawn toward the magnet. This phenomenon is called *induced magnetism*.

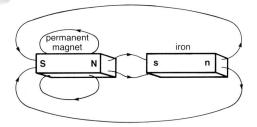

Figure 5-5
A permanent magnet induces N and S poles in a piece of iron.

The same effect takes place if we bring the S pole of the bar magnet near the block of iron. The magnetic domains will swing around 180 degrees. Consequently, the induced magnetic poles will reverse, and so the iron will again be attracted to the magnet. That is why a nail is attracted to either the N or S pole of a permanent magnet.

If the magnet is brought very close to the iron block, many more magnetic domains will line up, causing the induced magnetic poles to became stronger. Consequently, the force of attraction will become greater.

What happens when the permanent magnet is removed? Ideally, the magnetic domains will return to their random state and the induced magnetic poles will disappear. However, in practice, the magnetic domains do not all swing back to where they were before. As a result, the block of iron will have a weak N and S pole when the magnet is removed. This is known as *remanent magnetism*. In effect, the piece of iron becomes a weak permanent magnet after it has been subjected to a strong magnetic field.

Although we have chosen iron to illustrate the phenomenon of induced and remanent magnetism, the same effect takes place when any magnetic material is subjected to an external field.

5.6 Permanent magnets

Some materials possess magnetic domains that are very hard to turn around. Once they have taken a position, they tend to stay locked in place. If such a material is put in an intense magnetic field, all its magnetic domains will line up, causing all the N-poles to point in the same direction. However, when the external field is removed, the magnetic domains remain frozen in place. Consequently, the material continues to have very strong N and S poles, meaning that it has become a permanent magnet.

Powerful permanent magnets are made of alloys of iron, aluminum, cobalt, platinum, yttrium, oxygen, etc. Alnico V, for example, is a typical permanent magnet alloy made up of 51% iron, 8% aluminum, 14% nickel, 24% cobalt, and 3% copper.

Ceramic magnets form another class of permanent magnets. They are much lighter than metallic magnets and have resistivities as high as those of good insulators. Among other applications, they are used in the manufacture of door-sealing magnetic strips and magnetic tapes.

We sometimes want to demagnetize a permanent magnet. This can be done in three ways:

a) heating the magnet until it is red hot,

b) subjecting it to a powerful but gradually decreasing ac field, or

c) striking it repeatedly with sharp blows.

Each of these methods causes the magnetic domains to take a random position.

5.7 Magnetic field created by a current

If we bring a compass close to a current-carrying conductor, the needle always swings to a position at right angles to the conductor (Fig. 5-6). When the current stops flowing, the needle returns to its normal position in the earth's magnetic field. If the current is reversed, the compass needle also reverses, but it still remains at right angles to the conductor. This experiment shows that an electric current produces a magnetic field.

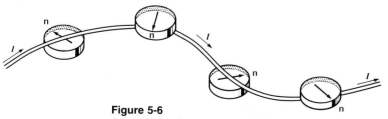

Figure 5-6
A current produces a magnetic field.

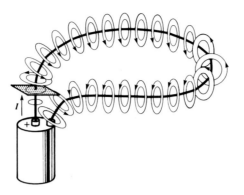

Figure 5-7
The magnetic field around a conductor is composed of circular loops.

Further experiments show that the lines of force exist in circular loops around the conductor (Fig. 5-7). The direction of the lines depends upon the direction of current flow. The so-called right-hand rule gives the relationship between the two:

If the conductor is grasped in the right hand so that the thumb points in the direction of current flow, the fingers indicate the direction of the lines of force (Fig. 5-8).

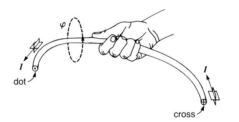

Figure 5-8
The right-hand rule indicates the direction of the flux lines around a current-carrying conductor.

Referring to Fig. 5-8, you will note that one end of the conductor bears a cross X, while the other bears a dot. The cross shows that current is flowing into the conductor and away from you. The dot indicates that current is flowing out of the conductor and toward you. The dot and cross may be thought of as being the tip and tail of a feathered arrow.

5.8 Magnetic field produced by a short coil

Figure 5-9 shows a coil having 100 turns and carrying a current of 5 A. Each turn produces a rather weak magnetic field because the current is small. However, with 100 turns, the field is 100 times stronger, and therefore many flux lines are produced. The magnetic field is similar to that produced by a disc-shaped permanent magnet (Fig. 5-10).

The coil also produces a N pole and a S pole. The N pole is where the flux comes out of the coil and the S pole is where it goes in (Fig. 5-9).

We can increase the strength of the field by raising the current above 5 A. If we double the current, the number of flux lines will double. If we increased the current

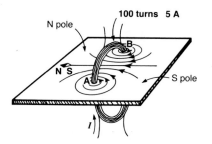

100 turns 5 A

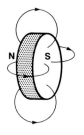

Figure 5-10
The field around a disc-shaped permanent magnet is similar to that produced by the coil in Fig. 5-9.

Figure 5-9
A circular coil produces a N and S pole.

to 50 A, the number of flux lines would increase ten times. However, a large current causes the wire to overheat, and the coil may burn out.

5.9 Magnetomotive force and flux density, the tesla (T)

The *magnetomotive force* (mmf) of a coil is equal to the number of turns it has, times the current it carries. Thus, a coil of 50 turns carrying a current of 8 A develops a mmf of $(50 \times 8) = 400$ ampere turns (400 At). The mmf of a coil is the driving force that produces the flux.

Returning to Fig. 5-9, if the mmf of the coil is increased, the number of lines of force inside the coil will increase. As a result, the lines inside the coil become more densely packed. In other words, the number of lines per square centimeter increases. This so-called *flux density* is a measure of the strength of a magnetic field. For example, in Fig. 5-2 the field is stronger near the poles because the flux density is greater there. The SI unit of flux density is the tesla, equal to one weber per square meter. The tesla (symbol T) is equivalent to 10 000 lines per square centimeter.*

5.10 Magnetic field of a long coil, electromagnets

The long coil, or solenoid, shown in Fig. 5-11 is equivalent to a series of short coils (Fig. 5-9) stacked side by side. The resulting magnetic field consists of flux lines

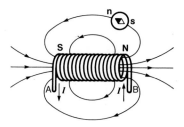

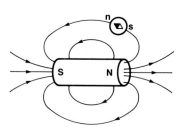

Figure 5-12
The magnetic field around the bar magnet is similar to that produced by the solenoid of Fig. 5-11.

Figure 5-11
Magnetic field created by a solenoid.

---

*   In the centimeter-gram-second system of units, the unit of flux density is the gauss, equal to 1 line per square centimeter.

that pass through the center of the coil. Each line forms a separate, closed loop. The direction of the lines inside the coil is easily found by applying the right-hand rule to one of the turns.

The magnetic field has the same shape as that produced by a bar magnet of equal size (Fig. 5-12). Furthermore, like a magnet, the coil produces a N pole at one end and a S pole at the other.

Unfortunately, the coil produces only a small amount of flux, even when the current is raised to the point at which the coil begins to overheat. But if we introduce a soft iron core inside the coil, the flux increases dramatically. The reason is that the mmf of the coil causes some of the magnetic domains in the iron to line up. The enormous flux produced by the magnetic domains adds therefore to the rather weak flux created by the coil alone. The result is a powerful electromagnet. Electromagnets produce a large flux with a moderate mmf.

Electromagnets act in a sense like variable permanent magnets because they enable us to vary the flux by simply varying the current. Furthermore, the magnetic polarity can be reversed by reversing the current flow. Because of this versatility, it is not surprising that electromagnets are found in almost every electric machine and device.

5.11 Typical applications of electromagnets

One of the most important uses of electromagnets is to create the field in dc motors and generators. In Fig. 5-13, the coils are wound on iron pole-pieces so as to produce a N and S pole. The flux lines follow the dotted path shown in the diagram. They flow across the two air gaps and return, in closed loops, by way of the iron frame.

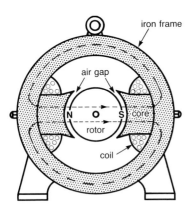

Figure 5-13
Magnetic field created by the coils in a dc generator.

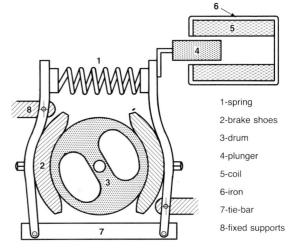

1-spring
2-brake shoes
3-drum
4-plunger
5-coil
6-iron
7-tie-bar
8-fixed supports

Figure 5-14
Electric brake to prevent an elevator motor from turning when the motor is stopped.

Figure 5-14 shows a magnetic brake for an elevator. When the elevator is not moving, a powerful spring causes brake shoes to clamp around a drum. The drum is directly coupled to the elevator motor. As soon as the motor is energized, current starts flowing in the coil. This attracts the iron plunger, causing it to move inside the coil. This releases the pressure of the brake shoes against the drum, and the elevator motor is free to turn. When the motor stops, the electromagnet is automatically de-energized, which releases the iron plunger. The brake shoes again come into action, clamping the drum and preventing further movement of the elevator.

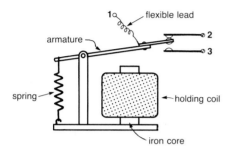
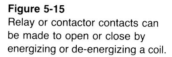

Figure 5-15
Relay or contactor contacts can be made to open or close by energizing or de-energizing a coil.

Figure 5-15 shows a magnetic contactor, which is also called a relay. When no current flows in the holding coil, contact is made between terminals 1 and 2. But when the coil is excited, a flux is produced in the iron core, which attracts the iron armature. This overcomes the tension of the spring, causing contacts 1 and 3 to close. The opening and closing of these contacts enable us to control a device from a distance.

5.12 Properties of magnetic circuits

Most electrical machines, such as motors, generators, transformers, and relays, contain coils that produce a magnetic field. The amount of flux produced depends upon the mmf of the coil, the quality of the iron, the length of the air gap, and the particular dimensions of the machine. Calculating the flux in such complicated magnetic circuits is a problem for designers. Nevertheless, we can gain some useful insights by examining the behavior of a few simple magnetic circuits. In Sections 5-13 to 5-15 we discuss three circuits: a non-magnetic core circuit, a magnetic-core circuit, and a magnetic-core circuit having an air gap.

5.13 Magnetic circuit with a non-magnetic core

Consider Fig. 5-16, in which a coil of 800 turns is wound on a non-magnetic core, such as wood. The core has a length of 1 m and a cross-section of 1 cm^2. A dc source enables us to raise the current gradually from zero to 4 A. A special fluxmeter (not shown) indicates the flux produced in the core.

When the current is 1 A, the mmf is 800 ampere-turns and the fluxmeter indicates 10 lines. If the current is doubled, the mmf becomes 1600 At and the flux

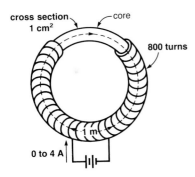

Figure 5-16
Coil wound on a non-magnetic core.

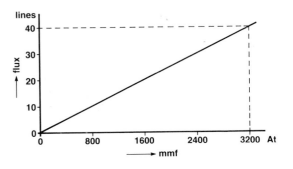

Figure 5-17
In a non-magnetic material, the flux is directly proportional to the mmf.

increases to 20 lines. Finally, with a current of 4 A, the mmf is 3200 At and the fluxmeter indicates 40 lines.

If we plot the number of lines of force versus the mmf of the coil, we obtain a straight line (Fig. 5-17). We conclude that the flux is directly proportional to the mmf when the magnetic circuit is composed of a non-magnetic material, such as wood. The same graph would be obtained if the core were made of air or cement, or even if it were a vacuum. Consequently, non-magnetic materials have the same magnetic properties.

5.14 Magnetic circuit with a magnetic core

Let us replace the wooden core of Fig. 5-16 by an iron core having the same dimensions. We again raise the current from zero to 4 A and observe the corresponding values of flux.

We discover that the same mmf produces much more flux than before. Thus, for a current of 1 A, the mmf is again 800 At, but the flux is 15 000 lines, instead of only 10 lines.

If we double the mmf to 1600 At, the flux rises to 18 000 lines. Finally, for a mmf of 3200 At, the flux is 19 000 lines. These results are plotted in Fig. 5-18. They produce an upward-sloping curve that gradually flattens as the mmf increases.

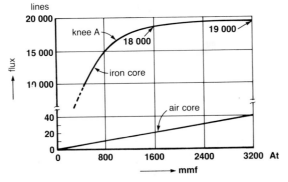

Figure 5-18
Saturation curve obtained when an iron core is used in Fig. 5-16.

This is called a *saturation curve*. More precisely, it is the saturation curve for this particular magnetic circuit.

In analyzing the saturation curve, we note that the flux increases rapidly up to the so-called "knee" of the curve (region A). Beyond the knee the flux increases only slightly, despite large increases in mmf. The reason is that most of the magnetic domains in the iron core line up for rather small values of mmf. They contribute an enormous amount of flux, and so the total flux in the core is high, even though the coil mmf is low. But as we raise the mmf, only a few more magnetic domains remain to be lined up, with the result that the flux cannot increase much more.

Eventually, when the coil mmf is high enough, all the magnetic domains will be lined up. When this point is reached, the iron core cannot produce any additional flux. It is said to be totally saturated. If we increase the mmf beyond this level, the flux will continue to rise, but the increase is then due to the coil alone. In iron, total saturation sets in at a flux density of about 2 T.

5.15 Magnetic circuit with an air gap

As a final experiment, let us cut an opening 1 mm long in the iron core of Fig. 5-16. This small air gap replaces the iron that was there before. Consequently, for a given mmf, we would expect the flux to lie somewhere between the values it had in Fig. 5-17 (non-magnetic core) and Fig. 5-18 (all-magnetic core).

If we repeat the test, we obtain a flux of 13 000 lines at 1600 At, and 18 000 lines at 3200 At. The corresponding saturation curve for this air gap circuit is shown in Fig. 5-19. Note that as the flux increases, the iron begins to saturate, and it takes a progressively larger mmf to increase the flux by a given amount. Thus, the saturation curve flattens off, just as it did when there was no air gap.

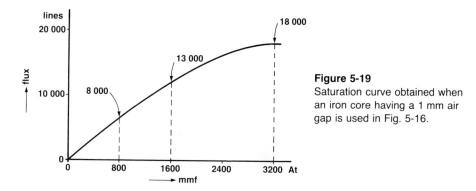

Figure 5-19
Saturation curve obtained when an iron core having a 1 mm air gap is used in Fig. 5-16.

We also note that to produce a given flux density, it takes a considerably higher mmf when the circuit contains an air gap. For example, to produce 15 000 lines, it takes 800 At with no air gap, but 2000 At with a 1 mm air gap. Thus, the longer the air gap, the more mmf that is required to produce a given flux.

5.16 Reluctance of a magnetic circuit

The flux created in a magnetic circuit depends directly upon the mmf of the coil and inversely upon the "magnetic resistance" of the circuit. The lower the magnetic resistance the higher the flux will be. For example, a closed iron core offers less resistance to the creation of magnetic flux than one having an air gap. The magnetic resistance of a circuit is called its *reluctance*. Circuits that require a large mmf to produce a given flux are said to have a high reluctance.

5.17 Relative permeability

We have seen that when the coil in Fig. 5-16 is excited, it produces far more flux in an iron core than in a non-magnetic core. The reason for this is sometimes attributed to the *permeability* of the core material. Thus, iron is considered to be much more permeable then air is. Some magnetic materials, such as permalloy, are even more permeable than iron. This means that for a given coil mmf, the flux in a sample of permalloy will exceed that in a similar sample of iron.

The *relative permeability* of a substance is the ratio of the flux produced in the substance to what it would be if the substance were air. The relative permeability is not constant, but varies with the mmf that is applied. We can therefore write:

$$\text{relative permeability} = \frac{\text{flux created in the magnetic substance at a given mmf}}{\text{flux created if the magnetic substance were replaced by air}} \qquad (5\text{-}1)$$

Iron that contains very little carbon is said to be soft. Soft iron has a high relative permeability. Cast iron contains considerable carbon and has a lower relative permeability.

Example 5-1:

Calculate the relative permeability of the iron in Fig. 5-18 at a mmf of 800 At.

Solution:

$$\text{relative permeability} = \frac{\text{flux in iron at 800 At}}{\text{flux in air at 800 At}} \qquad \text{Eq. 5-1}$$

$$= \frac{15\ 000 \text{ lines}}{10 \text{ lines}}$$

$$= 1500$$

5.18 Torque developed by a magnetic system

Motors and generators are essentially composed of a set of fixed and moveable magnets. The forces and torques developed between them are an important part of the electromechanical conversion process. In this section we examine the nature of these torques for a simple system of magnets.

Consider Fig. 5-20, in which two bar magnets A and B are fixed in space, and a third bar magnet C is free to pivot around its center 0. With C in the position shown, its respective N and S poles are attracted with considerable force to the opposite poles of the fixed magnets. However, the forces pull in opposite directions, and so they tend only to stretch magnet C.

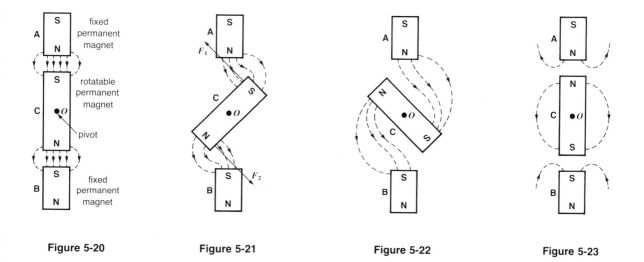

Figure 5-20   Figure 5-21   Figure 5-22   Figure 5-23

Let us rotate magnet C through an angle of 30 ° (Fig. 5-21). Because of the magnetic attraction between the S pole of C and the N pole of A, a force F_1 will be produced, acting in the direction shown. A similar force F_2 acts between the N pole of C and the S pole of B. These forces produce a twisting effect on C, tending to bring it back to its original position. Thus, magnet C is subjected to a counterclockwise (ccw) torque. The magnitude of the torque depends upon the angle of rotation. It increases progressively until the angle reaches 90°. Beyond this point (Fig. 5-22), the torque gradually decreases, eventually becoming zero at 180° (Fig. 5-23).

It is interesting to observe how the flux patterns change as the angle moves from zero to 180 degrees. In particular, you will note that the lines never cross each other.

If we rotate C beyond 180°, the torque reverses, becoming clockwise (cw). It rises to a maximum at 270° (Fig. 5-24), and then gradually falls to zero, as we come back to our starting point at 360°.

If we plot the torque versus the angle of rotation, we obtain the curve shown in Fig. 5-25. A positive torque means that the torque acting on C is trying to drive it ccw.

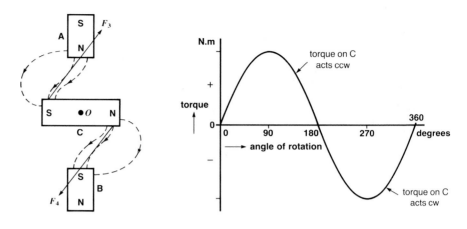

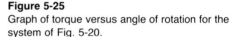

Figure 5-24
The torque on C now acts cw.

Figure 5-25
Graph of torque versus angle of rotation for the system of Fig. 5-20.

5.19 Reluctance torque

Let us replace magnet C by a soft iron bar D having exactly the same shape. At an angle of 0°, N and S poles will be induced in D but, although forces are produced, there is no torque (Fig. 5-26). If we rotate D by 30°, N and S poles are again induced, and a torque is developed that acts ccw (Fig. 5-27).

At an angle of 90° (Fig. 5-28), N and S poles are still induced, but they give rise to four equal forces F_2 that act in the directions shown. The net torque on D is therefore zero. If we rotate the bar beyond 90°, the torque reverses, becoming cw (Fig. 5-29).

A torque is exerted on the iron bar because the reluctance of the magnetic path

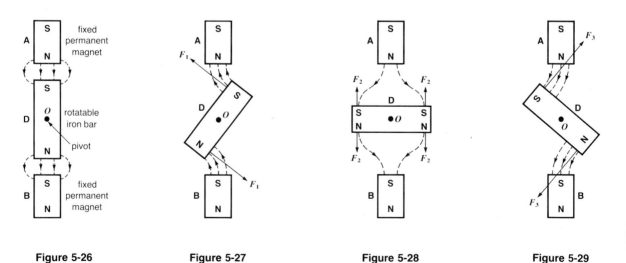

Figure 5-26   **Figure 5-27**   **Figure 5-28**   **Figure 5-29**

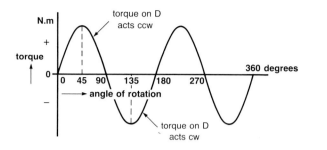

Figure 5-30
Graph of torque versus angle of rotation for the system of Fig. 5-26.

varies with the position of D. Thus, the reluctance is less in Fig. 5-26 than in Fig. 5-27 because the air gaps are shorter. Consequently, the torque is called a *reluctance torque*. It varies with the angle of rotation according to the curve of Fig. 5-30. The torque reaches a maximum at 45°, compared with 90° in the case of the permanent magnet.

Note that if D were round instead of oblong, the reluctance would be the same for every position because the air gaps would be constant. Consequently, the torque acting on D would always be zero.

5.20 Hysteresis losses

Consider Fig. 5-31, in which two coils are connected in series to a source of alternating current. Because the current reverses periodically (say at 60 Hz), the magnetic field ϕ will also reverse periodically. Let us introduce a piece of iron between the coils so that the ac flux passes through it (Fig. 5-32). The magnetic domains will tend to line up with the field, but because the field is alternating, the magnetic domains will also reverse periodically. But we recall that magnetic domains tend to stick in the position they happen to be in. Consequently, when they reverse, frictional effects take place, causing the iron to heat up. In other words, an alternating field produces losses in a magnetic material because of the back and forth motion of the magnetic domains. These losses are called *hysteresis losses*. The

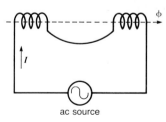

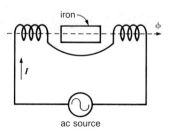

Figure 5-31
Alternating flux created by alternating current *I*.

Figure 5-32
Hysteresis losses in the iron bar are produced by the reversal of magnetic domains.

magnitude of the losses (in watts) depends directly upon the frequency of the ac flux and upon its strength.

5.21 Hysteresis losses due to rotation

Hysteresis losses also occur when a magnetic material rotates in a stationary magnetic field. Referring to Fig. 5-33, permanent magnets A and B produce a field that passes through a rotor made of iron. The rotor can be rotated by some external means, such as a hand crank. The magnetic field causes the magnetic domains in the rotor to line up. Note however, that they are oriented outward in region x and inward in region y. As we turn the rotor, region y will move to the bottom and occupy the position formerly held by region x. At the same time, region x will move to the top and occupy the position formerly held by y. Consequently, the orientation of the magnetic domains reverses 180° every time the rotor makes half a turn. If we

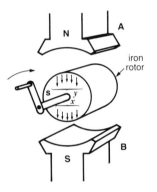

Figure 5-33
Rotation produces hysteresis losses in the rotor.

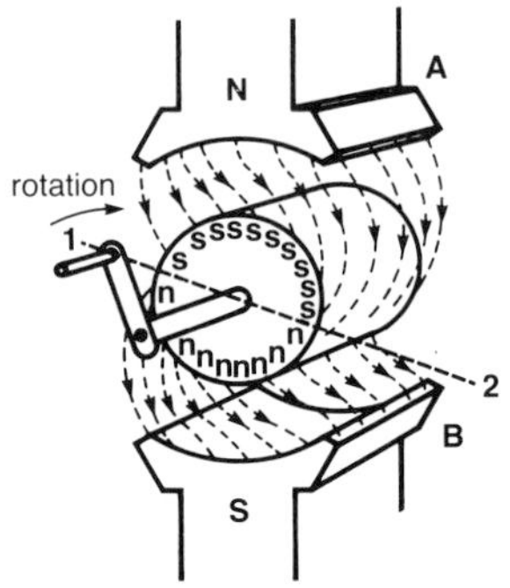

Figure 5-34
The hysteresis losses produce an opposing mechanical torque when the rotor turns.

continually rotate the rotor, the magnetic domains will reverse back and forth, just as they did in the case of an ac field. As a result, hysteresis losses will again occur, causing the rotor to heat up. The losses depend upon the speed of rotation and the strength of the magnetic field.

5.22 Torque produced by hysteresis

Suppose the rotor in Fig. 5-34 is being turned cw. The stationary N pole continually induces s poles in the rotor, and these s poles will exist until they come under the influence of the stationary S pole. The magnetic domains then reverse, and n poles are induced, as shown. These n poles also continue to exist as the rotor turns until they come under the influence of the stationary N pole. The resulting nn——n and ss——s poles are distributed around the rotor, as shown in Fig. 5-34. They are skewed, or twisted, with respect to the N, S poles of magnets A and B.

As a result, a ccw torque is exerted on the rotor. We must therefore do mechanical work to turn the rotor cw. It is precisely this mechanical work that furnishes the energy needed to reverse the magnetic domains. Thus, hysteresis losses in a rotor always produce a torque, or drag, that tends to oppose the rotation of the rotor.

5.23 Summary

In this chapter, we have covered all the basic principles of magnetism and electromagnets. We have learned that a current produces a magnetic field and that this field can be strengthened by shaping conductors into coils. The amount of flux can be increased even more by using magnetic materials inside the coil.

The theory of magnetic domains is very useful because it helps explain such phenomena as induced magnetism, residual magnetism, magnetic saturation, and hysteresis losses.

We were also able to draw a comparison between magnetic circuits and electrical circuits. Thus, the mmf of a coil is the driving force that produces a magnetic flux ϕ, in the same way as an emf produces a current I. Similarly, the reluctance of a magnetic circuit represents its "opposition" to the production of magnetic flux, in the same way as resistance represents the opposition to current flow.

The torque developed by a system of magnets shows how magnetic principles can produce mechanical effects. In the next chapter, we will find that forces and torques can also be produced when current-carrying conductors are placed in a magnetic field. These forces and torques are later put to practical use in the design of motors, generators, and other electrical devices.

TEST YOUR KNOWLEDGE

5-1   The SI unit of magnetic flux is the

a. maxwell   b. weber   c. tesla

5-2   The unit of magnetic flux density is the

a. Wb/m^2 b. Tesla (T)   c. tesla (T)

5-3   The shape of a coil having 60 turns and carrying a current of 17 A has an effect on the mmf it develops

☑ true   ☐ false

5-4   Flux lines cannot exist in a vacuum.

☐ true   ☐ false

5-5   A coil of 5000 turns carrying 1 mA produces a mmf of

a. 5 At   b. 5 Wb   c. 500 At

5-6   In Fig. 5-21, the flux lines within the magnets pass from the N to the S pole.

☑ true   ☐ false

5-7   A compass needle is made of

a. soft iron   b. a permanent magnet material
c. a permanent magnetic material that has
been magnetized

5-8   Hysteresis losses can be produced in copper.

☑ true   ☐ false

QUESTIONS AND PROBLEMS

5-9   Name the properties of lines of force.

5-10   How can you determine the magnetic polarity of a magnet?

5-11   The earth's magnetic field in the state of New York is directed toward the Arctic Circle. Explain.

5-12   Using the right-hand rule, determine the magnetic polarity of end A in the electromagnet of Fig. 5-35.

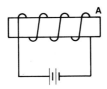

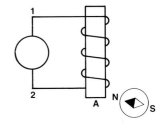

Figure 5-35
See Problem 5-12.

Figure 5-36
See Problem 5-14.

5-13   A coil having 250 turns is connected to a 50 V dc source. If the coil resistance is 10 Ω, calculate the mmf developed by the coil.

5-14   A compass needle is oriented toward an electromagnet, as shown in Fig. 5-36. Determine the direction of current flow and the electric polarity (+) (−) of terminal 1.

5-15   In Fig. 5-37, an electromagnet having an iron core is placed next to a permanent magnet.

    a)   Will there be attraction or repulsion between them?

    b)   Will there be attraction or repulsion if the terminals of the coil are reversed?

    c)   Will there be attraction or repulsion if the coil is disconnected from the source?

Figure 5-37
See Problem 5-15.

Figure 5-38
See Problem 5-16.

5-16   The electromagnets in Fig. 5-38 are connected to a 600 V source. Coil A has 2000 turns, and a resistance of 50 Ω. Coil B has 800 turns and a resistance of 70 Ω. Calculate a) the mmf produced by each coil and b) the power dissipated in each coil.

    c)   If the coils have the same size, which one will get hotter?

    d)   Will the magnets attract or repell each other?

    e)   What happens to the force of attraction (or repulsion) if the battery connections are reversed?

5-17   In Fig. 5-35, why does the mmf of the coil decrease as the coil heats up, even though the battery voltage is fixed?

5-18   A flux of 1.2 mWb exists in a magnet having a cross section of 30 cm^2. Calculate the flux density.

CHAPTER 6

Electromagnetic Forces

Introduction and Chapter Objectives

The operation of motors, generators, transformers, and many other electric devices is based upon two basic laws.

The first law states that if a current-carrying conductor is placed in a magnetic field, it will be subjected to a mechanical force.

The second law states that whenever the flux inside a coil varies, a voltage is induced across its terminals.

In this chapter we cover the first law, along with some of its applications. The second law, known as Faraday's law of electromagnetic induction, will be covered in Chapter 8.

6.1 Force on a straight conductor

In Chapter 5 we saw that a current-carrying conductor is surrounded by a magnetic field. If the current flows into the page, the lines of force appear as shown in Fig. 6-1. We also know the shape of the magnetic field between the N and S poles of two permanent magnets (Fig. 6-2). If the conductor is placed in this magnetic field

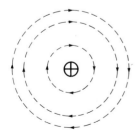

Figure 6-1
Magnetic field around a conductor when the current flows into the page.

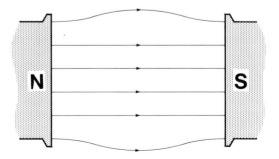

Figure 6-2
Magnetic field between the poles of a permanent magnet.

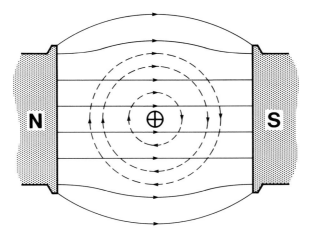

Figure 6-3
Superposition of the magnetic fields.

(Fig. 6-3), we observe that

1. The conductor is subjected to a force that tends to move it downward.
2. If we reverse the direction of current flow, the force acts upward.

To understand how this force is produced, let us examine the resulting magnetic field when the conductor is between the two poles. It cannot have the shape shown in Fig. 6-3 because lines of flux never cross each other. However, in the space above the conductor we note that the circular flux lines created by the current point in the same direction as the N-S magnetic field. Consequently, the field above the conductor is strengthened. By the same reasoning, the field below the conductor is weakened because the lines of force act in opposite directions. Furthermore, the lines of force produced by the N and S poles must remain the same whether the conductor is there or not. The net result is that the number of lines above the conductor is greater than the number below. The magnetic field must therefore have the shape illustrated in Fig. 6-4.

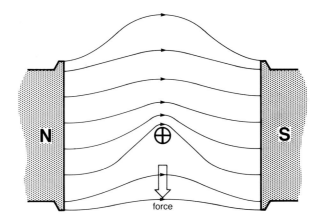

force

Figure 6-4
Resulting shape of the magnetic field, and the corresponding force on the conductor.

But we recall that lines of force behave like tight elastic bands that repel each other. When they are concentrated above the conductor, as they are in Fig. 6-4, the force of repulsion between them forces the conductor downward.

This method of determining the direction of the force can be used in other situations. In Fig. 6-5, for example, the force on the conductor acts toward the right.

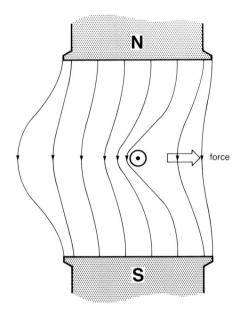

Figure 6-5
The direction of the force depends upon the direction of the magnetic field and the direction of current flow.

6.2 Magnitude of the force

The magnitude of the force on a conductor depends upon:

a) the current carried by the conductor: the greater the current the greater the force.
b) the flux density of the magnetic field in which the conductor is located: the higher the flux density, the greater the force.
c) the length of the conductor that is immersed in the magnetic field: the longer it is, the greater the force.
d) the way the conductor is oriented with respect to the magnetic field. The force is greatest when the conductor is perpendicular to the field (Fig. 6-6). It is zero when the conductor is parallel to the field (Fig. 6-7). Between those two extremes, the force has intermediate values. In practice, the conductors and the magnetic field are arranged to be perpendicular to each other.

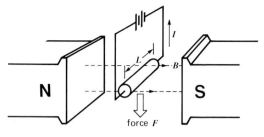

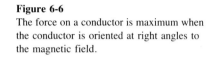

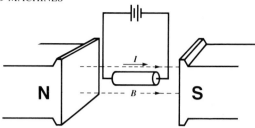

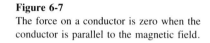

Figure 6-6
The force on a conductor is maximum when the conductor is oriented at right angles to the magnetic field.

Figure 6-7
The force on a conductor is zero when the conductor is parallel to the magnetic field.

The maximum value of the force is given by the equation:

$$F = BLI \qquad (6\text{-}1)$$

where

$F =$ force, in newtons (N)

$B =$ flux density of the magnetic field, in teslas (T)

$L =$ length of the conductor in the field, in meters (m)

$I =$ current in the conductor, in amperes (A)

Example 6-1:

A conductor 4 m long has 3 m of its length immersed in a magnetic field. The flux density of the field is 0.5 T, and the conductor carries a current of 200 A. Calculate the force acting on the conductor if it is perpendicular to the field, as shown in Fig. 6-6.

Solution:

$F = BLI$ Eq. 6-1
$\quad = 0.5 \times 3 \times 200$
$\quad = 300$ N, or about $300/4.448 = 67$ lbf.

6.3 Force between two conductors

When two current-carrying conductors are placed side by side, they either attract or repel each other. If the currents flow in the same direction, the conductors are attracted, otherwise they are repelled. These forces are a direct consequence of the basic law we have just described.

Consider, for example, Fig. 6-8, in which two parallel conductors, A and B, carry currents that flow in the same direction (into the page). Conductor B produces its own circular magnetic field, but it is also immersed in the field created by

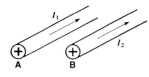

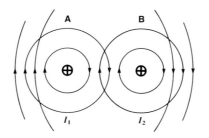

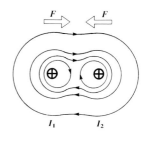

Figure 6-8
Two conductors carrying currents I_1 and I_2 that flow in the same direction.

Figure 6-9
Superposition of the magnetic fields created by currents I_1 and I_2.

Figure 6-10
Resulting magnetic field when currents flow into the page. The conductors are attracted.

conductor A. Referring to Fig. 6-9, it is clear that the resulting magnetic field to the right of B is strengthened. On the other hand, the magnetic field to the left is weakened because the fields of A and B are opposed in this region. The increased flux density on the right-hand side of B produces a force that pushes B to the left.

By the same reasoning, a force acts on A, but it is directed toward the right. It follows that conductors A and B are attracted to each other. The complete magnetic field around both conductors is shown in Fig. 6-10.

If the conductors carry currents that flow in opposite directions, as shown in Fig. 6-11, the conductors will repel each other. The corresponding shape of the magnetic field is given in Fig. 6-12.

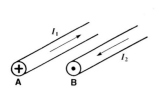

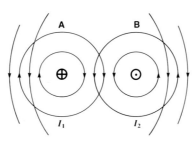

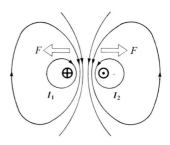

Figure 6-11
Two conductors carrying currents I_1 and I_2 that flow in opposite directions, and superposition of the magnetic field created by each.

Figure 6-12
Resulting magnetic field when currents flow in opposite directions. The conductors are repelled.

For a given length and spacing, the force between two parallel conductors depends upon the product of the currents they carry. Under normal conditions, the force is quite small. But if a short-circuit occurs, the line currents may become 100 times larger than usual. This produces a force that is 100 × 100, or ten thousand, times greater than normal. Heavy copper busbars have been bent by the action of these powerful short-circuit forces, which may reach peaks of several tons. In substations, special bracing is needed to prevent busbars from becoming deformed or being pulled off their supports.

6.4 Forces acting on a coil

The turns of a coil can be looked upon as parallel conductors that carry a current flowing in the same direction. As a result, the turns are attracted to each other, and the entire coil tends to become compressed under the action of these forces (Fig. 6-13).

At the same time, each turn tends to balloon outward when it carries a current. To understand why this takes place, let us consider a single turn on the coil (Fig. 6-14). We can imagine the turn consists of short sections, each of which is a short conductor. Sections that are diametrically opposite, such as **a** and **b**, carry currents that flow in opposite directions. Based on Fig. 6-12, these sections repel each other, and so the turn tends to expand.

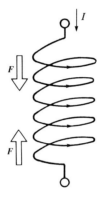

Figure 6-13
The turns of a coil are attracted to each other, tending to compress the coil.

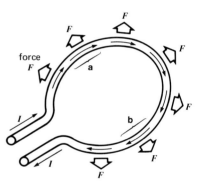

Figure 6-14
Opposite segments of a turn repel each other, and so the turn tends to expand.

Normally, the forces tending to compress a coil while causing it to balloon outward are quite small. But in special cases, such as a short-circuit on a big transformer, the forces may become so great that heavy coils are torn apart.

6.5 Blow-out coils

Some power circuits carry currents of several hundred amperes. Such circuits are often opened and closed by fast-acting switches called *circuit breakers*. When a circuit breaker opens, an intense arc continues to bridge the opening contacts. To extinguish the arc as quickly as possible, it is driven against a set of insulated plates so as to lengthen it, cool it, and chop it into smaller pieces. A jet of compressed air could do this and effectively blow out the arc. However, a pair of coils placed at right angles to the arc can accomplish the same result (Fig. 6-15). These so-called *blow-out coils* carry the current that flows in the circuit breaker. Furthermore, the magnetic field they create is arranged to intercept the arc. The arc is therefore subjected to a force, and because it has almost negligible mass, the force drives the arc at terrific speed against the insulated plates, thus extinguishing it.

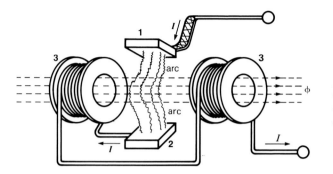

Figure 6-15
Blow-out coils can quickly extinguish an electric arc. As movable contact 1 separates from fixed contact 2, the arc is in the field created by blowout coils 3.

6.6 Torque produced by a rectangular coil

The torque exerted by an electric motor is produced by current-carrying coils that are located in a magnetic field. To show how such a torque is developed, consider a single coil ABCD placed between the poles of a magnet (Fig. 6-16). The coil carries a current, and consequently a force F_1 acts on side AB (Fig. 6-17). An equal force F_2 acts on side CD. No forces act on sides BC and AD because they lie outside the magnetic field.

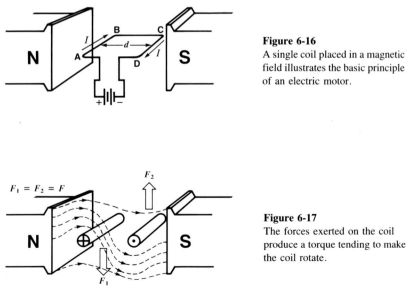

Figure 6-16
A single coil placed in a magnetic field illustrates the basic principle of an electric motor.

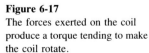

Figure 6-17
The forces exerted on the coil produce a torque tending to make the coil rotate.

Forces F_1 and F_2 act in opposite directions because the current flows in opposite directions in the two "active" sides of the coil. Consequently, the coil is subjected to a torque. The magnitude of the torque is given by

$$T = Fd \qquad (6\text{-}2)$$

where

T = torque, in newton meters (N.m)

F = force on one coil side, in newtons (N)

d = width of coil, in meters (m)

Example 6-2:

The rectangular coil in Fig. 6-16 has 40 turns and carries a current of 90 A. Coil side AB is 100 cm long and BC is 60 cm long. If the magnetic field created by the magnets has a flux density of 0.5 T, calculate the torque developed by the coil.

Solution:

1. The force exerted on each conductor is

$$
\begin{aligned}
F &= BLI & \text{Eq. 6-1}\\
&= 0.5 \times 0.1 \times 90\\
&= 4.5 \text{ newtons} = 4.5 \text{ N}
\end{aligned}
$$

2. The total force acting on side AB is

$$F = 40 \text{ conductors} \times 4.5 \text{ N} = 180 \text{ N}.$$

3. The torque produced by the coil is

$$T = Fd = 180 \times 0.6 = 108 \text{ newton meters} = 108 \text{ N.m} \qquad \text{Eq. 6-2}$$

6.7 Summary

A force will act on a current-carrying conductor if it is placed at right angles to a magnetic field. The force is directly proportional to the current and the flux density of the field in which the conductor is placed. Such electromagnetic forces can be very powerful.

In Chapter 5, we saw how magnets can produce forces and torques because of the attraction and repulsion of their magnetic poles. In this chapter, we learned that forces are also produced when a conductor is placed in a magnetic field. The force on a single conductor is given by the equation $F = BLI$.

We examined several cases in which forces are developed, including the force on an electric arc. We saw that the forces can produce a torque, such as when a rectangular coil is placed in a magnetic field.

Finally, one of the advantages of $F = BLI$ is that it enables us to look at the forces exerted on individual conductors. As we will see, this becomes particularly important in the study of dc and ac machines.

TEST YOUR KNOWLEDGE

6-1   If the N and S poles in Fig. 6-4 are interchanged, the force on the conductor will act upward.

    ☐ true   ☐ false

6-2   The flux density around the conductor of Fig. 6-1 gets weaker as the distance from the conductor increases.

    ☐ true   ☐ false

6-3   In Fig. 6-8, I_1 is 300 A and I_2 is 400 A, and the force of attraction is 10 N. If I_1 increases to 1200 A, the force becomes

  a. 160 N   b. 40 N   c. 22.86 N

6-4   In Fig. 6-10, if I_1 and I_2 both reverse, the conductor will be attracted, but the flux lines will reverse.

    ☐ true   ☐ false

6-5   In Fig. 6-13, if current I reverses, the forces F reverse.

    ☐ true   ☐ false

6-6   In Fig. 6-16, if the battery terminals are reversed, the coil will tend to rotate clockwise.

    ☐ true   ☐ false

6-7   In Fig. 6-13, if the current increases 20 times, the compression forces F will increase

  a. 4.472 times   b. 20 times   c. 400 times

QUESTIONS AND PROBLEMS

6-8   When a long coil carries a large current, the coil tends to get shorter. Explain.

6-9   Draw the resultant field between the magnets of Fig. 6-18 and show the direction of the force on the conductor.

6-10   Determine the magnetic polarity of the poles A and B in Fig. 6-19, knowing that the force on the conductor acts downward.

6-11   In which direction will conductor ab in Fig. 6-20 move?

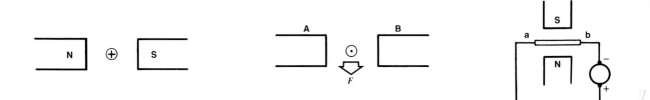

Figure 6-18
See Problem 6-9.

Figure 6-19
See Problem 6-10.

Figure 6-20
See Problem 6-11.

6-12   The electromagnet in Fig. 6-21 is connected to a dry cell having the polarity shown. In which direction does the conductor between the poles tend to move?

6-13   In Fig. 6-22, an electromagnet and rectangular coil are connected in series to a dc source having the polarity shown. (a) Will the coil turn cw or ccw? (b) If the polarity of the source is reversed, in which direction will the coil tend to rotate?

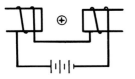

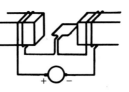

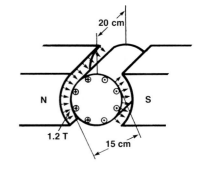

Figure 6-21
See Problem 6-12.

Figure 6-22
See Problem 6-13.

Figure 6-23
See Problem 6-15.

6-14   A conductor carrying 2000 A is placed in a field having a flux density of 0.5 T. Calculate the force exerted per inch of length.

6-15   Figure 6-23 is the schematic diagram of a motor. The rotor contains 80 conductors, each of which carries a current of 50 A. The poles produce a flux density of 1.2 T. Calculate (a) the force on each conductor and (b) the torque developed by the motor in newton meters, and in lbf.ft.

6-16   If the conductor in Fig. 6-3 carries an alternating current having a frequency of 60 Hz, the conductor will not move out of the field. Explain.

CHAPTER 7

AC Voltages and Currents

Introduction and Chapter Objectives

So far, we have used voltages and currents in simple dc circuits. However, most commercial circuits and machines carry alternating voltages and currents. In such circuits, the magnitude of the voltage and its polarity are constantly changing. Furthermore, the magnitude of the current, as well as its direction, are constantly changing. To show this alternating property, the voltages and currents are represented in a special way.

In this chapter, we will represent these voltages and currents by graphs, by phasors, and also in trigonometric form. We will also discuss the important concept of rate of change of a voltage or current. These simple rules and conventions are essential to an understanding of circuits and machines.

This chapter would be appropriately placed just before Chapter 13 on ac circuits. However, the chapters on inductors, capacitors, and electromagnetic induction also require a knowledge of the notation and concepts we are about to explain.

7.1 Positive and negative voltages

A voltage has two basic properties: a magnitude and a polarity. We are often inclined to emphasize its magnitude, but the polarity of a voltage is just as important. In this section we give a simple way to indicate both properties.

Consider Fig. 7-1, in which a generator produces 100 V, and terminal A is (+) and terminal B is (−). It is important to understand that the polarity of A is only positive with respect to B. By itself, terminal A has no polarity. Similarly, the polarity of terminal B is only negative with respect to that of terminal A; by itself, terminal B has no polarity. Using these facts, we now describe the generator voltage as either E_{AB} or E_{BA}.

The symbol E_{AB} stands for two things: (1) the magnitude of the voltage between terminals A and B, and (2) the polarity of A with respect to B.

Similarly, the symbol E_{BA} stands for the magnitude of the voltage between terminals A and B, and the polarity of B with respect to A.

Figure 7-1
Terminal A is positive with respect to terminal B. Similarly, terminal B is negative with respect to terminal A.

Figure 7-2
In reference to this figure, it is known that $E_{21} = -300$ V.

Figure 7-3
This figure shows the actual voltage and polarities referred to in Fig. 7-2.

Referring to Fig. 7-1, and recognizing that A is positive ($+$) with respect to B, we can write:

$$E_{AB} = +100 \text{ V}$$

Because B is negative ($-$) with respect to A, we could equally well write:

$$E_{BA} = -100 \text{ V}$$

Thus, $E_{AB} = +100$ V and $E_{BA} = -100$ V convey exactly the same information.

As a further example, suppose that in Fig. 7-2, $E_{21} = -300$ V. This tells us that (1) the voltage between the terminals is 300 V, and (2) terminal 2 is negative with respect to terminal 1. Figure 7-3 shows the actual condition.

This way of designating voltages is known as the double-subscript notation.

7.2 Graph of a variable voltage

The generator G in Fig. 7-4 produces an alternating voltage E_{12} whose magnitude and polarity change with time. The output voltage is represented graphically over a 30-second period in Fig. 7-5. To interpret this graph, let us take a few representative points. We select t = 0, 3, 9, and 17 seconds. Table 7A lists the corresponding values of E_{12}, which are simply read off the graph. The third column gives the meaning of these readings. For even greater clarity, the actual generator voltages and polarities are shown on the graph.

7.3 Rate of change of voltage

So far, we have considered the magnitude of a voltage and its polarity. However, a third property of a voltage is also important. It is the rate at which the voltage is changing. Thus, to completely describe a voltage, we must know three things: its magnitude, its polarity, and its rate of change.

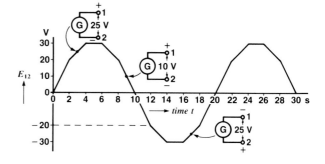

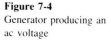

Figure 7-4
Generator producing an
ac voltage

Figure 7-5
Graph of voltage generated by G in Figure 7-4. The instantaneous
voltages and polarities are shown at t = 3, 9, and 17 s.

TABLE 7A   INTERPRETING AN AC VOLTAGE

Instant	Voltage	Interpretation
t = 0	$E_{12} = 0$	no voltage exists between the terminals
t = 3 s	$E_{12} = +25$ V	terminal 1 is (+) and voltage is 25 V
t = 9 s	$E_{12} = +10$ V	terminal 1 is (+) and voltage is 10 V
t = 17 s	$E_{12} = -25$ V	terminal 1 is (−) and voltage is 25 V

Referring to Fig. 7-5, we note that the voltage increases from zero to $+20$ V during the interval between 0 and 2 s. It continues to increase from $+20$ V to $+30$ V between 2 s and 4 s. The rate of change of voltage is said to be *positive* during these two intervals. However, between 4 s and 6 s, the voltage does not change at all; it remains fixed at $+30$ V. Its rate of change during this interval is therefore zero.

From 6 s to 8 s, the voltage decreases from $+30$ V to $+20$ V, and so the rate of change is negative. Then, between 10 s and 12 s, the voltage increases negatively from zero to -20 V. This again represents a negative rate of change. We can now generalize as follows:

a) When a voltage-time curve slopes upward, the rate of change of voltage is positive.

b) When the voltage-time curve slopes downward, the rate of change of voltage is negative.

c) The steeper the slope, the greater is the rate of change. Thus, the rate of change during the interval from 0 to 2 s is greater than that between 2 s and 4 s.

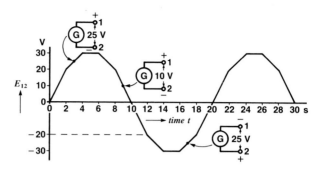

Figure 7-5

d) When the voltage-time curve slopes neither upward nor downward, the rate of change of voltage is zero. The slope of the curve is then horizontal.

e) Strange as it may seem, a voltage that is zero can still have a high rate of change. Thus, at $t = 10$ s, the voltage is momentarily zero, but it nevertheless has a negative rate of change because the curve slopes downward. Again, at $t = 20$ s, the voltage is zero but the rate of change is positive because the curve slopes upward.

Using these rules, we see that the rate of change of voltage is negative between 6 s and 14 s. Between 14 s and 16 s the rate of change is zero. Finally, between 16 s and 24 s, the rate of change is positive.

It is now clear that a positive voltage can have either a positive or negative rate of change. It depends upon whether the voltage is in the process of increasing or decreasing.

Example 7-1:

In Fig. 7-6, determine whether the rate of change of voltage is positive or negative at instant t_a, t_b, t_c and t_d. At what instants are the rates of change zero?

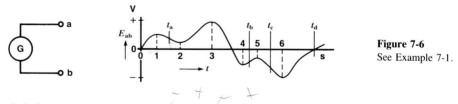

Figure 7-6
See Example 7-1.

Solution:

1. At instants t_a, t_b, t_c and t_d, the rate of change is respectively $(-)$, $(+)$, $(-)$, and $(+)$.

2. The rate of change is zero at every instant at which the slope of the curve is horizontal. This condition occurs at instants 1, 2, 3, 4, 5, and 6.

7.4 Calculating the rate of change of voltage

The rate of change of voltage is represented by the symbol $\Delta E/\Delta t$ which reads "delta *E* over delta *t*", or simply "delta *E*, delta *t*". ΔE is the change in voltage that occurs during an interval Δt. The interval can have any value — long or short — as long as the slope of the voltage-time curve is reasonably constant during the interval.

Returning to Fig. 7-5, let us calculate the rate of change of voltage between zero and 2 s. The slope of the curve does not change during this 2-second interval, consequently, the rate of change is the same at every instant. The voltage increases by 20 V during 2 s, and so the rate of change is:

$$\frac{\Delta E}{\Delta t} = \frac{\text{change in voltage}}{\text{time interval}} = \frac{20 \text{ V}}{2 \text{ s}} = 20 \text{ V/s}$$

Next, consider the interval between 16 s and 18 s. The voltage changes from -30 V to -20 V, which represents a change of 10 V. Thus,

$$\frac{\Delta E}{\Delta t} = \frac{\text{change in voltage}}{\text{time}} = \frac{10 \text{ V}}{2 \text{ s}} = 5 \text{ V/s}$$

The rate of change of voltage is $+5$ V/s during this interval because the voltage curve slopes upward.

7.5 Positive and negative currents

A current, like a voltage, has two basic properties: a magnitude and a direction. We can use positive and negative signs to indicate the direction of current flow. For example, the current in a resistor (Fig. 7-7) may flow from X to Y or from Y to X. If one of these directions is considered to be positive ($+$), the other is negative ($-$).

Figure 7-7
Current may flow from X to Y or from Y to X.

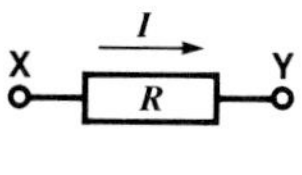

Figure 7-8
The arrow shows the direction of current that is arbitrarily considered to be positive.

The positive direction is shown arbitrarily by means of an arrow (Fig. 7-8). Thus, if a current of 2 A actually flows from X to Y, it flows in the positive direction and is designated by the symbol $+2$ A. Conversely, if the current actually flows from Y to X (direction opposite to that of the arrow), it is designated by the symbol -2 A.

A current whose direction is continually changing is called an *alternating current*, or ac current.

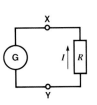

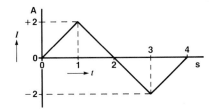

Figure 7-9
See Example 7-2.

Figure 7-10
See Example 7-2.

Example 7-2:

The circuit of Fig. 7-9 shows a current *I* flowing in a resistor R. The current varies according to the graph in Fig. 7-10. Interpret the physical meaning of this graph.

Solution:

According to the graph, the current increases from zero to $+2$ A during the interval from 0 to 1 s. Because the current is positive, it is actually flowing from Y to X in the resistor (direction of the arrow). During the interval from 1 s to 2 s, the current decreases from $+2$ A to zero, but it still circulates from Y to X in the resistor.

Between 2 s and 3 s, the current increases from zero to -2 A. Because the current is negative, it flows in a direction opposite to the arrow, that is, from X to Y in the resistor.

This example shows that a graph must always be associated with a corresponding circuit diagram in order to determine the actual direction of current flow.

7.6 Rate of change of current

The rate of change of current is calculated the same way as the rate of change of voltage. Thus, in Fig. 7-10, the current changes at a rate of $+2$ A/s in the interval between 0 and 1 s. Then, between 1 s and 3 s, the rate of change is -2 A/s.

7.7 Properties of a sinusoidal waveshape

In Chapter 1, we covered the basic principles of trigonometry. We also saw that a sine wave can be generated by a revolving rod. The reader should review these principles (covered in Sections 1-16 to 1-24) because they form the basis for describing commercial ac voltages and currents. Most alternating voltages and currents are sinusoidal, that is, they vary according to the properties of a sine wave.

Figure 7-11 shows a typical ac current that varies sinusoidally. It flows in a resistor R. The current reaches peak values of $+20$ A and -20 A at regular intervals of time. The interval between successive positive peaks is called a *cycle*. A cycle is also equal to the interval between two successive currents having the same magnitude, the same sign and the same rate of change. See, for example, instants t_a and t_b. In Fig. 7-11, the duration of one cycle is 1 s.

Figure 7-12 shows a sinusoidal voltage having a peak value of 170 V. It is

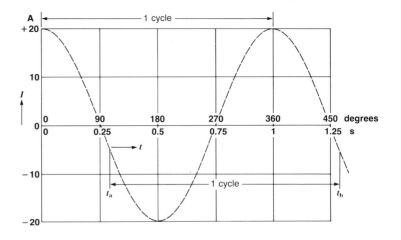

Figure 7-11
Graph of a sinusoidal ac current having a peak value of 20 A and a frequency of 1 Hz.

produced by a source G. As in the case of an alternating current, the interval between two voltages having the same magnitude, the same polarity and the same rate of change is called a cycle. Thus, the voltage completes one cycle between instants t_1 and t_2. However, it is always easier to measure the duration of one cycle by using two successive positive peaks. In this figure, the duration of one cycle is 1/60 s.

The *frequency* of an alternating current or voltage is equal to the number of cycles in one second. In Fig. 7-11, the frequency of the current is 1 cycle per second, or 1 hertz (symbol Hz). On the other hand, the frequency of the voltage in Fig. 7-12 is 60 Hz because one cycle is completed every 1/60 of a second.

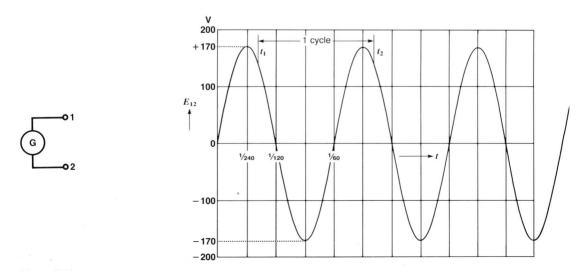

Figure 7-12
Graph of a sinusoidal ac voltage having a peak value of 170 V and a frequency of 60 Hz.

7.8 Rate of change of a sinusoidal voltage

Referring to Fig. 7-12, it is obvious that the rate of change of voltage varies from one instant to the next. Thus, it is momentarily zero whenever the voltage is at its positive or negative peak. However, the rate of change is maximum when the voltage is zero because at these moments the curve is steepest. This maximum rate of change is given by the equation:

$$\left(\frac{\Delta E}{\Delta t}\right)_{max} = 2\pi f E_m \tag{7-1}$$

where

$(\Delta E/\Delta t)_{max}$ = maximum rate of change of voltage, in volts per second (V/s)

E_m = peak value of the sinusoidal voltage, in volts (V)

f = frequency in hertz (Hz)

2π = constant to take care of units (approximate value = 6.28)

The maximum rate of change of the voltage in Fig. 7-12 is therefore

$$2\pi f E_m = 6.28 \times 170 \times 60 = 64\ 056 \text{ V/s}.$$

This appears to be a very fast rate of change, but compared with other industrial voltages it is quite moderate. For example, on some high-voltage transmission lines the maximum rate of change is as great as 3 million volts per second. Later, we will see that such high rates of change produce important effects in electric circuits.

7.9 Rate of change of a sinusoidal current

When a current varies sinusoidally, its maximum rate of change also occurs when the current is passing through zero. It is given by the equation:

$$\left(\frac{\Delta I}{\Delta t}\right)_{max} = 2\pi f I_m \tag{7-2}$$

where

$(\Delta I/\Delta t)_{max}$ = maximum rate of change of current, in amperes per second (A/s)

I_m = peak value of the sinusoidal current, in amperes (A)

f = frequency in hertz (Hz)

2π = constant (approximate value = 6.28)

Going back to Fig. 7-11, the maximum rate of change of current is

$2\pi f I_m = 2 \times 3.14 \times 1 \times 20 = 125.6$ amperes per second $= 125.6$ A/s

7.10 Representing sinusoidal voltages and currents by phasors

The sine waves of Figs. 7-11 and 7-12 are similar to the sine wave generated by the revolving rod of Fig. 1-23, Chapter 1. This similarity enables us to represent an ac voltage in a very simple way. In effect, just as a revolving rod can generate a sine wave, we can imagine that a given sine wave is due to a revolving rod. In the case of a sinusoidal voltage, the imaginary "rod" has a length equal to the peak value of the voltage, and its speed of rotation is equal to the frequency.

The voltage in Fig. 7-12 has a peak value of 170 V and a frequency of 60 Hz. It can therefore be represented by the slender rod OA, having a length of say 170 mm and rotating at a speed of 60 revolutions per second (Fig. 7-13). We show the rod in three different positions (0A, 0A′ and 0A″) to illustrate the correspondence between it and the sine wave. However, we really only need to show its position at time

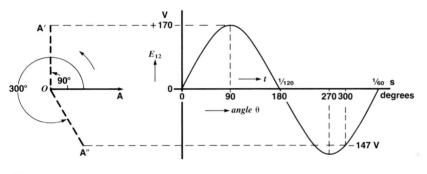

Figure 7-13
Phasor OA revolving at 60 revolutions per second can represent the ac voltage of Fig. 7-12.

$t = 0$. Thus, the single horizontal "rod" OA represents the sine wave completely.

This dramatic simplification has far-reaching effects in all ac circuit theory, and we will use it many times.

Because of its importance, this "rod" is called a *phasor*. The phasor is a straight line that carries an arrowhead at the end opposite to the point of rotation.

Alternating currents can also be represented by phasors. Thus, the sine wave of Fig. 7-11 can be represented by phasor OB (Fig. 7-14). In this case, the phasor is drawn in the vertical position because at $t = 0$ the current is at its maximum positive value (Fig. 7-11). The length of phasor OB may be 20 mm or any other convenient length that represents a current of 20 A.

7.11 Degree scale versus time scale

Because a sinusoidal voltage or current can be represented by a revolving phasor, it is usually more convenient to draw the sine wave with the horizontal axis calibrated

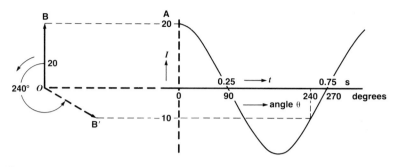

Figure 7-14
Phasor OB revolving at 1 revolution per second can represent the ac current of Fig. 7-11.

in degrees of rotation rather than in seconds or milliseconds. This angular scale is used in Figs. 7-13 and 7-14. To avoid confusion with mechanical degrees, the angles in sine waves and phasor diagrams are called *electrical degrees*. An interval of 360° corresponds to one complete revolution of a phasor, which, in turn, corresponds to the duration of one cycle. We can therefore convert any interval in degrees to an interval of time, and vice versa.

Example 7-3:

In Fig. 7-13, calculate the time interval that corresponds to an angle of 30 °.

Solution:

1. The duration of one cycle $= 1/f = 1/60$ s $= 0.0167$ s $= 16.67$ ms

2. One cycle is equivalent to 360 electrical degrees

3. One electrical degree corresponds therefore to:
    16.67 ms/360 $= 0.0463$ ms

4. Thirty electrical degrees corresponds to $0.0463 \times 30 = 1.389$ ms

7.12 Relationship between two sinusoidal waveshapes

In a given ac circuit, the sinusoidal voltages and currents all have the same frequency, but they usually do not reach their peak values at the same instant. Thus, in Fig. 7-15, voltage E_{12} and current I in a device Z attain their peak positive values at instants t_1 and t_2 respectively. The corresponding angles are θ_1 and θ_2 degrees. The current peak occurs after the voltage peak, and so the current is said to *lag* behind the voltage. Alternatively, the voltage is said to *lead* the current. The amount of lag (or lead) is expressed as the difference between θ_1 and θ_2. Thus, if $\theta_1 = 90°$ and $\theta_2 = 127°$, current I lags voltage E_{12} by an angle $\theta = (127 - 90) = 37°$.

If the positive peaks of voltage and current occur at the same instant, the voltage and current are said to be *in phase*. This condition is illustrated in Fig. 7-16. If the positive peak of voltage occurs at the same instant as the *negative* peak of current,

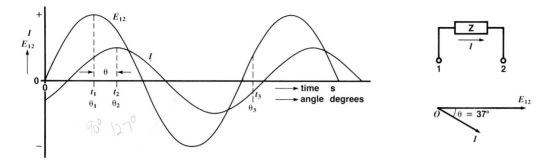

Figure 7-15
Current I lags behind voltage E_{12} by $(t_2 - t_1)$ seconds or $(\theta_2 - \theta_1)$ electrical degrees.

the voltage and current are said to be 180° out of phase (Fig. 7-17). We could also say that the current lags behind the voltage by 180°. And, in this special case, we could even state that the current leads the voltage by 180°.

Whenever we specify the lag or lead of one sine wave with respect to another, the angle must not exceed 180°.

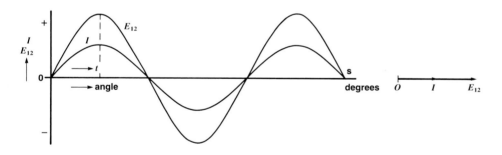

Figure 7-16
Voltage E_{12} and current I are in phase.

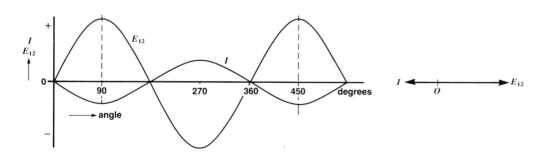

Figure 7-17
Voltage E_{12} and current I are 180° out of phase.

7.13 Typical phasor diagrams

The sinusoidal voltages and currents of Figs. 7-15 to 7-17 can be represented by phasors. The so-called *phasor diagrams* are shown beside the E_{12} and I sine waves. Note how much simpler they are than the corresponding curves.

We shall again encounter phasor diagrams when we study ac circuits. This introduction to phasors and phasor diagrams will then be useful.

7.14 Instantaneous values

We sometimes want to know the value of a voltage or current at a particular instant. For example, in Fig. 7-15, we may want to know the value of E_{12} and I at instant t_3. We can make such calculations by expressing the voltages and currents in trigonometric form.

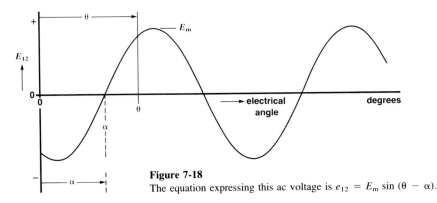

Figure 7-18
The equation expressing this ac voltage is $e_{12} = E_m \sin (\theta - \alpha)$.

Figure 7-18 shows a sinusoidal voltage E_{12} having a peak value E_m. To describe it in trigonometric form, we first identify the angle α where (1) the voltage is zero and (2) its slope is positive. We select a value of α that is less than 360°. The instantaneous value E_{12} corresponding to any angle θ is then given by the equation:

$$E_{12} = E_m \sin (\theta - \alpha) \qquad (7\text{-}3)$$

where θ is the number of electrical degrees that have elapsed since the beginning of the sine wave (see Fig. 7-18). We can calculate the corresponding time t by the equation:

$$t = \theta / (360 f) \qquad (7\text{-}4)$$

Using a hand calculator, we can easily calculate the value of the voltage at any angle θ (or any instant t).

Example 7-4:

Express the sine wave of Fig. 7-11 in trigonometric form. Calculate the current at $t = 35.6$ s.

Solution:

1. The current is zero and the slope is positive at $t = 0.75$ s. But measured from $t = 0$, this corresponds to an angle α of 270°.

2. The peak current is 20 A, and so the instantaneous current is
$$I = 20 \sin (\theta - 270) \qquad \text{Eq. 7-3}$$

3. The frequency is 1 Hz, and so
$$t = \theta/360f = \theta/360 \text{ or } \theta = 360t \qquad \text{Eq. 7-4}$$

4. The angle θ corresponding to 35.6 s is
$$\theta = 360 \, t = 360 \times 35.6 = 12\ 816°$$

Consequently,

$$
\begin{aligned}
I &= 20 \sin (\theta - 270) \\
&= 20 \sin (12\ 816 - 270) \\
&= 20 \sin 12\ 546° \\
&= -16.18
\end{aligned}
$$

5. The current is 16.18 A, and it flows opposite to the arrow shown in Fig. 7-11

Example 7-5:

Referring to Fig. 7-19, the voltage has a peak value of 170 V, and the current has a peak of 8 A. The frequency is 60 Hz, and the current lags 37° behind the voltage. Calculate the values of E_{12} and I at an angle of 390° and the corresponding time t.

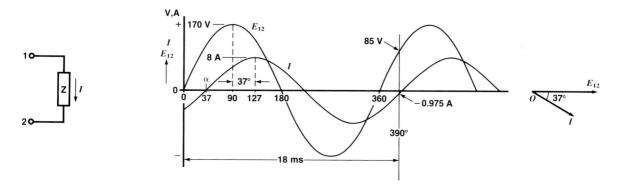

Figure 7-19
See Example 7-5.

Solution:

1. The expression for E_{12} is
$$
\begin{aligned}
E_{12} &= E_\text{m} \sin(\theta - \alpha) \qquad \text{Eq. 7-3} \\
&= 170 \sin (\theta - 0) \\
&= 170 \sin \theta
\end{aligned}
$$

2. The expression for I is
$$I = I_m \sin(\theta - \alpha)$$
$$= 8 \sin(\theta - 37)$$

3. at $\theta = 390°$
$$E_{12} = 170 \sin 390 = 85 \text{ V}$$
$$I = 8 \sin(390 - 37) = -0.975 \text{ A}$$

4. $\qquad t = \theta/(360f)$ Eq. 7-4
$$= 390/(360 \times 60)$$
$$= 0.018 \text{ s, or } 18 \text{ ms}$$

7.15 Summary

In this chapter, we learned how to interpret positive and negative voltages and currents. We also discovered that an ac voltage has three important properties: a magnitude, a polarity, and a rate of change. Similarly, an ac current has a magnitude, a direction, and a rate of change.

We learned that a voltage or current can be changing very rapidly, even when its value is zero. In the case of an ac sine wave, the voltage or current changes most quickly when it is zero.

Voltages and currents in an ac circuit do not always reach their peak values at the same time. This has given rise to terms such as lag, lead, and out-of-phase. A current is said to lag behind a voltage when the positive current peak occurs after the positive voltage peak.

The concept of a rotating rod that generates a sine wave was introduced in Chapter 1. We now see that it has practical applications because it enables us to represent sine waves of voltage and current by means of phasors. Phasors are much easier to draw than sine waves, but in later chapters we will find they have other useful properties.

We also learned how to express a voltage or current in trigonometric form. This enables us to calculate the exact value at any instant. However, we seldom need to know these instantaneous values.

Finally, we found that positive and negative voltages imply a change of polarity. Similarly, positive and negative currents imply a change in the direction of flow.

TEST YOUR KNOWLEDGE

7-1   The equation $E_{78} = -45$ V means that the voltage is 45 V and

    a. terminal 78 is negative   b. terminal 8 is positive
    c. terminal 7 is negative with respect to terminal 8

7-2   The equation $I = -2$ A is meaningless unless

    a.   we have a graph of the current
    b.   we have a circuit diagram showing the positive direction of current flow
    c.   both (a) and (b)

7-3   The arrow showing current flow in a circuit is

    a.   the direction in which the current is actually flowing
    b.   the direction in which the current should flow
    c.   the direction in which the current is assumed to flow

7-4   In Fig. 7-5, at $t = 19.5$ s, the following is true:

    a. $E_{21} = 5$ V   b. $E_{12} = 5$ V   c. $E = -5$ V

7-5   In any device having terminals A and B, the sum $E_{AB} + E_{BA}$ is always zero.

    ☐ true   ☐ false

7-6   In Fig. 7-11, the instant $t = 0.5$ s corresponds to

    a. 180°(electrical)   b. 90°   c. 180° (mechanical)

7-7   In Fig. 7-11, the interval between 0.27 s and 0.52 s corresponds to an electrical angle of

    a. 60°   b. 180°   c. 90°

7-8   Fig. 7-12 shows, within the boundaries of the graph, a total of 5.5 cycles.

    ☐ true   ☐ false

7-9   In Fig. 7-12, 3 ms corresponds to

    a. 0.0648°   b. 0.018°   c. 64.8°

7-10 In Fig. 7-12, one second corresponds to

    a. 360°   b. 21 600°   c. 43 200°

7-11   An angle of 3476° is equivalent to an angle of

      a. 236°   b. 9.655°   c. 0.655°

7-12   An angle of 340° is equivalent to an angle of

      a. 680°   b. 160°   c. $-20°$

7-13   A positive voltage is more important than a negative voltage.

      ☐ true   ☐ false

QUESTIONS AND PROBLEMS

7-14   In Fig. 7-10, is there any moment when the rate of change of current is zero? Explain.

7-15   In Fig. 7-10, if $t = 1.5$ s, is E_{XY} positive or negative? Explain.

7-16   Using the same scales as in Fig. 7-11, draw the sine wave of a current having a peak value of 10 A and a frequency of 2 Hz. Assume the current is zero (with positive slope) at $t = 0$.

7-17   Using graph paper scaled off at 30° intervals, draw a sine wave of E_{12}, so that the peak value is 200 V and the frequency is 50 Hz. The voltage starts at zero, with a positive slope, and lasts for 30 ms.

7-18   In Fig. 7-11, superpose the curve of a sine wave of voltage E_{12} having an amplitude of 10 V and a frequency of 1 Hz with

      a) voltage in phase with the current

      b) voltage lagging 90° behind the current

      c) voltage leading the current by 90°.

7-19   Draw the graph of $E_{12} = 50 \sin (\theta - 30)$, starting from $\theta = 0$.

7-20   Draw the graph of $E_{12} = 50 \sin (\theta + 30)$.

7-21   In Fig. 7-11, starting from $t = 0$ calculate the value of I after a period of 2 h, 3 min, 25.2 s.

7-22   In Fig. 7-12, what is the angle corresponding to $E_{12} = -100$ V if the slope is positive?

CHAPTER 8

Electromagnetic Induction

Introduction and Chapter Objectives

Faraday's law of electromagnetic induction, the last basic law we have to study, has hundreds of useful applications. In this chapter we will describe how a voltage is generated and what determines its magnitude and polarity. Then we will show how the concept of cutting lines of flux also leads to a generated voltage. Finally, we will discuss eddy currents and the reason for laminating iron cores.

8.1 Voltage induced in a coil, Faraday's law of induction

Consider the coil in Fig. 8-1, wound on a non-magnetic core and connected to a sensitive voltmeter. The flux from a permanent magnet passes through the interior of the coil. When the coil and magnet are both stationary, no voltage is observed.

However, if the magnet is suddenly pulled downwards, a voltage appears across the terminals of the coil. The voltage exists whenever the magnet is moving. But as soon as the magnet stops, the voltage falls to zero. When the magnet moves downward, the voltage has the polarity shown in Fig. 8-2.

If the magnet is moved upward, a voltage is again produced as long as the magnet is moving. However, its polarity is the reverse of its previous polarity (Fig. 8-3).

The voltage generated by the coil is said to be an *induced voltage*. Furthermore, the flux that passes through the coil is said to be "linked" with the coil.

Michael Faraday showed that the induced voltage is due to the *change* in flux that passes through the coil. Thus, whenever the flux inside a coil changes, a voltage is induced across its terminals. This phenomenon is called *electromagnetic induction*.

In the above example, we used a moving magnet to change the flux inside the coil. But Faraday showed that no matter how the flux is produced and no matter what causes it to vary, a voltage is always induced when the flux linked by a coil varies.

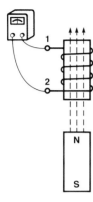

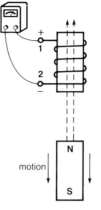

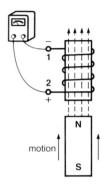

Figure 8-1
When the permanent magnet is stationary, no voltage appears across the terminals of the coil.

Figure 8-2
When the magnet is moved downwards, a voltage E_{12} is induced in the coil. E_{12} is positive.

Figure 8-3
When the magnet is moved upwards, a voltage E_{12} is induced in the coil. E_{12} is negative.

8.2 Magnitude of the induced voltage

The magnet in Fig. 8-4 is arranged so that it can move only from position 1 to position 2. This ensures that the initial and final flux inside the coil are always the same. Thus, whether the magnet moves quickly or slowly, the change of flux inside the coil is constant. As before, a voltmeter indicates the voltage induced in the coil.

If the magnet is moved slowly, we find the induced voltage is only a few millivolts. But if the magnet moves quickly, the induced voltage may be several volts. We conclude that the magnitude of the induced voltage does not depend upon the change in flux, but on the *rate* at which it is changing. According to Faraday's law, the voltage induced in a coil is given by the equation:

$$E = N \times \text{rate of change of flux} \qquad (8\text{-}1)$$

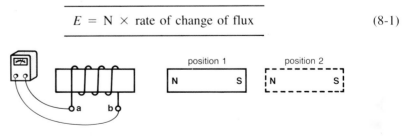

Figure 8-4
The magnitude of the induced voltage depends upon the rate of change of flux inside the coil. See Example 8-1.

which may also be expressed as

$$E = N \frac{\Delta\phi}{\Delta t}$$ (8-2)

where

E = voltage induced in the coil, in volts (V)

N = number of turns on the coil

$\Delta\phi$ = change of flux inside the coil, in webers (Wb)

Δt = duration of the change, in seconds (s)

The rate of change of flux is expressed in webers per second.

Example 8-1:

A coil having 2000 turns surrounds 50 mWb of flux produced by a large permanent magnet (Fig. 8-4). When the magnet is moved to position 2, the flux inside the coil drops to 10 mWb. Calculate the average voltage induced in the coil if the magnet is moved in 0.1 s.

Solution:

1. The change in flux linking the coil is:
   $\Delta\phi$ = 50 − 10 = 40 milliwebers = 40 mWb

2. The interval Δt during which the change takes place = 0.1 s

3. The rate of change of flux is
   $\Delta\phi / \Delta t$ = 40 × 10^{-3}/0.1 = 0.4 webers per second = 0.4 Wb/s

4. From Faraday's law we have:
   $E = N \Delta\phi / \Delta t$ Eq. 8-2
   E = 2000 × 0.4 = 800 V

8.3 Induced current

Let us connect a resistor across the terminals of a coil having N turns (Fig. 8-5). If we bring a magnet near the coil, a voltage is again induced. However, the voltage now causes a current I to flow in the resistor, and therefore in the coil. This current is called an *induced current*.

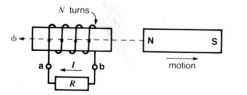

Figure 8-5
A current is induced in the circuit whenever the flux inside the coil is changing.

It is important to note that the induced current produces a mmf *NI* while it is flowing in the coil.

An induced current will also circulate if the terminals of the coil are short-circuited. Although there no external voltage can be measured, a voltage is still induced in the coil. The situation is similar to short-circuiting the terminals of a battery. A large current flows because the battery still develops a voltage, despite the fact it cannot be measured externally.

8.4 Direction of the induced current

Consider Fig. 8-6, in which a coil is connected to a resistor. A flux produced by an external device (not shown) links with the coil. The flux is directed toward the left. Suppose that the flux increases. A voltage will be induced which, in turn, produces a current. The question is, in what direction will the induced current flow? It has been found that the current always flows in a direction so that the magnetomotive force it creates *opposes* the change of flux inside the coil. If the flux increases, the mmf will tend to oppose the increase. On the other hand, if the flux decreases, the mmf will tend to oppose the decrease.

To illustrate what happens, suppose the flux in Fig. 8-6 increases. To oppose this increase, the mmf of the coil must be directed toward the right (Fig. 8-7). Consequently, the current in the coil must flow in the direction shown.

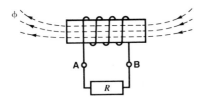

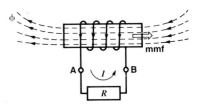

Figure 8-6
The coil tries to oppose any change in the flux which threads through the coil.

Figure 8-7
The external flux ϕ is increasing, and the resulting induced current opposes the increase.

As another example, suppose the flux in Fig. 8-8 is decreasing. To oppose the decrease, the mmf must be directed to the right. Consequently, by applying the right-hand rule, the induced current must flow in the direction shown.

Figure 8-8
The external flux ϕ is decreasing and the resulting induced current opposes the decrease.

⟵ Does induced current always flow to right!

8.5 Polarity of the induced voltage, Lenz's law

In the previous examples we were able to determine the direction of the induced current. But knowing this, we can determine the polarity of the terminals of the coil because current always flows from $(+)$ to $(-)$ in a resistor. Thus, in Fig. 8-7, terminal A is positive and terminal B is negative. Similarly, in Fig. 8-8, terminal C is positive and D is negative.

The polarity of the terminals is independent of the value of the resistor connected across the coil. Thus, even if the resistance is infinite (meaning that the terminals are on open circuit), the polarities will still be the same.

We therefore have a method of determining the polarity of the terminals. We simply assume a resistor is connected across the terminals and find the resulting direction of current flow, and the problem is solved. This method is known as Lenz's law, which may be stated as follows:

The polarity of the induced voltage across a coil is such that it tends to circulate a current that will oppose the change of flux inside the coil.

8.6 Methods of inducing a voltage

The following examples will help us see some of the ways in which a voltage can be induced in a coil. None of them represents a practical, commercial device, but each illustrates a basic principle that is used in the manufacture of motors, generators, transformers, and relays.

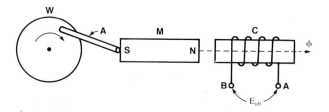
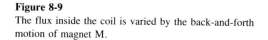

Figure 8-9
The flux inside the coil is varied by the back-and-forth motion of magnet M.

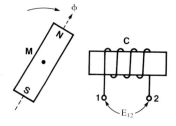
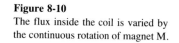

Figure 8-10
The flux inside the coil is varied by the continuous rotation of magnet M.

In Fig. 8-9, a rotating wheel W connected to a crank arm A causes magnet M to move back and forth. This motion causes the flux inside coil C to vary periodically. When the magnet moves to the right, terminal A is $(+)$; when it moves to the left, terminal A is $(-)$. Consequently, an ac voltage is induced whose frequency depends upon the speed of rotation of the wheel.

In Fig. 8-10, a rotating magnet M causes the flux inside coil C to vary periodically. Consequently, an ac voltage E_{12} is induced whose frequency depends upon the speed of rotation of the magnet.

In Fig. 8-11, a stationary magnet M produces a constant flux ϕ. A coil mounted on a soft iron core rotates clockwise, as shown. Because the coil rotates, the flux inside it varies periodically between a maximum and zero. An ac voltage E_{34} is induced whose frequency depends upon the speed of rotation. Note that the terminals rotate with the coil. If we want to connect a stationary load to the terminals, sliding electrical contacts must be used between the terminals and the load.

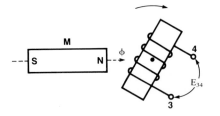

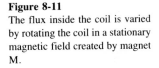

Figure 8-11
The flux inside the coil is varied by rotating the coil in a stationary magnetic field created by magnet M.

In Fig. 8-12, a rotating piece of iron F is placed between stationary coil C and fixed magnet M. The reluctance of the magnetic circuit varies with the position of F. Thus, the flux in the coil is maximum when F is horizontal. An instant later, the flux is minimum when F is vertical. An ac voltage E_{56} is again induced whose frequency depends upon the speed of rotation of F.

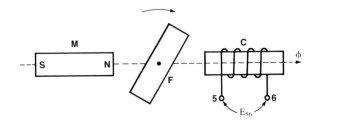

Figure 8-12
The flux inside the coil is varied by rotating an iron bar between the coil and magnet M.

In Fig. 8-13, a rotating rectangular coil C is placed between two magnets. The flux inside the coil varies between zero and maximum. In the position shown, the flux linking the coil is zero. An ac voltage E_{78} is induced whose frequency depends upon the speed of rotation of the coil.

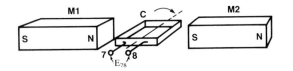

Figure 8-13
The flux inside the coil is varied by rotating the coil in the stationary magnetic field created by magnets M1 and M2.

In Fig. 8-14, a battery B is connected to coil A by means of a switch. The switch opens and closes periodically. When the switch is open, the flux linked by coil C is zero. But when the switch is closed, current I in coil A produces a flux ϕ that links coil C. The flux increases very quickly when the switch closes. As a result, a high voltage E_{12} is induced. Then, when the switch is opened, the flux collapses very quickly to zero, and a high voltage is again induced. E_{12} is an alternating voltage and its frequency depends upon the number of times the switch opens and closes per second.

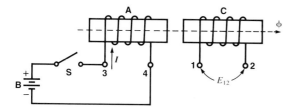

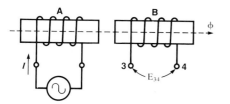

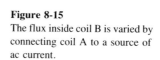

Figure 8-14
The flux inside coil C is varied by opening and closing switch S.

In Fig. 8-15, an ac source is connected to coil A, so that an alternating current I flows in the winding. This produces a flux ϕ that continually changes in magnitude and direction. Thus, an ac voltage E_{34} is induced in coil B despite the fact that both coils are stationary. The frequency of E_{34} depends upon the frequency of I.

Figure 8-15
The flux inside coil B is varied by connecting coil A to a source of ac current.

In Fig. 8-16, the flux ϕ inside a coil is assumed to increase continually without ever reaching a maximum. Under these conditions, the polarity of the terminals never changes, and terminal **a** is always negative. This means that E_{ab} is a dc voltage, just like that produced by a battery. Unfortunately, it is impossible to create a flux that increases indefinitely. Although we can induce a dc voltage for brief periods (as shown), the flux must eventually reach an upper limit. The value of E_{ab} then abruptly falls to zero because the flux is no longer changing.

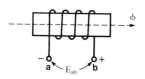

Figure 8-16
If it were possible for flux ϕ to increase indefinitely, the polarity of terminals a and b would never reverse, and we would have a dc source.

8.7 Relationship between an ac flux and the induced ac voltage

The previous examples have shown that when the flux varies periodically inside a coil, an ac voltage is induced. Let us now look more closely at this relationship between the voltage and the flux.

Figure 8-17 shows a stationary coil that is linked by an alternating flux ϕ. The flux is assumed to be positive when it is directed to the right, as shown. It is negative when it is directed to the left. Thus, the flux is positive when it enters face A of the coil and negative when it comes out of face A.

Let us assume the flux is varying sinusoidally, as shown in Fig. 8-18. The question is, what kind of voltage will appear across the terminals?

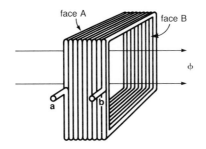

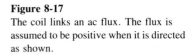

Figure 8-17
The coil links an ac flux. The flux is assumed to be positive when it is directed as shown.

First, we know that the voltage depends only upon the rate of change of flux. The rate of change is zero at instants 2, 4, and 6, because at these moments the flux is momentarily not changing. In effect, the slope of the flux wave is zero.

Second, we know that the voltage is maximum when the flux is changing the most rapidly. This occurs at instants 1, 3, 5, and 7 because at these moments the slope of the flux wave is steepest.

Third, the rate of change of flux is positive whenever the slope of the flux wave is positive. Looking at the curve, this occurs between instants 1 and 2, and again between instants 4 and 6. The induced voltage is positive during these intervals. Thus, if terminal **a** is positive during these intervals, E_{ab} is (+). Conversely, whenever the slope of the flux wave is negative, the induced voltage E_{ab} will be negative.

Thus, we conclude that the induced voltage must have the general shape shown in Fig. 8-19. In effect, the voltage is maximum when the flux through the coil is zero. Then, the voltage is zero when the flux in the coil is maximum. Clearly, the flux and the voltage are out of step. The two are said to be *out of phase* because they do not reach their maximum positive values at the same time.

8.8 Coil revolving in a stationary field

Referring to Fig. 8-17, an ac voltage is induced in the coil because the flux through it is varying with time. We can produce exactly the same flux variations inside the coil by rotating the coil in a stationary magnetic field (Fig. 8-20). In the position shown, the flux is entering face A. Furthermore, it is decreasing because the coil is turning ccw. It follows that E_{ab} is negative at this moment.

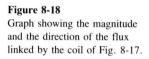

Figure 8-18
Graph showing the magnitude
and the direction of the flux
linked by the coil of Fig. 8-17.

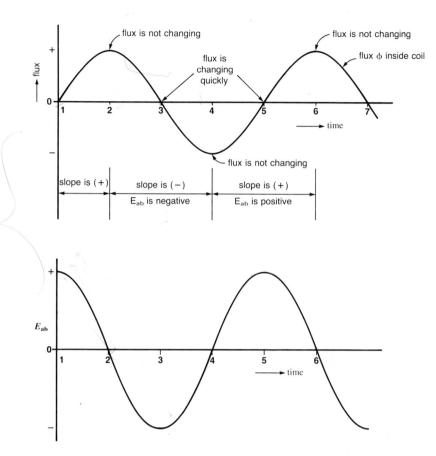

Figure 8-19
Graph of the voltage induced in
the coil of Fig. 8-17.

When the coil is momentarily horizontal, the flux through it is zero. However, E_{ab} will be maximum because the flux through the coil is changing very quickly. The waveshape of the voltage induced in Fig. 8-20 is identical to that shown in Fig. 8-19. One complete cycle is made every time the coil completes one revolution. The frequency of the voltage is therefore equal to the number of revolutions per second. Thus, if the coil turns at 1800 r/min, the frequency is 1800/60 = 30 Hz.

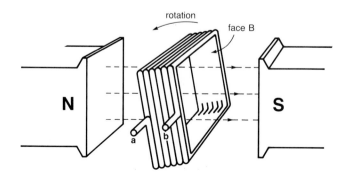

Figure 8-20
A practical way to generate an ac
voltage is to rotate a coil in a stationary magnetic field.

The peak value of E_{ab} depends directly upon the number of turns on the coil and on the speed of rotation. If we double the number of turns (or the speed), the peak voltage will double. The peak voltage also depends upon the flux produced by the magnets. If the flux is doubled, the peak voltage will double.

8.9 Cutting lines of flux

Consider a single turn of wire ABCD, having terminals X and Y (Fig. 8-21). A stationary magnet, located below conductor AB, produces a flux density B. The turn of wire is moving rapidly to the right. It is obvious that the flux enclosed by the single turn is in the process of decreasing. Consequently, an induced voltage E_{XY} appears between terminals X and Y.

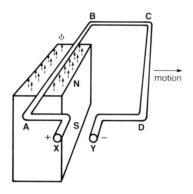

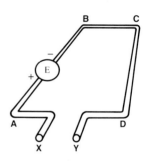

Figure 8-21
The voltage induced between terminals X and Y can be considered to be due to conductor AB cutting the flux lines.

Figure 8-22
The flux-cutting concept means that a voltage E is induced in conductor AB.

It is sometimes convenient to imagine that the voltage is induced because conductor AB is "cutting" the lines of force produced by the magnet. The voltage is assumed to be induced in conductor AB alone, while conductors BC, CD, AX, and DY merely act as connecting wires. The electrical condition is shown in Fig. 8-22. With this flux-cutting approach, the same voltage E appears between terminals X and Y as with the standard flux-linkage method. However, the conductor must cut the flux at right angles. The magnitude of the induced voltage is given by the equation:

$$E = BLv \qquad (8\text{-}3)$$

where

E = voltage induced in the conductor, in volts (V)

B = flux density that is being "cut" by the conductor, in teslas (T)

v = speed of the conductor, in meters per second (m/s)

L = length of the conductor that is cutting the flux, in meters. (m)

Example 8-2:

Conductor AB in Fig. 8-21 has a length of 0.35 m, but only 0.3 m are actually cutting flux lines. The magnet produces a density of 0.4 T, and the conductor is moving at 120 km/h. Calculate the voltage induced in the conductor, and the voltage that appears between terminals X and Y.

Solution:

$$E = 0.4 \times 0.3 \times \frac{120\ 000}{3600} = 4\ V \qquad\qquad Eq.\ 8\text{-}3$$

The voltage induced in conductor AB is 4 V, and the same voltage appears across terminals X and Y.

We use this alternative method of calculating the induced voltage because in many practical cases we know the value of the flux density and the speed of the conductor.

In some machines, the coil is stationary and the flux is moving. This does not change the induced voltage because v in Eq. 8-3 is the *relative* speed between the conductor and the magnetic field.

8.10 Polarity of the induced voltage in a straight conductor

We can predict the polarity of the voltage induced in a straight conductor by means of a simple rule, called Fleming's right-hand rule. It works as follows:

1. Extend the thumb, forefinger, and middle finger so they are at right angles to each other.

2. Point the thumb in the direction the conductor is moving.

3. Point the forefinger in the direction of the flux.

4. The middle finger then points to the positive end of the conductor.

To illustrate the application of this rule, consider Fig. 8-23, in which a conductor is moving upwards, and cutting across the magnetic field produced by two magnets. By extending the fingers according to the above rule, we find that extremity A is (+) and B is therefore (−).

Some students find Fleming's rule awkward to apply, particularly when the flux and motion are directed in ways that require unnatural positions of the hand. Furthermore, the relative motion of the conductor has to be used when the conductor is stationary and the flux is cutting across it. The following alternative rule may be used:

1. Stretch the right hand out flat, with the thumb at right angles to the fingers.

2. Let the fingers point in the direction of the flux.

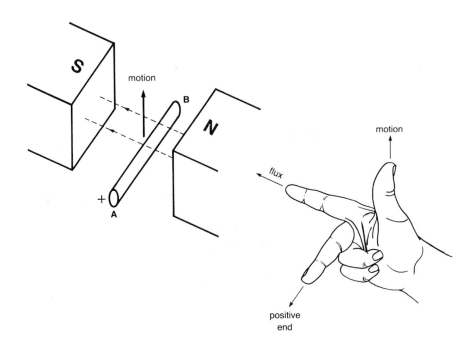

Figure 8-23
The polarity of the induced voltage can be determined by Fleming's right-hand rule.

3. Press the palm of the hand against the edge of the conductor that is cutting the flux.

4. The thumb then points toward the positive end of the conductor.

Figure 8-24 illustrates the application of this rule.

8.11 Eddy currents

Consider an ac flux ϕ that links a rectangular-shaped conductor (Fig. 8-25). According to Faraday's law, an ac voltage E_{ab} is induced across its terminals.

If the conductor is short-circuited, a large ac current I_1 will flow, causing the conductor to heat up. If a second conductor is placed inside the first, a smaller voltage is induced because it links a smaller flux. Consequently, the short-circuit current I_2 is less than I_1 and so, too, is the dissipated power. Figure 8-26 shows four such concentric conductors carrying currents I_1, I_2, I_3, and I_4.

In Figure 8-27, the ac flux passes through a solid metal plate. The plate is equivalent to a set of rectangular conductors that touch each other. Currents swirl back and forth inside the plate, following the paths shown in Fig. 8-27. These

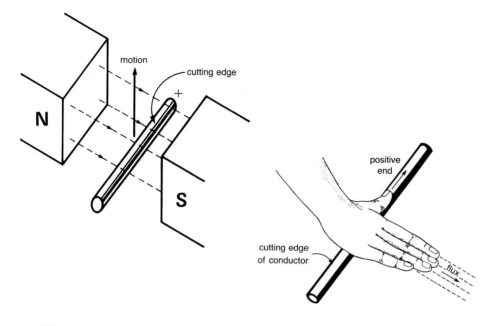

Figure 8-24
The polarity of the induced voltage can also be determined by an alternative method.

so-called *eddy currents* (or Foucault currents) can be very large because of the low resistance of the plate. Consequently, a metal plate penetrated by an ac flux can become very hot.

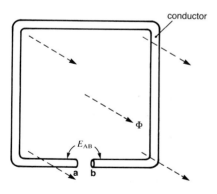

Figure 8-25
An ac flux ϕ induces an ac voltage in the solid conductor.

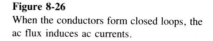

Figure 8-26
When the conductors form closed loops, the ac flux induces ac currents.

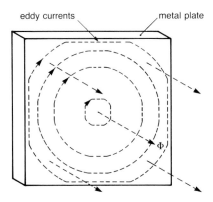

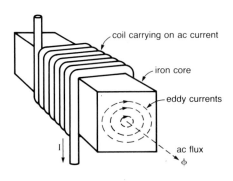

Figure 8-27
When an ac flux penetrates a metal plate, the induced currents are called eddy currents.

Figure 8-28
Eddy currents flowing in an iron core can produce large losses and much heat.

8.12 Eddy currents in a stationary iron core

The eddy current problem becomes crucially important when iron has to carry an ac flux. Figure 8-28 shows a coil carrying an ac current that produces an ac flux in a solid iron core. Eddy currents are set up as shown, and they swirl back and forth throughout the length of the core. A large core could eventually become red hot (even at a frequency of 60 Hz) because of the I^2R losses.

We can reduce the losses by splitting the core along its length, taking care to insulate the two sections from each other (Fig. 8-29). The voltage induced in each section is one-half of what it was before, with the result that the eddy currents — and the corresponding losses — are considerably reduced.

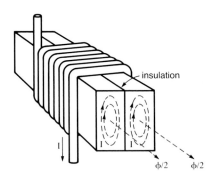

Figure 8-29
The losses can be reduced by subdividing the core into smaller sections, and insulating the sections from each other.

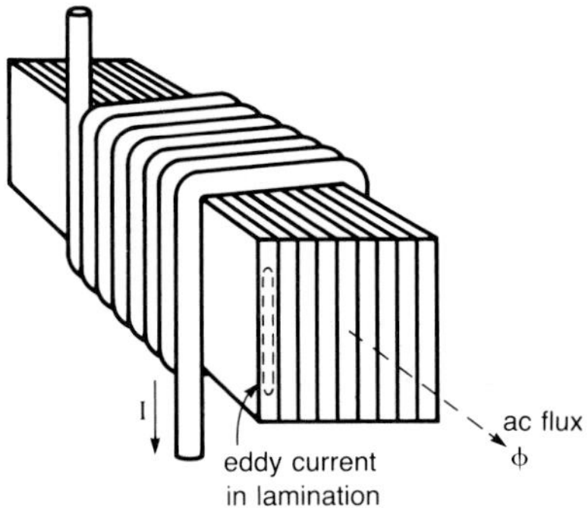

Figure 8-30
Laminating the iron core can greatly reduce the core losses.

If we continue to subdivide the core, we find that the losses decrease progressively. In practice, the core is composed of stacked laminations, usually a fraction of a millimeter thick (Fig. 8-30).

The cores of ac motors and generators are always laminated. A thin coating of insulation covers each lamination to prevent electrical contact between the laminations. They are stacked on top of each other and are held tightly in place by bolts and end-pieces.

8.13 Eddy current losses in a revolving core

Eddy currents are also produced when a rotor turns in a constant magnetic flux. Consider, for example, a solid iron rotor that revolves between the poles of a magnet (Fig. 8-31). As it turns, the rotor cuts flux lines and, according to Faraday's law, a voltage is induced along its length. Because of this voltage, large eddy currents flow in the rotor because its resistance is very low. These eddy currents produce I^2R losses which are immediately converted into heat.

To reduce these losses, we laminate the armature. The laminations are tightly stacked, with the flat sides running parallel to the flux lines (Fig. 8-32).

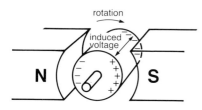

Figure 8-31
Eddy currents are induced when a solid iron rotor turns in a stationary magnetic field.

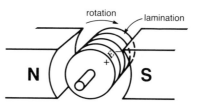

Figure 8-32
Laminating the iron core reduces the eddy current losses.

8.14 Summary

In this chapter, we covered the last physical law that is needed to explain the behavior and properties of any electromechanical machine and device. Consequently, we are now prepared to undertake the study of rotating machines, transformers, and circuit components.

In particular, we have learned that a voltage can be induced in two ways: 1) by varying the flux inside a coil, and 2) by moving a conductor so that it cuts across a magnetic field. Both methods are based upon Faraday's law of electromagnetic induction. The induced voltage will produce an induced current, but only if the coil (or the conductor) is part of a closed loop.

In studying some ingenious ways in which a voltage can be induced, we found that the voltage is always ac. As we will see in succeeding chapters, the most important method of inducing a voltage is to rotate a coil in a stationary magnetic field. Alternatively, a revolving field can move across a stationary coil.

One of the consequences of Faraday's law is that eddy currents are induced in a block of metal when it is subjected to an alternating flux. Thus, to keep the eddy currents and losses within bounds, it is essential to laminate iron cores whenever they carry an ac flux. For the same reason, we must laminate iron rotors that turn in a stationary field.

TEST YOUR KNOWLEDGE

8-1   A coil of 2000 turns surrounds a constant flux of 0.2 Wb. The voltage induced is

a. 400 V   b. zero   c. cannot be
                                                calculated

8-2   When the flux changes inside a coil that is on open-circuit

a. a voltage is induced   b. a current is induced
c. a voltage and a current are induced

8-3   A voltage of 8 V is induced in a coil when the flux changes from 800 mWb to 600 mWb in 2 s. What is the magnitude of the voltage when the flux changes from 8000 mWb to 8200 mWb in 4 s?

a. 16 V   b. 4 V   c. 40 V

8-4   Eddy currents are always produced when an ac flux passes through a solid metal plate.

☐ true   ☐ false

8-5   In Fig. 8-25, if the flux ϕ is increasing E_{ab} is

a. positive   b. negative   c. zero

8-6   In Fig. 8-21, the induced voltage is 6 V when conductor AB moves to the right at 2 m/s. Calculate the voltage when it moves vertically upward at 20 m/s.

a. 60 V   b. zero   c. -60 V

8-7   In Fig. 8-10, E_{12} is momentarily

a. positive   b. zero   c. negative

QUESTIONS AND PROBLEMS

8-8   A conductor that is 2 m long moves at 60 km/h across a field whose flux density is 0.6 T. Calculate the magnitude of the induced voltage.

8-9   In Fig. 8-9, (a) is the flux in the coil increasing or decreasing and (b) what is the polarity of A with respect to B at this moment?

8-10   A coil having 200 turns is linked by a flux of 3 mWb. If the flux decreases to 1.3 mWb in 12 ms, calculate the average value of the induced voltage.

8-11   Why do the iron cores of ac magnets have to be laminated and those of dc magnets do not?

8-12   A dc voltage can be induced in a coil for only a short time. Explain.

8-13   In Fig. 8-6, the coil has 1000 turns and is linked by a flux of 180 mWb. If the flux decreases to zero at a uniform rate in a period of two hours, calculate the magnitude of the induced voltage and determine its polarity. Is the induced voltage dc or ac?

CHAPTER 9

Inductors and Inductance

Introduction and Chapter Objectives

In addition to rotating machines and transformers, industrial power circuits contain three important components: resistors, inductors, and capacitors. We already know what a resistor is. The object of this chapter is to explain the properties and behavior of inductors. We will also discuss mutual inductance, which is the beginning of transformer theory. Capacitors will be covered in Chapter 10, and transformers in Chapter 16.

An inductor is simply a coil consisting of one or more turns. Its most important property is its ability to store energy. This enables it to act, for short periods, as either a source or a load. We will also examine the behavior of an inductor that carries a sinusoidal ac current. This preliminary approach serves as an introduction to Chapter 13 on ac circuits. Finally, we will see that a knowledge of inductors helps greatly in the study of machines, transformers, and transmission lines.

9.1 Mutual induction

Consider two coils A and B wound on a common iron core (Fig. 9-1). Coil A and rheostat R are connected in series across the terminals of a battery. By changing the position of the rheostat, current I_a can be varied. This causes the mmf of coil A to vary, and hence the flux ϕ in the core. The magnitude of ϕ will therefore increase and decrease with I_a. But according to Faraday's law, any change in ϕ will induce a voltage in coil B. This means that a change in the current flowing in coil A will induce a voltage E_b in coil B.

The phenomenon by which a change of current in one coil induces a voltage in another is called *mutual induction*.

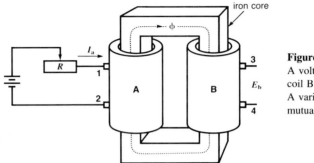

Figure 9-1
A voltage E_b is induced in coil B when current I_a in coil A varies. This is called mutual induction.

If a resistor is connected across the terminals of coil B, the induced voltage E_b will cause a current to flow, and so the resistor will heat up. This thermal energy must come from energy that is supplied to coil A. Thus, the mutual induction between two coils enables energy to be transmitted from one coil to another without any physical connection between them. The coils are said to be *coupled* magnetically by the common flux ϕ that links them.

9.2 Mutual inductance; the henry (H)

In Fig. 9-1, the magnitude of the induced voltage E_b depends upon the rate at which the flux ϕ is changing. But the rate of change of flux depends upon the rate of change of the mmf developed by coil A. This in turn depends upon the rate at which current I_a is changing. It follows that the induced voltage E_b is proportional to the rate of change of current I_a. We can therefore write

$$E_b = M \times \text{rate of change of current } I_a \qquad (9\text{-}1)$$

or

$$E_b = M\,\frac{\Delta I_a}{\Delta t} \qquad (9\text{-}2)$$

where

$$E_b = \text{voltage induced in coil B, in volts (V)}$$

$$\Delta I_a/\Delta t = \text{rate of change of current in coil A, in amperes per second (A/s)}$$

$$M = \text{a constant, whose magnitude depends upon the number of turns on the two coils and the coupling between them}$$

Figure 9-1A
This large inductor having a rating of 0.5 H
and able to carry a dc current of 1800 A, is
used to smooth the dc current in a 300 kV dc
transmission line. Because an inductor
opposes any change of current that flows
through it, it effectively suppresses the ac
currents in the dc line. The coil is immersed
in an oil-filled tank and the external radiators
are fan-cooled during the summer. The two
HV bushings lead the dc current into and out
of the inductor. (*Courtesy Manitoba Hydro*)

The constant *M* in Eq. 9-1 is called the *mutual inductance* of the two coils. It is simply a convenient multiplier that relates the rate of change of current in one coil to the voltage induced in the other. The mutual inductance of two coils is a property they share that is just as real as their resistance.

What happens if we interchange the connections in Fig. 9-1 so that coil B carries a variable current I_b and coil A is on open-circuit? A voltage E_a will be induced in coil A, and its value is given by the equation:

$$E_a = M \frac{\Delta I_b}{\Delta t} \tag{9-3}$$

where *M* has the same value as in Eq. 9-2. The mutual inductance between two coils is therefore the same, no matter which coil is being excited.

The SI unit of mutual inductance is the henry (symbol H). The mutual inductance between two coils is said to be 1 henry when a rate of change of 1 ampere per second in one coil induces 1 volt in the other.

Example 9-1:

The mutual inductance between coils A and B of Fig. 9-1 is 3.7 H. Calculate the voltage induced in coil B if the current in coil A changes from 7A to 2A in 2 seconds.

Solution:

1. The change in current is: $\Delta I_a = 7 - 2 = 5$A

2. The duration of the change is: $\Delta t = 2$ s

3. The rate of change of current is: $\Delta I_a / \Delta t = 5/2 = 2.5$ A/s.

    $E_b = 3.7 \times 2.5 = 9.25$ V   Eq. 9-2

9.3 Self-induction and self-inductance

Consider Fig. 9-2, in which a coil A and rheostat R are connected in series across a battery. As before, current I_a can be varied by means of the rheostat, and this produces a change in flux ϕ. But, by Faraday's law, any change of flux inside a coil induces a voltage across its terminals. Therefore, a change of current in coil A induces a voltage across its own terminals. This phenomenon is called *self-induction*.

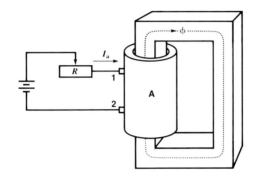

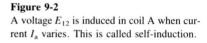

Figure 9-2
A voltage E_{12} is induced in coil A when current I_a varies. This is called self-induction.

By analogy with Eq. 9-1, the self-induced voltage is given by the equation :

$$E_a = L \times \text{rate of change of current } I_a \qquad (9\text{-}4)$$

or

$$E_a = L \, \frac{\Delta I_a}{\Delta t} \qquad (9\text{-}5)$$

where

$E_a =$ voltage induced in coil A, in volts (V)

$\Delta I_a/\Delta t =$ rate of change of current the coil A, in amperes per second (A/s)

$L =$ a constant whose magnitude depends upon the number of turns on the coil and the reluctance of the magnetic circuit

The constant L is called the *self-inductance* of the coil. It is simply a convenient multiplier that relates the self-induced voltage to the rate of change of current. Nevertheless, the self-inductance of a coil is one of its basic properties that is just as real as its resistance.

For example, coil A and coil B in Fig. 9-1 each has a certain self-inductance, in addition to their mutual inductance. The self-inductances are usually quite different, depending mainly upon the number of turns. Thus, if coil A has more turns than coil B, its inductance will be greater. (The term "inductance" is often used instead of "self-inductance".)

The SI unit of self-inductance is also the henry. A coil is said to have a self-inductance of 1 H when a rate of change of current of 1 A/s induces 1 V across its own terminals.

Example 9-2:

A coil having a self-inductance of 2 H is connected across a 3 V battery (Fig. 9-3). The current is 5 A, and it can be interrupted in 0.1 s using switch S. What is the average value of the induced voltage while the current is falling to zero?

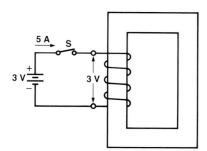

Figure 9-3
Coil connected to a 3 V battery.

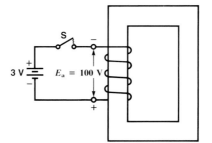

Figure 9-4
The induced voltage (100 V) is much higher that the battery voltage when the circuit is interrupted.

Solution:

$$E_a = L \frac{\Delta I_a}{\Delta t} = 2 \times \frac{5 - 0}{0.1} = 100 \text{ V} \qquad \text{Eq. 9-5}$$

Note that the induced voltage (100 V) in Fig. 9-4 is much higher than the battery voltage (3 V) in Fig. 9-3.

9.4 Energy stored in a coil

A magnetic field is a form of stored energy. The energy stored in the magnetic field of a coil is given by the equation:

$$W = \frac{1}{2} LI^2 \qquad (9\text{-}6)$$

where

$W =$ energy stored in the magnetic field, in joules (J)

$L =$ self-inductance of the coil, in henries (H)

$I =$ current in the coil, in amperes (A)

The stored energy increases with the square of the current. Thus, if the current in a coil doubles, the energy increases four times.

Example 9-3:

A coil having an inductance of 4 H carries a current of 10 A. Calculate the energy stored in its magnetic field. How much energy does the coil release when the current falls from 10 A to 5 A ?

Solution:

1. The initial energy in the coil is

$$W = \frac{1}{2} LI^2$$ Eq. 9-6

$$W_1 = \frac{1}{2} \times 4 \times 10^2 = 200 \text{ joules} = 200 \text{ J}$$

2. The final energy in the coil is

$$W_2 = \frac{1}{2} LI^2 = \frac{1}{2} \times 4 \times (5)^2 = 50 \text{ J}$$

3. The energy released by the coil is

$$W = W_1 - W_2 = 200 - 50 = 150 \text{ J}$$

Equation 9-6 tells us that the stored energy increases as the current increases. Thus, if the current in a coil is increasing, the stored energy is increasing. This means that the coil is absorbing energy from the circuit, and the coil therefore acts as a *load*. On the other hand, when the current decreases, the stored energy decreases. During such periods the coil releases energy, and it then acts as a *source*. The coil can therefore be either a load or a source, depending upon whether the current is increasing or decreasing.

Knowing that a coil behaves this way enables us to determine the polarity of the voltage across its terminals. The coil in Fig. 9-5 is part of an electrical circuit (not shown). It carries a current I that flows into terminal 1. The coil is assumed to have negligible resistance. Consequently, the IR drop across its terminals is always zero.

If the current is constant, the flux is not changing, and so the induced voltage is zero. As a result, E_{12} in Fig. 9-5 is zero. But if the current is increasing (Fig. 9-6), the flux is increasing, and so a voltage E_{12} will be induced. Because the coil acts as a load, the polarity of terminal 1 is positive (see Section 2-12).

If the current is decreasing (Fig. 9-7), the energy in the field is dropping and so the coil returns energy to the electrical circuit. Consequently, the coil now acts as a source, and terminal 2 is positive. Note that the polarity is independent of how the coil is wound.

This analysis brings out one of the important properties of an inductance: it tends to oppose the increase or decrease of current in a circuit. For example, in Fig. 9-6, the increasing current causes the coil to generate a voltage E_{12} that attempts to drive a current in the opposite direction to I. Conversely, in Fig. 9-7, the decreasing current causes the coil to generate a voltage that attempts to circulate a current in the same direction as I.

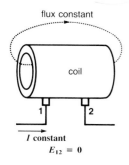

flux constant

coil

1   2

I constant

$E_{12} = 0$

Figure 9-5
The induced voltage is zero because the current is not changing.

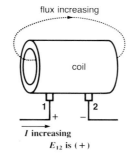

flux increasing

coil

1   2

I increasing

E_{12} is (+)

Figure 9-6
When the current is increasing, the coil absorbs energy, and so it acts as a load. Therefore terminal 1 is positive.

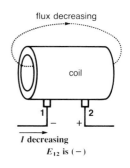

flux decreasing

coil

1   2

I decreasing

E_{12} is (−)

Figure 9-7
When the current is decreasing, the coil releases energy, and so it acts as a source. Therefore terminal 1 is negative.

9.5 Equivalent circuit of a coil

An inductor is a coil that has zero resistance. It is represented by the wiring diagram symbol shown in Fig. 9-8. If the coil has resistance (as all coils do), the equivalent circuit is given by Fig. 9-9. When the coil carries a variable current, the IR drop

L

A   B

Figure 9-8
Circuit diagram of a perfect inductor having an inductance L.

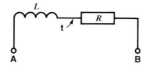

L R

t

A   B

Figure 9-9
Circuit diagram of an inductor having an inductance L and resistance R.

occurs across the coil resistance R, and any induced voltage appears across the inductance L. The voltage across the terminals of the coil is equal to the sum of these two voltages. The terminals are obviously A and B. Point **t** is not accessible; it exists only in the equivalent circuit. Energy stored in the coil is associated with L.

9.6 Build-up of current in a coil

When a coil is connected across a dc source (Fig. 9-10) the current does not immediately reach its final value because the induced voltage opposes the increase in current. Consequently, the current increases gradually, as shown in Fig. 9-11. When the final value of current is reached, the flux is constant, and so the induced voltage is zero. The current is then limited only by the resistance of the coil.

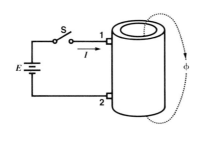

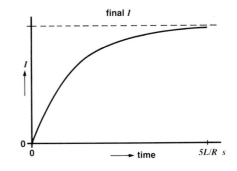

Figure 9-10
When a coil is connected to a dc source, the induced voltage opposes the build-up of current I.

Figure 9-11
The current reaches its final value after 5L/R seconds.

The time to reach this final state depends upon the ratio L/R, which is called the *time constant T* of the coil. For all practical purposes, the current reaches its final value about $5L/R$ seconds after the switch is closed.

Example 9-4:

A coil having an inductance of 4 H and a resistance of 2 Ω is connected across a 20 V dc source. Calculate

1. The time constant of the coil

2. The final current in the coil

3. The time to reach the final current

4. The energy stored in the coil

Solution:

1. The time constant of the coil is
    $$T = L/R = 4/2 = 2 \text{ s}$$

2. The final current is
    $$I = E/R = 20/2 = 10 \text{ A}$$

3. The time to reach 10 A is
    $$t = 5L/R = 5 \times 4/2 = 10 \text{ s}$$

4. The energy stored in the coil is
    $$W = 1/2 \ LI^2 = 1/2 \times 4 \times (10)^2 = 200 \text{ J} \qquad \text{Eq. 9-6}$$

9.7 Opening the circuit of a coil

Whenever we open the circuit of a coil by means of a switch, several things happen. As soon as the switch opens, the current in the coil is supposed to fall to zero. But if it did, an extremely high voltage E_a would be induced across the coil. The polarity of E_a is shown in Fig. 9-12. The voltage across the switch contacts is therefore $(V_s + E_a)$, where V_s is the voltage of the source. This very high voltage causes the

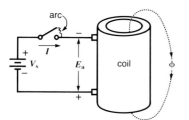

Figure 9-12
When a switch opens, the high induced voltage E_a produces an arc across the switch contacts.

air between the switch contacts to break down, and an arc is established between them. Thus, although the switch is partly open, the coil is not yet disconnected because the arc completes the circuit.

The high voltage may also break down the coil insulation, causing a short-circuit between turns. Furthermore, it is a shock hazard.

In addition, most of the energy previously stored in the magnetic field is dissipated in the electric arc. The resulting heat may damage and even melt the switch contacts.

The induced voltage can be limited by connecting a resistor across the coil terminals (Fig. 9-13). The resistor also absorbs an important part of the stored magnetic energy, thus reducing the heat at the switch contacts. Unfortunately, the resistor consumes power while the coil is in operation.

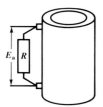

Figure 9-13
The magnitude of the induced voltage can be reduced by connecting a resistor across the coil.

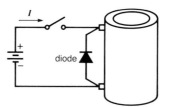

Figure 9-14
The induced voltage appearing across the coil can be reduced to less than 2 V by using a diode.

Another effective method of suppressing both the high voltage and the arc is to connect a diode across the coil (Fig. 9-14).* The induced voltage across the terminals can never exceed the conduction voltage of the diode (about 2 V), and all the stored energy is dissipated in the coil resistance. The drawback to this method is

* The diode must be connected as shown; if it is reversed, it will short-circuit the source.

that the flux decays slowly because current continues to circulate in the coil (and diode) even after the line current *I* is zero.

The most practical way to limit the high voltage is to use a voltage-sensitive resistor (varistor) in parallel with the coil. A varistor has a high resistance at the normal operating voltage, but when the voltage exceeds a certain level, its resistance automatically becomes much lower.

9.8 Typical inductance of coils

Coils that have air cores usually have inductances that are measured in microhenries or millihenries. On the other hand, if the coil is wound on an iron core, the inductance may be several henries. For example, the coil in Fig. 9-15 has an inductance, by itself, of 80 mH. But if it is mounted on a 12 × 12 cm soft iron core having an air gap of 1 mm, the inductance increases to about 18 H (Fig. 9-16).

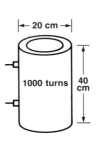

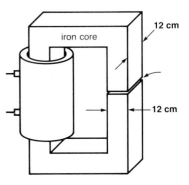

Figure 9-15
This coil having an air core
has an inductance of 80 mH.

Figure 9-16
The same coil mounted on an iron core
having an air gap has an inductance of
18 H.

9.9 Inductance of a transmission line

In discussing a transmission line, we are usually concerned with its resistance. However, a transmission line also has inductance. When a line carries current, it behaves like a huge, single turn of wire (Fig. 9-17). In effect, the current in the line produces a magnetic field, and so the line behaves like an inductor. The inductance

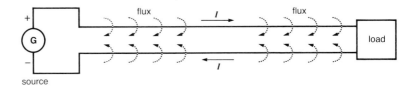

Figure 9-17
A two-conductor transmission line has inductance because it acts like a long turn of wire.

depends upon the length of the line, the spacing between the conductors, and their diameter. For example, a line 10 km long composed of two conductors having a diameter of 25 mm and spaced 3 m apart has an inductance of about 20 mH. If the length of the line were doubled, its inductance would double.

9.10 Inductor carrying an ac current

Alternating current is almost exclusively used on power distribution systems. Accidental short-circuits can occur on such systems, and the resulting currents can be enormous. To limit their magnitude, inductors are sometimes connected in series with the line. Inductors are also used to stabilize the voltage on transmission lines. Because of these practical applications, it is important to understand the voltage-current relationship when an inductor carries a sinusoidal ac current.

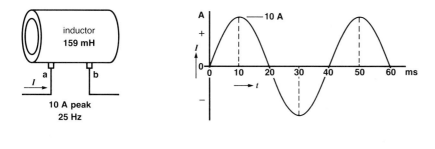

Figure 9-18
Perfect inductor carrying a sinusoidal ac current.

Figure 9-19
Waveshape of the current flowing in the inductor of Fig. 9-18.

Figure 9-18 shows a coil having an inductance of 159 mH, connected in series with a 25 Hz ac line. It carries a sinusoidal current I having a peak value of 10 A. The duration of one cycle is 1/25 s, or 40 ms (Fig. 9-19).

A voltage E_{ab} is induced across the terminals of the coil, and its magnitude is given by Eq. 9-5:

$$E_{ab} = L\ \frac{\Delta I}{\Delta t}$$

where L is the inductance of the coil.

Let us try to get a general idea of the waveshape of the induced voltage. First, the rate of change of current is zero at instants corresponding to 10, 30, and 50 ms. Consequently, the voltage is zero at these instants.

Second, the current is positive between zero and 10 ms. It is therefore flowing in the direction of the arrow (Fig. 9-18). But because the current is increasing, terminal **a** is $(+)$, and so E_{ab} is positive during this interval (Fig. 9-20). By the same reasoning, E_{ab} is negative between 10 ms and 20 ms.

We can generalize by stating that E_{ab} is $(+)$ when the slope of I is $(+)$.

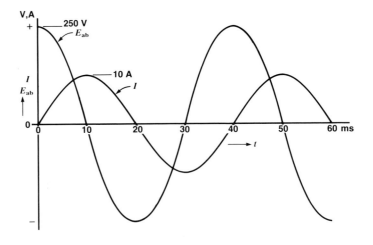

Figure 9-20
Waveshape of the ac voltage induced across the terminals of the inductor in Fig. 9-18.

Third, the rate of change of current is greatest at 0, 20, 40, and 60 ms. The induced voltage is therefore greatest at these instants. Recalling Eq. 7-2 in Section 7-10, the maximum rate of change of current is:

$$(\Delta I / \Delta t)_{max} = 2\pi f I_m = 2 \times 3.1416 \times 25 \times 10 = 1570.8 \text{ A/s}$$

The maximum induced voltage is therefore

$$E_{ab} = L \frac{\Delta I}{\Delta t} = 0.159 \times 1570.8 = 250 \text{ V}$$

Based on this line of reasoning, E_{ab} has the waveshape given in Fig. 9-20. It so happens that E_{ab} is also sinusoidal, just as I is. But E_{ab} and I are out of phase because when the one is maximum, the other is zero, and vice versa. Based on the discussion of Section 7-12, the current lags 90° behind the induced voltage. This is one of the fundamental properties of an inductor. We will examine some of its other properties when we reach Chapter 13 on ac circuits.

9.11 Summary

We learned that when the flux of one coil links with a second coil, mutual induction takes place between them. Mutual induction enables us to transfer energy from one coil to another without the need of an electrical connection.

We also discovered that a coil always has self-inductance (measured in henrys), in addition to its resistance. Self-inductance produces some important effects: (a) it opposes the increase (or decrease) of current in a coil, (b) it causes a high voltage to appear whenever the circuit of a coil is opened, and (c) it causes the current to lag 90° behind the voltage in a sinusoidal ac circuit.

An inductor also has the ability to store energy, which is why it can act either as a source or a load. It acts as a source whenever the current is decreasing and as a load when the current is increasing.

This chapter has made use of the important concepts explained in Chapter 7. Thus, we were able to make use of the maximum rate of change of a sinusoidal current and to observe the phase angle between an ac voltage and current.

The vast majority of machines and devices contain one or more coils. Consequently, the basic principles we have covered in this chapter will be particularly useful in subsequent studies.

TEST YOUR KNOWLEDGE

9-1   If we increase the number of turns on a coil, its inductance increases.

☐ true   ☐ false

9-2   In Fig. 9-1, if $E_{34} = -20$ V, there is an induced voltage in coil A.

☐ true   ☐ false

9-3   In Fig. 9-3, the coil has a resistance of 1 Ω. If the switch is closed, the initial current is

a. 0   b. 3 A   c. 1.5 A

9-4   When the current in a coil is constant for 1 minute, the induced voltage in the coil is zero during that period.

☐ true   ☐ false

9-5   When the current in a coil is constant for one tenth of a microsecond, the induced voltage during that short period is

a. very high   b. zero   c. dependent upon the coil inductance

9-6   The energy stored in a coil is constant during one second. This means that the current in the coil is

a. zero   b. constant   c. increasing

9-7   The current in a coil is momentarily 5 A, and the voltage is 60 V. If the current is increasing and the coil resistance is zero, the coil

a. absorbs 300 J   b. absorbs 300 W
c. delivers 300 W to the circuit

9-8   When an inductor carries a sinusoidal current, the voltage leads the current by 90°.

☐   true   ☐   false

9-9   A coil has a self-inductance of 7 H. If the instantaneous current is 5 A, the induced voltage is

a. unknown   b. 35 V   c. 0 V

9-10   A sinusoidal voltage has a frequency of 400 Hz and a peak value of 200 V. The maximum rate of change is

a. 80 kV/s   b. 502.6 kV/s   c. 160 kV/s

QUESTIONS AND PROBLEMS

9-11   The current in a coil changes from 5 A to 1 A in 0.4 s. If the induced voltage is 40 V, calculate the self-inductance of the coil.

9-12   In Fig. 9-1, $E_b = +40$ V when I_a changes at the rate of 2 A/s. Calculate a) the value of E_b when $\Delta I_a/\Delta t = 90$ A/min and b) the value of the mutual inductance M.

9-13   In Problem 9-12, the connections are interchanged so that coil B is excited, and coil A is on open circuit. If I_b changes at the rate of 2 A/s, what is the value of E_a?

9-14   In Fig. 9-1, what is the polarity of terminal 1 if I_a is positive and decreasing?

9-15   A coil having an inductance of 3 H carries a current of 40 A. What is the energy stored in the magnetic field?

9-16   A large inductor used in atomic research has an inductance of 10 mH and a resistance of 4 mΩ. If it carries a constant dc current of 6000 A, calculate a) the energy stored in the coil, b) the voltage across the terminals, and c) the time constant of the coil.

9-17   In Fig. 9-10, the coil has an inductance of 4 H and a resistance of 2 Ω. The supply voltage E is 12 V. When the switch is closed, current I increases gradually from zero to its final value. Knowing that the equivalent circuit of the coil is given by Fig. 9-9, calculate:

a. the final value of I

b. the IR drop when $I = 5$ A

c. the induced voltage when $I = 5$ A

d. the rate at which the current is changing when $I = 5$ A

e. the energy stored in the coil when $I = 5$ A

9-18   A 12 V battery is connected to an inductor of 2 H. If the inductor resistance is negligible, what is the current in the coil after 0.1 s? after 8 s? after 1 min? after 1 h?

9-19   If the coil in Fig. 9-18 carries a current having a peak value of 10 A and a frequency of 50 Hz, what is the peak value of the induced voltage?

9-20   A coil whose inductance is 10 H carries a sinusoidal current having an effective value of 2 A and a frequency of 60 Hz. Calculate the peak value of the induced voltage.

CHAPTER 10

Capacitors and Capacitance

Introduction and Chapter Objectives

Capacitors are simple devices composed of a set of plates separated by a dielectric. Like inductors, they are able to store energy. An inductor stores energy when current flows in a coil but a capacitor stores it when a voltage exists across its plates.

We will describe how a charge transfer between the plates takes place. This leads us to a definition of the farad, which is the unit of capacitance. Then, after studying the charge and discharge of a capacitor, we will use a numerical example to illustrate its basic behavior when it is connected to an ac line.

10.1 Free electrons in a metal

In Section 2-3, we learned that an enormous number of free electrons exist inside a conductor. It is estimated that in metals there are about 6×10^{19} per cubic millimeter. Because 1 coulomb corresponds to 6.24×10^{18} electrons, the free electron charge is equivalent to 10 coulombs per cubic millimeter.

The electrons are "free" in the sense that they are not attached to any particular atom. However, they cannot escape the block of metal, despite their high speed. The boundaries of the metal act like walls which prevent the electrons from flying out. However, if external electrons touch the metal surface, they are immediately captured and themselves become prisoners inside the "walls".

Normally, the negative charge carried by the free electrons is exactly counterbalanced by the equivalent positive charge on the stationary nuclei of the atoms. Thus, a conductor is normally neutral.

10.2 Transfer of charge and potential difference

Consider two blocks of metal A and B perfectly insulated from each other and both electrically neutral (Fig. 10-1). Suppose that by some means we are able to remove electrons from block A and deposit them on block B. The resulting lack of electrons on A gives it a positive charge, and the corresponding surplus of electrons on B gives it a negative charge. This difference in charge produces a difference of potential between the two blocks. If we connected a voltmeter between them, we would measure a voltage.

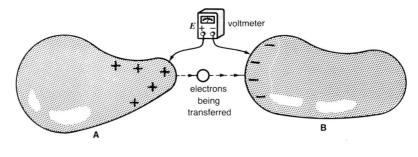

Figure 10-1
Transferring electrons from one body to another creates a difference of potential between them.

The more electrons we carry from A to B, the greater the voltage becomes. The magnitude of the voltage depends therefore upon the number of coulombs that have been transferred. However, it also depends upon the shape of the bodies, the distance between them, and the kind of dielectric (air, oil, paper, etc.) that separates them.

Only a very small fraction of the free electrons in a body has to be transported to produce a high voltage. For example, suppose the bodies are two identical plates having an area of 1 cm^2 and a thickness of 0.01 mm and separated by a distance of 1 cm (Fig. 10-2). Each plate has a volume of 1 mm^3 and contains about 10 coulombs of free electrons. If we transfer only 0.000 000 001 coulombs from one plate to the other, the resulting voltage between them will reach about 10 000 volts (Fig. 10-3)!

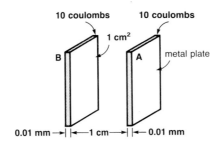

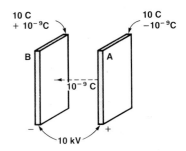

Figure 10-2
Each metal plate contains about 10 coulombs of free electrons.

Figure 10-3
If only 10^{-9} coulombs are transferred, the resulting voltage is 10 kV.

Obviously, in making such transfers, we don't have to worry about running short of available electrons.

10.3 Energy stored in an electric field

Suppose that a substantial electric charge is transferred from block A to block B in Fig. 10-4, making A' strongly positive and B strongly negative. If we now carry an

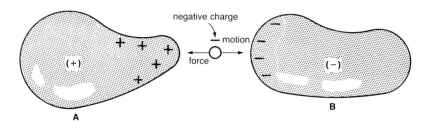

Figure 10-4
Mechanical work is needed to transfer the electric charge. This is converted into electric energy which is stored in the electric field.

additional negative charge from A to B, it will be repelled by the negative charge on B and attracted by the positive charge on A. Thus, a force will act to the left as we move the charge to the right. Consequently, we have to use energy and do real mechanical work whenever we transfer charges from A to B.

What happens to this energy? It is stored in the so-called *electric field* between the two blocks. Energy can therefore be stored in an electric field just as it can be stored in a magnetic field.

The energy stored in the electric field is given by the equation:

$$W = \frac{1}{2} QE \tag{10-1}$$

where

W = energy stored, in joules (J)

Q = electric charge transferred between the two bodies, in coulombs (C)

E = voltage between the bodies, in volts (V)

10.4 Capacitance; the farad (*F*)

Returning to Fig. 10-1, we recall that the voltage E between the charged bodies increases in direct proportion to the charge Q that is transferred. It so happens that for any pair of bodies that are fixed in relation to each other, the ratio Q/E is a constant. This constant is called the *capacitance* of the two bodies. The equation is:

$$C = Q/E \qquad\qquad (10\text{-}2)$$

where

C = capacitance of the two bodies, in farads (F)

Q = charge transferred, in coulombs (C)

E = voltage between the bodies, in volts (V)

The SI unit of capacitance is the farad (symbol F). The capacitance of two bodies is equal to 1 farad if the transfer of 1 coulomb between them produces a difference of potential of 1 volt. In practice, the farad is a much larger capacitance than we ordinarily encounter. For this reason, the microfarad, equal to one millionth of a farad, is a more commonly used unit.

The capacitance of two bodies is a basic property they have, that is just as real as their physical size.

Example 10-1:

In Fig. 10-4, electrons equivalent to 0.2 C are transferred from A to B, and the resulting voltage between the bodies is 2000 V. Calculate

1. the work done in transferring the electrons

2. the energy stored in the electric field

3. the capacitance of the two bodies

Solution:

1. The work done is

   $W = 1/2\ QE$ Eq. 10-1

   $= 1/2 \times 0.2 \times 2000$

   $= 200$ J

2. The energy in the field is equal to the work done in transferring the charges, or 200 J

3. The capacitance of the two bodies is

   $C = Q/E$ Eq. 10-2

   $= 0.2/2000 = 0.0001$ F $= 100$ microfarads $= 100\ \mu$F

10.5 Capacitors

Capacitors are commercial devices that can store an electric charge. They are composed of two metal plates separated by a thin insulator (Fig. 10-5). The insulator (called dielectric) may be air, mica, or paper, among other things, and the plates are usually made of thin aluminum foil.

metal plates

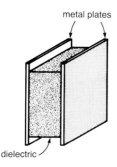

dielectric

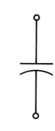

Figure 10-5
A capacitor consists of two metal plates separated by a dielectric.

Figure 10-6
Circuit diagram symbol for a capacitor.

The capacitance of such a parallel-plate capacitor depends on three factors:

1. the area of the plates
2. the distance between the plates
3. the type of dielectric between the plates.

The capacitance is given by the equation:

$$C \;=\; 8.85 \;\times\; 10^{-12} \, \frac{kA}{d} \tag{10-3}$$

where

$C =$ capacitance, in farads (F)

$A =$ area of the plates, in square meters (m^2)

$d =$ distance between the plates, in meters (m)

$k =$ dielectric constant of the insulation, a simple number

$8.85 \times 10^{-12} =$ constant to take care of units

The dielectric constant of air or a vacuum is equal to 1, and that of oil and paper is about 3. Consequently, for a given capacitor size (A and d), the capacitance is increased by a factor of 3 when paper, instead of air, is used to separate the plates.

According to Eq. 10-3 the capacitance doubles when the area of the plates is doubled. But it also doubles if the distance between them is halved. As a result, in manufacturing a capacitor of a given value, we tend to minimize the distance between the plates, in order to reduce its size and cost.

The circuit diagram symbol for a capacitor is shown in Fig. 10-6.

10.6 Energy stored in a capacitor

By combining Eqs. 10-1 and 10-2, we obtain another useful equation for the energy stored in a capacitor:

$$W = \frac{1}{2} CE^2 \qquad (10\text{-}4)$$

where

W = energy, in joules (J)

C = capacitance, in farads (F)

E = voltage across the capacitor, in volts (V)

It is interesting to recall that the energy stored in a magnetic field is given by a similar equation (Eq. 9-6):

$$W = \frac{1}{2} LI^2$$

Example 10-2:

Two aluminum plates each having an area of 2 m^2 are separated by a sheet of paper 0.08 mm thick. The capacitor is charged to 1000 V. If the dielectric constant of paper is 2.5, calculate:

1. the capacitance, in microfarads

2. the charge on the negative plate, in coulombs

3. the energy stored in the capacitor, in joules

Solution:

1. $C = 8.85 \times 10^{-12} \dfrac{kA}{d}$ 　　　　　　　　　　　　　Eq. 10-3

$$C = 8.85 \times 10^{-12} \times \frac{2.5 \times 2}{0.08 \times 10^{-3}}$$

$\qquad = 0.000\ 000\ 553$ farads $= 0.553 \times 10^{-6}$ F

$\qquad = 0.553$ microfarads $= 0.553\ \mu F$

2. $Q = CE$ 　　　　　　　　　　　　　　　　　　　　　Eq. 10-2

$Q = 0.553 \times 10^{-6} \times 1000 = 0.553 \times 10^{-3}$ coulombs

$\qquad = 553$ microcoulombs $= 553\ \mu C$

3. $\quad W = \dfrac{1}{2}\ CE^2$ Eq. 10-4

$$= \dfrac{1}{2} \times 0.553 \times 10^{-6} \times (1000)^2$$

$$= 0.276 \text{ joules } = 0.276 \text{ J}$$

10.7 Using a source to transfer a charge

The simplest way to transfer electrons from one capacitor plate to another is to connect the plates to the terminals of a battery (Fig. 10-7). Electrons are immediately removed from the plate connected to the ($+$) battery terminal and deposited on the plate connected to the ($-$) battery terminal. When an electric charge is transferred this way, the voltage between the capacitor plates builds up very quickly. However, as soon as the potential difference between the plates is equal to the battery emf, the transfer of electrons ceases. We can then disconnect the battery, and the voltage across the plates will be maintained forever, at least in theory (Fig. 10-8).

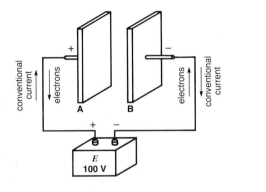

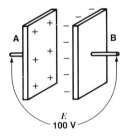

Figure 10-7
Electrons can be transferred from one plate to another by using a battery.

Figure 10-8
A perfect capacitor remains charged forever.

   Let us connect a resistor across the capacitor (Fig. 10-9). The difference of potential between the plates will cause electrons to flow from the negative plate, through the resistor, and into the positive plate. Consequently, the capacitor will gradually discharge, and the voltage between the plates will eventually fall to zero. When this happens, the two plates will again be electrically neutral.

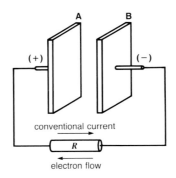

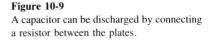

Figure 10-9
A capacitor can be discharged by connecting a resistor between the plates.

10.8 Capacitor as a source or load

Referring again to Fig. 10-7, we recall that conventional current flow is opposite to electron flow, Consequently, when the capacitor is charging, conventional current flows out of the positive battery terminal. Thus, the battery is a source, delivering energy to the capacitor. This is to be expected because the electric field between the plates is building up. The capacitor is acting as a load because current is flowing into its (+) terminal.

In Fig. 10-9, the capacitor is discharging its energy into a resistor. As a result, conventional current is flowing out of its (+) terminal. Consequently, the capacitor is now acting as a source.

A capacitor can therefore behave either as a source or load, which is similar to the behavior of a coil.

10.9 Commercial capacitors

Commercial capacitors are made of two extremely thin sheets of metal separated by equally thin sheets of paper or plastic. The metal and plastic sheets are rolled into a cylinder which is placed inside a protective can (Fig. 10-10). High-voltage capacitors are often impregnated with oil to provide additional security against breakdown. The question of insulation breakdown is important. As we have seen, the capacitance can be increased by reducing the thickness of the dielectric between the plates. But if the dielectric is too thin, it may no longer be able to withstand the voltage between the plates. In this case, an arc will leap between the plates, puncturing and carbonizing the dielectric and putting the capacitor out of service.

A capacitor designed to operate at 1000 V must therefore have a thicker dielectric than one operating at only 100 V. To attain the same capacitance, the plates of a 1000 V capacitor must therefore have a larger area. It follows that a 1000 V capacitor is more bulky than a 100 V capacitor of equal microfarad rating.

The rated voltage of a capacitor is usually indicated on the nameplate. When capacitors of several thousand volts are needed, they are made by connecting lower-voltage units in series. Similarly, when a very large capacitance is required, the practice is to connect several identical capacitors in parallel.

Figure 10-10
Typical construction of a commercial capacitor.

Table 10A gives an idea of the approximate size of commercial capacitors.

TABLE 10 A   TYPICAL CAPACITOR SIZES

Capacitance (mF)	Operating voltage (V)	Approximate dimensions (mm)	Approximate dimensions (in)
40	440 (ac)	140 × 115 × 75	5.5 × 4.5 × 3
5	30 000 (dc)	610 × 360 × 180	24 × 14 × 7
1.3	14 400 (ac)	460 × 340 × 115	18 × 13 × 4.5

In the interest of safety, large commercial capacitors contain discharge resistors permanently connected across the capacitor terminals. The resistors are usually inside the protective casing of the capacitor. Capacitors rated 600 V and lower are designed to discharge within 1 minute after being disconnected from the line. Capacitors rated above 600 V become discharged after a 5-minute interval. However, it is always good practice to short-circuit the terminals of a capacitor before touching them.

10.10 Capacitors in parallel and in series

When capacitors are connected in parallel, the total capacitance C_T is equal to the sum of the individual capacitances.

$$C_T = C_1 + C_2 + C_3 + \ldots + C_n \qquad (10\text{-}5)$$

When a group of capacitors is connected in series, the total capacitance C_T is given by the equation:

$$\frac{1}{C_T} = \frac{1}{C_1} + \frac{1}{C_2} + \frac{1}{C_3} + \ldots + \frac{1}{C_n} \qquad (10\text{-}6)$$

Example 10-3

Three capacitors of 8, 42, and 56 microfarads are connected in series. What is the resulting capacitance of the group?

Solution:

$$\frac{1}{C_T} = \frac{1}{8} + \frac{1}{42} + \frac{1}{56} \qquad \text{Eq. 10-6}$$

$$= 0.1250 + 0.0238 + 0.0178 = 0.1666$$

$$C_T = \frac{1}{0.1666} = 6 \ \mu F$$

10.11 Charging a capacitor; time constant

When a capacitor charges, how do the voltage and current vary with time? Consider the circuit of Fig. 10-11, in which a capacitor C and resistor R are connected to a battery by means of a switch. The battery produces a voltage E_b. The capacitor is initially discharged, and so the voltage across its terminals is zero. However, as soon as the switch is closed, current begins to flow in the circuit, and the capacitor charges. As a result, voltage E_c across the capacitor increases gradually with time. But as E_c increases, voltage E_R across the resistor decreases because $E_r = E_b - E_c$ a smaller E_r means that current I decreases as the capacitor charges. Eventually, the current becomes zero when $E_c = E_b$.

If we plot the graph of E_c versus time, we get the curve shown in Fig. 10-12. The rate at which the capacitor charges up depends upon the so-called *time constant* of the circuit. The time constant is given by the equation:

$$T = RC \qquad (10\text{-}7)$$

where

$T = $ time constant, in seconds (s)

$R = $ resistance of the circuit, in ohms (Ω)

$C = $ capacitance, in farads (F)

For all practical purposes, we can assume that the capacitor is completely charged after a period equal to 5 T seconds.

Note that the initial current in the circuit is given by $I = E_b / R$. Thus, if the series resistance is low, the initial charging current is very large. A large initial current (called inrush current) will not harm the capacitor, but it may burn the switch contacts at the moment the switch closes.

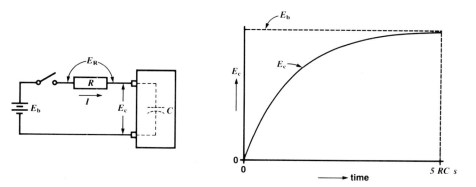

Figure 10-11
The voltage across a capacitor builds up
gradually when the capacitor is charged
in series with a resistor.

Figure 10-12
The capacitor is fully charged after a
period of 5*RC* seconds.

Example 10-4:

A 100 μF capacitor is connected in series with a 20 Ω resistor to a 128 V dc source.
Calculate

1. the time constant of the circuit

2. the approximate time to charge the capacitor

3. the initial current in the circuit

Solution:

1. The time constant is
    $$T = RC = 20 \times 100 \times 10^{-6} = 2 \times 10^{-3} = 2 \text{ ms} \qquad \text{Eq. 10-7}$$

2. The time to charge the capacitor is
    $$T = 5\,RC = 5 \times 2 = 10 \text{ milliseconds} = 10 \text{ ms}$$

3. The initial current is
    $$I = E_b / R = 128/20 = 6.4 \text{ A}$$

10.12 Discharging a capacitor

We can discharge a capacitor by connecting a resistor across its terminals. If a low
resistance is used, the discharge current can be very high. Even a small capacitor
can deliver a current of several hundred amperes. This is an important feature of a
capacitor: it can deliver very large currents for brief periods.

Figure 10-13 shows a capacitor that is in the process of discharging. Note that
current is flowing out of the (+) terminal, and so the capacitor is acting as a source.
The energy dissipated in the resistor is obtained at the expense of the decreasing
electric field between the plates. The voltage across the capacitor falls rapidly at
first and then much more slowly (Fig. 10-14). We can assume that a capacitor is
completely discharged after a period equal to 5 *T* seconds.

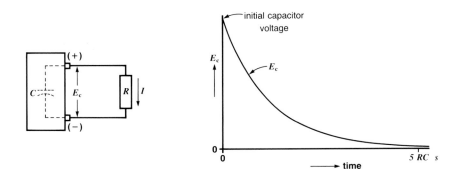

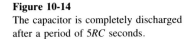

Figure 10-13
The voltage across the capacitor
decreases with time.

Figure 10-14
The capacitor is completely discharged
after a period of 5RC seconds.

10.13 Basic capacitor equation

When the voltage across a capacitor is constant, the capacitor is neither charging nor discharging, and so the capacitor current is zero (Fig. 10-15). But if the capacitor voltage is rising, the capacitor is charging some more, and so a current flows (Fig. 10-16). Conversely, if the voltage is falling, the capacitor must be discharging, and a current again flows, but in the opposite direction (Fig. 10-17).

The current in a capacitor does not depend upon the magnitude of the applied voltage, but only on the rate at which it is changing. In effect, the current depends directly upon the capacitance and the rate of change of voltage. This is expressed by the equation:

$$I = C \times \text{rate of change of voltage} \qquad (10\text{-}8)$$

or

$$I = C \, \frac{\Delta E}{\Delta t} \qquad (10\text{-}9)$$

where

I = capacitor current, in amperes (A)

C = capacitance, in farads (F)

ΔE = change in the capacitor voltage, in volts (V)

Δt = duration of the change, in seconds (s)

Example 10-5:

The voltage across the terminals of a 100 μF capacitor changes from 500 V to 200 V in 0.001 s. What is the current during this interval?

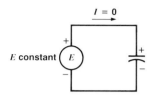

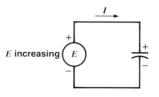

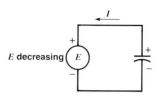

Figure 10-15
The current in a capacitor is zero whenever the voltage across it is constant.

Figure 10-16
A current flows whenever the voltage across a capacitor is *increasing*.

Figure 10-17
A current flows whenever the voltage across a capacitor is *decreasing*.

Solution:

1. The change in voltage is: $\Delta E = 500 - 200 = 300$ V

2. Using Eq. 10-9 we have:

$$I = C \frac{\Delta E}{\Delta t} = \frac{100 \times 10^{-6} \times 300}{0.0001} = 30 \text{ A}$$

Equation 10-8 tells us that for a given capacitance, the faster the voltage changes, the larger the current is.

The voltage changes very quickly when an uncharged capacitor is suddenly connected to a source E_s, with a switch. The capacitor voltage leaps from zero to E_s, and so the rate of change of voltage is theoretically infinite. The resulting current is also theoretically infinite. But in practice, the current is limited by the resistance and inductance of the source. Nevertheless, it can easily reach several hundred amperes, and the switch must be designed to carry this heavy current without damage. Otherwise, current-limiting resistors or inductors must be used to keep the inrush current within the rating of the switch.

Small capacitors are sometimes used to suppress high-voltage spikes on industrial power circuits. Such spikes are caused by lightning or switching disturbances. The capacitors (usually rated at about 0.5 μF) are placed across the terminals of electric equipment that has to be protected. When a spike of several thousand volts comes in over the line, the sudden change of voltage causes the capacitor to draw a large current. This absorbs the energy in the spike and reduces its amplitude to safe limits, as far as the equipment is concerned.

This application highlights another important property of capacitors: they oppose any change in voltage across the terminals.

10.14 Capacitance of transmission lines

A transmission line has capacitance because the two parallel conductors act like the plates of a capacitor (Fig. 10-18). There is also a capacitance between each line and ground because the ground acts like an enormous flat plate. The total capacitance depends upon the length of the line, the diameter of the conductors, and the spacing between them. For example, a line 10 km long composed of two conductors having a diameter of 25 mm and spaced 3 m apart has a capacitance of about 0.05 μF.

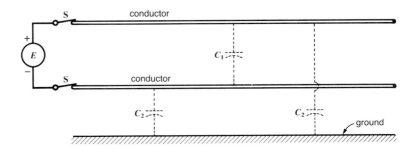

Figure 10-18
The two conductors of a transmission line act like the plates of a capacitor.

The existence of line capacitance raises certain dangers. Suppose the transmission line in Fig. 10-18 is connected to a 69 000 V source, and the switch at the source end is then opened. Although the line is completely disconnected, it will remain charged, and substantial, dangerous voltages will remain between the conductors and ground. If linemen are doing maintenance work, it is essential to short-circuit the lines to ground before beginning repairs, so that the lines are discharged.

10.15 Capacitor connected to an ac voltage

We have already learned that industrial line voltages vary sinusoidally. It is therefore useful to see how the current in a capacitor varies under these ac voltage conditions.

The capacitor in Fig. 10-19 is connected to a 25 Hz ac source that generates a sinusoidal voltage E_{ab} having a peak value of 250 V. The capacitor has a rating of 255 μF. Because the frequency is 25 Hz, the duration of one cycle is 1/25 s, or 40 ms. To determine the waveshape of the current we reason as follows:

First, according to Eq. 10-8, the current is zero when the rate of change of voltage is zero. This occurs at 10, 30, and 50 ms because the slope of E_{ab} is zero at these instants (Fig. 10-19).

Second, E_{ab} is positive and increasing between zero and 10 ms. The capacitor is charging, and so it is receiving energy from the source and acting as a load. Because terminal **a** is (+) during this interval, the actual current must flow in the direction of the arrow. The current is therefore (+) between 0 and 10 ms. By the same reasoning, I is (−) between 10 ms and 20 ms. Thus, I is (+) whenever the slope of E_{ab} is (+).

Third, the current is greatest when the voltage is changing most rapidly. This occurs at 0, 20, 40, and 60 ms. Thus, the current is maximum when the capacitor

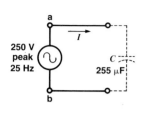

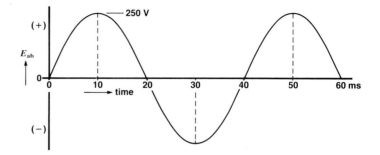

Figure 10-19
A 250 V (peak), 25 Hz voltage is applied to a 255 μF capacitor.

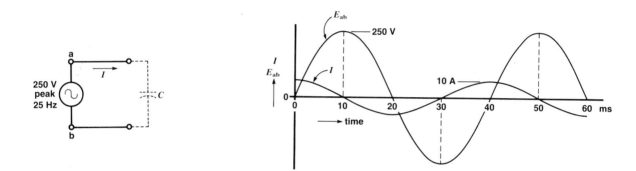

Figure 10-20
The resulting current is 10 A (peak), and it leads the voltage by 90°.

voltage is zero. This may be surprising, but we must remember that the current in a capacitor does not depend upon the *magnitude* of the voltage across its terminals, but only upon the *rate* at which the voltage is changing.

Based on Eq. 7-1, Section 7-9, we recall that the maximum rate of change of voltage is given by

$$(\Delta E / \Delta t) = 2\pi f E_m$$

$$= 2 \times 3.1416 \times 25 \times 250 = 39\ 270 \text{ V/s}$$

The peak current is therefore

$$I = C \frac{\Delta E}{\Delta t} = 255 \times 10^{-6} \times 39\ 270 = 10 \text{ A}$$

The waveshape of the current is shown in Fig. 10-20.

It so happens that the current also varies sinusoidally. Clearly, E_{ab} and I are out of phase. Based on Section 7-12, the current is 90° ahead of the voltage. This is one of the important properties of a capacitor when it is connected to an ac source. You will recall that in the case of an inductor, the current lagged 90° *behind* the voltage. Thus, a capacitor and inductor behave somewhat like complementary twins.

10.16 Summary

This chapter showed us that a capacitor is able to store electrical energy, and that is why it can act as a source or a load. In many ways, the properties of a capacitor are similar to those of an inductor. However, the energy in an inductor is proportional to the square of the current, whereas in a capacitor it is proportional to the square of the voltage. Furthermore, in an inductor, the rate of change of current produces a voltage, whereas in a capacitor, the rate of change of voltage produces a current. In both devices, we can see the great importance of the rate of change of a voltage or current.

We learned that when a capacitor is connected to a sinusoidal ac voltage, it draws a sinusoidal ac current. The current leads the voltage by 90°, which is again the opposite to an inductor, in which the current lags behind the voltage. In later chapters we will see how these complementary properties of capacitors and inductors can be put to practical use.

TEST YOUR KNOWLEDGE

10-1   In Fig. 10-2, if electrons are transferred from plate B to plate A, the plates will be attracted to each other.

☐ true   ☐ false

10-2   In Fig. 10-2, if a thick sheet of paper is placed between the plates the capacitance of the plates will

a. increase   b. decrease   c. remain the same

10-3   In Fig. 10-5, the voltage between the plates is 1000 V. If the voltage drops to 200 V, this will affect the capacitance of the plates.

☐ true   ☐ false

10-4   In Fig. 10-11 $R = 10\ M\Omega$ and $C = 30\ \mu F$. The time constant of the circuit is:

a. 0.3 s   b. 5 min   c. 3 s

10-5   A capacitor that is initially charged to 600 V is connected across a 5 Ω resistor. The initial discharge current is

a. 120 A   b. apparently 120 A, but it depends upon the size of the capacitor

10-6   A capacitor that is charged to 600 V stores 100 J of energy. If the voltage increases to 900 V, the stored energy will be:

a. 150 J   b. 225 J   c. 44.4 J

10-7   A 50 μF and a 80 μF capacitor are connected in parallel; the resultant capacitance is

a. 130 μF   b. 30.77 μF   c. 94.34 μF

10-8   When two unequal capacitors are connected in series, the instantaneous current is the same in both

☐ true   ☐ false

10-9   In Fig. 10-20, if the frequency of the voltage is doubled, the peak current will become

a. 20 A   b. 10 A   c. 40 A

10-10   The current in a capacitor lags 270° behind the voltage

☐ true   ☐ false

10-11   The voltage in a capacitor lags 90° behind the current.

☐ true   ☐ false

QUESTIONS AND PROBLEMS

10-12   Why are industrial capacitors equipped with discharge resistors?

10-13   A 60 kV, 20 μF capacitor is equipped with a discharge resistor of 3 MΩ.

a) How long will it take the capacitor to discharge?

b) How much energy is dissipated as heat in the resistor?

10-14   A 100 μF capacitor is connected across a 60 Hz line having a peak voltage of 170 V. Calculate (a) the maximum rate of change of voltage and (b) the peak current in the capacitor.

10-15   Twelve 60 μF capacitors are connected in series across a 36 kV dc source. Calculate (a) the voltage across each capacitor, (b) the energy stored in each capacitor and (c) the total capacitance of the capacitor bank.

10-16   A special dc source supplies a constant current of 1 mA to a 50 μF capacitor. How long will it take for the voltage to increase from 20 V to 500 V?

10-17   A 100 μF capacitor carries a charge of 0.004 C. Calculate the voltage across the terminals.

10-18   A 10 μF capacitor rated at 600 V is smaller than a 10 μF capacitor rated at 60 kV. Explain.

10-19   We wish to assemble a capacitor bank of 300 μF, rated at 1200 V. A large number of 100 μF capacitors rated at 600 V are available. How many capacitors are needed, and how should they be connected?

10-20   The capacitor shown in Fig. 10-5 is composed of two plates each having an area of 2 m². The dielectric is a piece of plate glass 0.5 mm thick and having a dielectric constant of 7. The dielectric strength of the glass is 80 kV/mm. Calculate:

a)   the capacitance of the capacitor

b)   the maximum voltage that can be applied to the plates

c)   the maximum energy that can be stored without causing the dielectric to break down.

10-21   A 12 V car battery can supply a current of 15 A for a period of 2 h.

a)   How much energy does this represent in joules?

b)   How much energy is stored in the 5 μF capacitor listed in Table 10A?

c)   How many such capacitors are needed to store the same amount of energy as that in the car battery?

CHAPTER 11

Direct Current Generators

Introduction and Chapter Objectives

Whenever we use mechanical rotation to generate electricity, the output is always an ac voltage. This is true for any conceivable arrangement of coils, flux, and type of rotation.

The only way to obtain a dc voltage is to periodically reverse the terminals of the coil in which the voltage is induced. Thus, whenever the polarity is about to change, we reverse the terminals so that the output is always positive. This reversing action is called *commutation*. The device that achieves the reversing is called a *commutator*.

Today, electronic rectifiers that convert ac power into dc are often used instead of dc generators. These rectifiers are more efficient, and in addition, they have no moving parts. Why, then, do we study dc generators? The reason is that many dc motors operate briefly as generators. Furthermore, dc motors and generators are built the same way and share the same properties.

In this chapter, we will explain the basic method by which a generator produces a dc voltage. We then discuss the construction and properties of shunt and compound generators. The effect of armature reaction and its influence on sparking at the brushes are also covered. Finally, we will explain the use of commutating poles to improve machine performance.

11.1 Generating a dc voltage

In Section 8-8, we saw that an ac voltage is induced in a coil when it rotates between the poles of a permanent magnet. This simple generator is again shown in Fig. 11-1. However, we have added slip rings and brushes so that the ac voltage is available

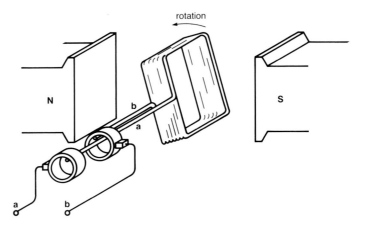

Figure 11-1
Simple ac generator producing an output voltage between terminals a and b.

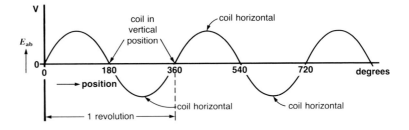

Figure 11-2
Waveshape of the ac output voltage and corresponding position of the coil
in Fig. 11-1.

between the stationary terminals a and b. The output voltage E_{ab} has the general shape given in Fig. 11-2.

In order to obtain a dc output, the negative half of the ac voltage must be rectified. Thus, when E_{ab} is rectified, we obtain the pulsating voltage E_{xy} shown in Fig. 11-3. We could attain this result by connecting a reversing switch to terminals a and b (Fig. 11-4). When E_{ab} is positive, terminal x is connected to a, and terminal y to b. But as soon as the polarity reverses, the switch is transferred so that terminal x is connected to b and terminal y to a. Because the polarity of E_{ab} changes every time the coil is in the vertical position, the switching action must be synchronized with the position of the coil.

A simpler way to rectify the voltage is to use a ring that is cut in half, so as to form two segments. The split ring is mounted on the shaft but insulated from it. The segments are directly connected to the two ends of the coil, as shown in Fig. 11-5. Two brushes enable us to bring the voltage to stationary terminals x and y. The

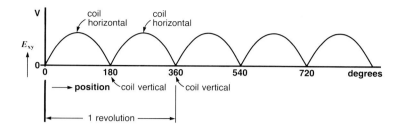

Figure 11-3
Waveshape of E_{ab} in Fig. 11-2 after rectification.

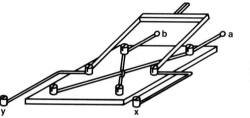

Figure 11-4
A manual reversing switch could rectify the ac output voltage in Fig. 11-1.

combination of the split ring and brushes is equivalent to the previous reversing switch. Note that the split ring is mounted in such a way that the switching action takes place when the coil is in the vertical position. The split ring is called a commutator.*

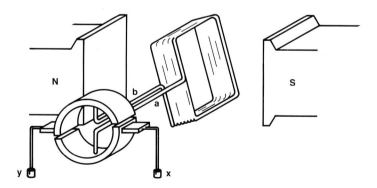

Figure 11-5
A split ring commutator acts as a reversing switch to rectify the ac voltage induced in the coil.

---

* The word commutator is the technical term for a switch.

11.2 Armature of a dc generator

We can obtain a higher dc voltage, and one that is much smoother, by increasing the number of coils and the number of segments on the commutator. The coils are identical and evenly spaced around a cylindrical iron core (Fig. 11-6). They are embedded in slots and insulated from the core by appropriate slot insulation. The coils are then connected to adjacent segments, and so there are as many segments as there are coils. A detailed explanation of the coil and commutator connections is given in Section 11-13.

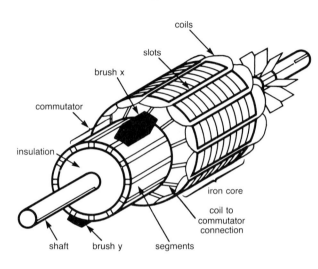

Figure 11-6
Components and construction of a dc armature.

Figure 11-7
Armature core is made of stacked laminations insulated from each other.

The assembly composed of the commutator, coils, and iron core is called the *armature*. The core is made of thin laminations stacked together to form a single block (Fig. 11-7). The laminated core reduces the eddy current losses when the armature rotates, as was explained in Section 8-13.

The dc output voltage is obtained by means of carbon brushes that ride on the commutator surface. In two-pole generators, the brushes are spaced 180° apart.

To summarize, when the armature of Fig. 11-6 rotates, the armature coils cut across the flux lines created by two stationary magnets (not shown). This action induces an ac voltage in each coil. The sum of the ac voltages induced in the coils is rectified by means of a commutator. The resulting dc voltage is picked off by means of two brushes riding on the commutator. The output voltage fluctuates slightly (Fig. 11-8), therefore it is not as steady as the voltage produced by a battery. However, this so-called "commutator ripple" is unimportant in most industrial applications.

Figure 11-9 and Fig. 11-10 show the armature of a commercial dc generator. The much larger armature of a 500 kW generator is shown in Fig. 11-11.

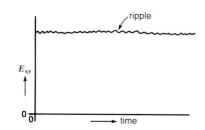

Figure 11-8
The dc output voltage of a multi-segment commutator contains only a small ripple.

Figure 11-9
Armature of a commercial dc generator showing coils, commutator, iron core, cooling fan, and bearings. (*Courtesy General Electric Company, USA*).

Figure 11-10
Closeup of the armature in Fig. 11-9, showing slot insulation, laminated core and wedges that hold the coils in the slots. The steel band on the right firmly holds the coils against the centrifugal forces. (*Courtesy General Electric Company, USA*).

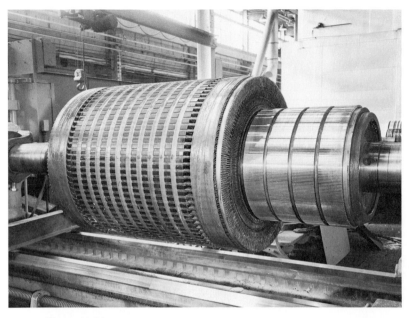

Figure 11-11
Armature of an 1150 kW, 750 V, 1000 r/min dc generator. (*Courtesy General Electric Company, USA*).

11.3 Field of a dc generator

The magnetic flux in a generator is created by its "field". Here, the term field means all the components needed to create the flux in the machine. Figure 11-12 shows the typical construction of the field of a two-pole dc generator. A circular cast-steel frame supports the salient poles and also provides a low-reluctance path for the magnetic flux. Mounting feet and lifting lugs are usually welded to the frame.

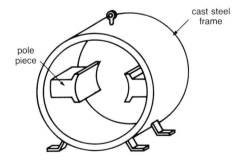

Figure 11-12
Field structure of a two-pole dc generator.

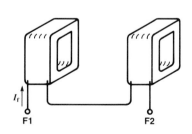

Figure 11-13
Shunt field coils are mounted on the pole-pieces and connected in series.

Two identical pre-wound coils are slipped over the poles and are then connected in series. These so-called *shunt* coils usually have many turns of small wire, and produce the mmf that creates the flux in the machine. The two field leads are brought out to terminals F1 and F2 (Fig. 11-13).

11.4 Assembly of a dc generator

Figure 11-14 and Fig. 11-15 show cutaway views of a dc generator when it is assembled. The armature fits between the salient poles, leaving two small air gaps. The field flux flows across the air gaps and armature and then divides in two parts as it returns by way of the cast steel frame. Because the armature core is made of iron, the reluctance of the magnetic circuit is reduced. Consequently, the flux is much greater than it otherwise would be.

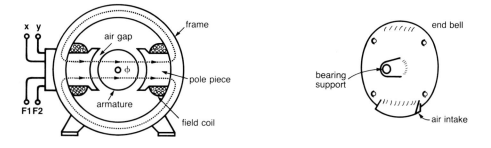

Figure 11-14
Cross section of a dc generator showing the flux path, and field terminals F1 and F2, and armature terminals X and Y.

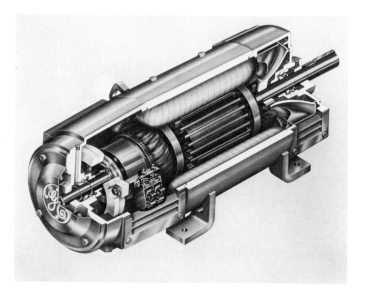

Figure 11-15
Cutaway view of a two-pole dc generator. (*Courtesy General Electric Company, USA*).

The armature shaft is supported by bearings that are mounted inside the two end-bells. The end-bells and frame usually have openings for the intake and exhaust of cooling air. A fan, mounted on the armature shaft, provides the required ventilation.

When the machine is being assembled, it is important to adjust the position of the brushes. They can be shifted as a unit, usually as much as 45° (Fig. 11-16). They are locked in the so-called *neutral* position, which corresponds to the spot where the dc output voltage at no-load is maximum.

Figure 11-16
Brush assembly of a four pole dc generator. The brushes are spring-loaded so they press uniformly against the commutator. (*Courtesy General Electric Company, USA*).

Another way to locate the neutral position is to connect the armature terminals x and y to a 60 Hz ac source. An ac voltmeter is connected across the field terminals F1, F2 and the brushes are shifted, as a unit, until the voltmeter reads zero. The brush yoke is then locked in this position. This method of finding the neutral position is based upon the principle of mutual induction discussed in Section 9-1. In effect, when the brushes are in the neutral position, the mutual inductance between the armature coils and field coils is zero.

11.5 Schematic diagram of a dc generator

A schematic diagram of the generator is given in Fig. 11-17. It shows the field flux ϕ created by field current I_f. Only one pole is shown, but it is understood that it

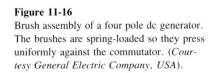

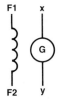

Figure 11-18
Schematic diagram of a separately excited dc generator using graphic symbols recommended by the Institute of Electrical and Electronic Engineers.

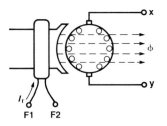

Figure 11-17
Schematic diagram of a dc generator. The single pole represents two or more poles.

represents both poles. The circles inside the armature represent the conductors that are lodged in the slots. The diagram can be further simplified to that shown in Fig. 11-18.

11.6 Current in the armature; armature reaction

Before studying the generator, let us see what happens when the armature carries a current. Referring back to Fig. 11-6, when current flows into brush y and out of brush x, it flows in all the armature coils. The armature then behaves like an electromagnet producing a mmf and corresponding N and S poles (Fig. 11-19). The armature mmf is almost as strong as the field mmf when the armature is carrying its rated current.

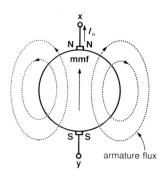

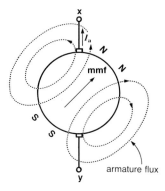

Figure 11-19
The armature produces a substantial mmf and flux when it carries its rated current. In this figure, the mmf acts along the axis of the brushes.

Figure 11-20
In some dc machines the armature mmf acts at an angle to the brush axis.

The direction of the armature mmf depends upon the position of the brushes. If the brushes are shifted 30°, the mmf shifts 30°. We are assuming that the mmf acts along the axis of the brushes (Fig. 11-19). In practice, it could be at any angle to the brush axis, as shown in Fig. 11-20. The direction of the mmf with respect to the brushes depends upon how the armature coils are connected to the commutator segments.

What happens to the mmf when the armature begins to rotate? It remains unchanged as long as the armature current I_a is the same. Thus, the mmf remains fixed in space whether or not the armature is turning. However, if the armature current reverses, the mmf reverses, as well as the N, S poles of the armature.

When the armature is mounted inside the generator, its mmf will have important magnetic effects. These effects are grouped under the general term *armature reaction*.

In Fig. 11-21, for example, the field mmf acts from left to right, but the direction of the armature mmf depends upon the brush setting. It so happens that when the

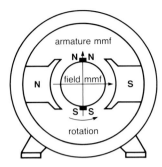
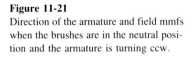

Figure 11-21
Direction of the armature and field mmfs when the brushes are in the neutral position and the armature is turning ccw.

brushes are in the neutral position, the armature mmf in a two-pole machine *always* acts at right angles to the field mmf. In a multi-pole machine the axis of the armature mmf is midway between adjacent poles.

11.7 Separately excited dc generator; no-load saturation curve

The flux in a dc machine is created by the field current I_f. When this current is supplied by an independent dc source, such as a battery, the generator is said to be *separately excited*. Figure 11-22 shows the schematic diagram of such a machine. The field current can be varied from zero to maximum by moving the potentiometer wiper from point A to point B. We assume the armature is driven by an internal combustion engine or electric motor.

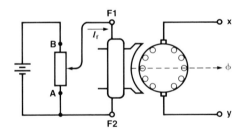

Figure 11-22
Separately excited generator using a potentiometer in order to vary the field current from zero to maximum.

When such a generator operates at no-load — with the armature terminals x and y on open circuit — the armature voltage E_{xy} depends upon the speed of rotation and the field current I_f.

If the speed of the driving motor is increased, the armature voltage increases in direct proportion to the armature speed. In effect, the armature conductors cut the flux lines more quickly, which raises the induced voltage in every conductor, according to the equation $e = BLv$. At no-load, the total induced voltage E_o appears as E_{xy} across the armature terminals.

On the other hand, if we increase the field current while keeping the speed constant, the flux ϕ will increase. The increase in flux will produce a proportional increase in the flux density B, and hence in the induced voltage E_o. However, as we continue to raise I_f, the iron in the magnetic circuit begins to saturate. This means that for a large increase in I_f, the flux will increase only slightly. Consequently, the induced voltage E_o and the corresponding output voltage E_{xy} will tend to flatten with increasing I_f.

If we plot E_{xy} versus I_f, we obtain a curve such as the one shown in Fig. 11-23. For low values of I_f, E_{xy} increases linearly with I_f. The reason is that the reluctance of the magnetic circuit is then mainly due to the air gaps. In Section 6-16, we saw that the flux in air is directly proportional to the mmf. But, as I_f is increased and the

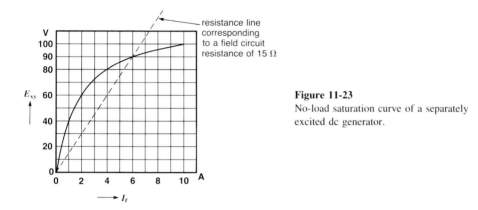

Figure 11-23

No-load saturation curve of a separately excited dc generator.

iron begins to saturate, the corresponding increase in flux is small, causing the curve to level off. This is called the *no-load saturation curve* of the generator.

If we reverse the direction of I_f, the flux will reverse, and this will reverse the polarity of E_{xy}.

Finally, if we reverse the direction of rotation, the polarity of E_{xy} will reverse.

11.8 Value of the induced voltage

The voltage induced in a dc generator is given by the equation

$$E_o = Kn\phi \qquad (11\text{-}1)$$

where

$E_o = $ dc voltage induced in the armature, in volts (V)

$n = $ speed of rotation, in revolutions per minute (r/min)

$\phi = $ flux per pole, in webers (Wb)

$K = $ a constant that depends upon the winding on the armature

The flux ϕ is that produced by the field coils. However, when the generator is under load, the value of ϕ can also be affected by the presence of the armature mmf.

At no-load, the magnitude of E_o can be measured by connecting a voltmeter across the armature terminals x and y.

Example 11-1:

The armature of a six-pole dc generator produces a voltage of 240 V when it turns at 800 r/min. The flux per pole is 0.04 Wb. Calculate the value of K for this machine.

Solution:
From $\qquad E_o = Kn\phi$ $\qquad\qquad\qquad\qquad$ Eq. 11-1

we have $\qquad 240 = K \times 800 \times 0.04$

therefore $\qquad K = 7.5$

11.9 Shunt generator at no-load

A *self-excited shunt generator* is one that provides its own field excitation. The shunt field is connected to the armature terminals, usually in series with a field rheostat R (Fig. 11-24). By varying the position of the rheostat, current I_f varies, which varies flux ϕ and the output voltage E_{xy}.

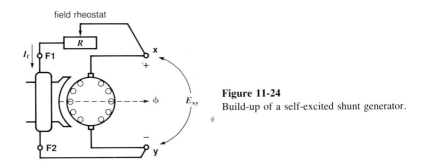

Figure 11-24
Build-up of a self-excited shunt generator.

Once the generator is in operation and generating a voltage, there is no problem in maintaining the field excitation. The question is, what happens when the generator is initially at rest and then put into service? If the flux from the poles is initially zero, E_{xy} will also be zero, no matter what the speed is. And if E_{xy} is zero, current I_f is zero, and so there is no excitation.

Fortunately, residual magnetism comes to the rescue. If the field poles are excited only once (such as by connecting them to a battery), some residual magnetism will remain. Consequently, when the generator runs at rated speed, a low voltage will be induced between terminals x and y. This will cause a weak current I_f to flow in the windings (Fig. 11-24). This increases the flux slightly above its residual value, which in turn increases E_{xy}, which increases I_f, which increases ϕ, which increases E_{xy} ... and so forth. Voltage E_{xy} will build up, and it seems it will become greater and greater without limit.

However, E_{xy} reaches a final value that depends upon the no-load saturation curve and the total resistance R_f of the shunt field circuit. If we draw a straight line corresponding to R_f, and superimpose it on the no-load saturation curve, the point of intersection gives the limiting value of E_{xy}. For example, given the saturation curve of Fig. 11-23, if $R_f = 15\ \Omega$, the resistance line intersects the curve at $E_{xy} = 90$ V. Because of the inductance of the field coils, it may take several seconds for the voltage to reach this final value (see Section 9-6).

If the resistance of the field rheostat is increased, the resistance line becomes steeper and a lower E_{xy} is obtained. If the resistance is too high the resistance line will no longer intersect the saturation curve, and the armature voltage will not build up.

Note that the armature voltage cannot build up if we reverse the direction of rotation. The reason is that the polarity of E_{xy} will reverse, causing the initial I_f to reverse, thus *reducing* the flux below its residual value. To remedy the situation, we must interchange the field connections F1, F2.

11.10 Separately excited generator under load

When a dc generator delivers power to a load (Fig. 11-25), several factors must be considered.

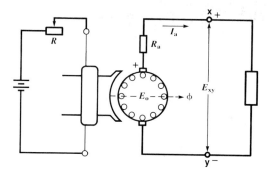

Figure 11-25
Separately excited dc generator under load.

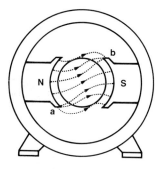

Figure 11-26
The flux crossing the armature becomes distorted and smaller because of armature reaction.

1. The flux ϕ induces a voltage E_o in the armature.

2. The armature has a low, but not negligible, resistance R_a. It causes an internal voltage drop $I_a R_a$, similar to that in a battery. Consequently, the voltage E_{xy} across the generator terminals is less than the induced voltage E_o.

3. The armature current I_a produces an armature mmf that acts at right angles to the field mmf. The resultant mmf is therefore inclined at an angle, as shown in Fig. 11-26. This causes the flux to move diagonally across the armature. Unfortunately, this increases the flux density in pole tips a and b, and so they become saturated. This, in turn, reduces the total flux ϕ that crosses the armature. As a result, the induced voltage E_o is less at full-load than at no-load.

The overall effect of armature resistance and armature reaction is that the armature voltage E_{xy} decreases with increasing load current, as shown in Fig. 11-27.

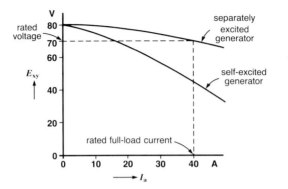

Figure 11-27
Voltage-regulation curves of separately excited and self-excited dc generators starting from the same no-load voltages. To obtain rated voltage at rated current, the no-load excitation of the self-excited generator must be raised.

The same voltage drop takes place in a self-excited shunt generator, but is accentuated. The reason is that as the armature voltage decreases (due to the effects just mentioned), there is a corresponding drop in field current. Therefore, the flux ϕ decreases even more with increasing load than it did before. Thus, the voltage regulation curve of a self-excited generator drops more steeply than that of a separately excited generator (Fig. 11-27).

11.11 Compound generator

One way to avoid the drop in armature voltage is to increase the field flux as the load increases. This can be accomplished by placing an extra winding around each field pole (Fig. 11-28). The winding carries the load current, and the mmf it produces adds to that of the shunt field. When this so-called *series* winding is added, we obtain a *compound generator*. Figure 11-29 and Fig. 11-30 show the construction of a 4-pole compound generator.

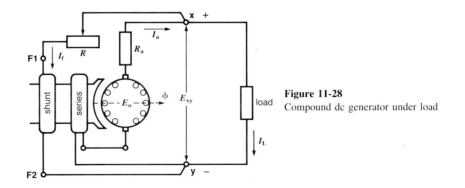

Figure 11-28
Compound dc generator under load

Figure 11-29
Construction of a four-pole compound generator showing the main poles and four intermediate commutating poles. (*Courtesy General Electric Company, USA*)

Figure 11-30
Construction of the field winding of the generator in Fig. 11-29. The two heavy terminals are the series field and the two small wires are the shunt field. (*Courtesy General Electric Company, USA*)

The series winding is composed of a few turns of heavy wire to carry the armature current. Depending upon the number of turns the terminal voltage under load can be made equal to, or greater than, the no-load voltage.

Typical voltage-regulation curves are shown in Fig. 11-31. The *flat-compound* generator has zero voltage regulation because the no-load voltage is equal to the full-load voltage. The *overcompound* generator has negative voltage regulation because the no-load voltage is less than the full-load voltage.

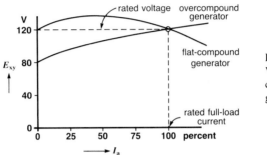

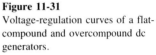

Figure 11-31

Voltage-regulation curves of a flat-compound and overcompound dc generators.

How much mechanical power is needed to drive the compound generator of Fig. 11-28? First, all the electric power that is consumed by the load, the shunt field, the series field, the armature resistance, and the shunt field rheostat must be equal to the total generated power, which is equal to $E_o I_a$. Second, this generated power is entirely produced by the "internal" mechanical power P_{mi} that is needed to drive the armature. Thus, we can write

$$P_{mi} = E_o I_a \tag{11-2}$$

where

P_{mi} = mechanical power needed to produce the electrical output, in watts (W)

E_o = induced voltage, in volts (V)

I_a = armature current, in amperes (A)

It takes slightly more mechanical power than P_{mi} to drive the generator because of the windage and friction losses, and the hysteresis and eddy-current losses in the armature core. Note that Eq. 11-2 applies to any dc generator — compound, shunt or series.

Example 11-2:

The following information is given for the two-pole compound generator shown in Fig. 11-28a.

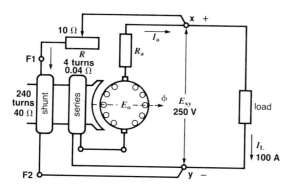

Figure 11-28a
See Example 11-2.

shunt field resistance = 40Ω
series field resistance = 0.04 Ω
armature resistance = 0.16 Ω
field rheostat resistance = 10 Ω
terminal voltage E_{xy} = 250 V
load current I_L = 90 A

windage and friction losses = 200 W
iron losses in the armature = 300 W
turns on the shunt field = 240
turns on the series field = 4

Calculate:

1. the shunt field current I_f

2. the voltage across the shunt field coils

3. the power loss in the series field

4. the value of the induced voltage E_o

5. the mechanical power needed to drive the generator

6. the total mmf of the shunt and series windings, per pole

Solution:

1. a) The total resistance R_t of the shunt field circuit is equal to the sum of the shunt field resistance (40 Ω) and the rheostat resistance (10 Ω),
        R_t = 40 + 10 = 50 Ω

    b) The shunt field current is
        $I_f = E_{xy}/R_t$ = 250/50 = 5 A

2. The voltage across the shunt field is
    $E = IR$ = 5 × 40 = 200 V

3. a) The current I_a in the armature circuit is
        $I_a = I_L + I_F$ = 90 + 5 = 95 A

    b) The loss in the series field is
        $P = I_a^2 R_s$ = 95² × 0.04 = 361 W

4. a) The voltage drop in the armature resistance is
$$v_a = I_a R_a = 95 \times 0.16 = 15.2 \text{ V}$$

b) The voltage drop across the series field
$$v_s = I_a R_s = 95 \times 0.04 = 3.8 \text{ V}$$

c) The total voltage drop in the armature circuit
$$v_t = 15.2 + 3.8 = 19 \text{ V}$$

d) The induced voltage
$$E_o = E_{xy} + 19 = 250 + 19 = 269 \text{ V}$$

5. a) The mechanical power to generate the electricity
$$P_{mi} = E_o I_a = 269 \times 95 = 25\ 555 \text{ W}$$

b) The mechanical power needed to overcome the windage, friction, and iron losses
$$P_f = 200 + 300 = 500 \text{ W}$$

c) Total mechanical power needed to drive the generator
$$P_{mi} + P_f = 25\ 555 + 500 = 26\ 055 \text{ W}$$

6. a) The mmf of the shunt field $NI_f = 240 \times 5 = 1200 \text{ At}$

b) The mmf of the series field $NI_a = 4 \times 95 = 380 \text{ At}$

c) The total mmf per pole $NI_f + NI_a = 1200 + 380 = 1580 \text{ At}$

11.12 Rating and voltage regulation of a dc generator

A dc generator is designed to produce a definite power output at a given terminal voltage and speed. These three quantities, together with the temperature rise, make up what is called the *rating* of the generator. For example, a generator may be rated at 120 kW, 240 V, 1750 r/min, with a 50°C temperature rise. The corresponding rated current is:

$$I = P/E = 120\ 000/240 = 500 \text{ A}$$

Another important characteristic of a generator is its voltage regulation, expressed in percent. It can be determined in the following way:

1. The generator is driven at rated speed and loaded up so that it delivers rated current;

2. The field excitation is adjusted so that the generator produces its rated voltage E_{FL};

3. Without touching the field rheostat or changing the speed, the load is disconnected, and the no-load armature voltage E_{NL} is measured;

4. the percent regulation is given by the equation:

$$\% \text{ regulation} = 100 \times \frac{E_{NL} - E_{FL}}{E_{FL}} \qquad (11\text{-}3)$$

where E_{FL} and E_{NL} have the meaning given above.

For example, in Fig. 11-27, the rated voltage of the separately excited generator is 70 V and the rated armature current is 40 A. The voltage regulation is therefore:

$$\% \text{ regulation} = 100 \times \frac{E_{NL} - E_{FL}}{E_{FL}}$$

$$= 100 \times \frac{(80 - 70)}{70} = 14.3\%$$

11.13 Actual construction of a dc armature

To understand some of the special aspects of a dc machine, it is useful to know how the armature is actually constructed. Consider Fig. 11-32, in which a single coil revolves ccw between the poles of a permanent magnet. Let us begin with the coil in the vertical position. Voltage E_{ab} is zero because conductors 1, 2 and 3, 4 are moving parallel to the flux lines. However, when the coil has moved through 90° (Fig. 11-33), the conductors are cutting across the flux lines and E_{ab} is maximum. By applying Fleming's right-hand rule, we can show that E_{ab} is positive. Let us suppose it is $+10$ V.

After another 90° (Fig. 11-34), the coil will again be vertical and $E_{ab} = 0$. This

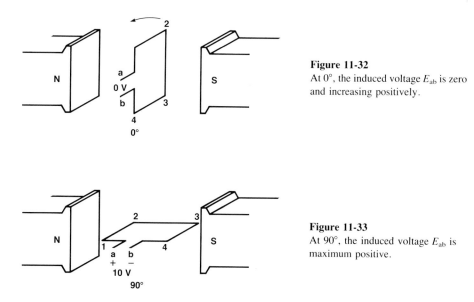

Figure 11-32
At 0°, the induced voltage E_{ab} is zero and increasing positively.

Figure 11-33
At 90°, the induced voltage E_{ab} is maximum positive.

means that between zero and 180°, E_{ab} is always positive. But if we go beyond 180°, E_{ab} is negative and remains so until the coil has reached 360°. For example, at an angle of (180 + 45) or 225°, $E_{ab} = -7$ V (Fig. 11-35).

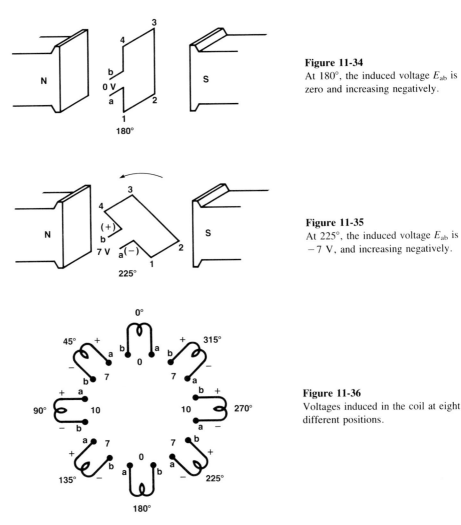

Figure 11-34
At 180°, the induced voltage E_{ab} is zero and increasing negatively.

Figure 11-35
At 225°, the induced voltage E_{ab} is -7 V, and increasing negatively.

Figure 11-36
Voltages induced in the coil at eight different positions.

We show these successive positions of the coil, and the corresponding voltages and polarities in Fig. 11-36. (To simplify the drawing, we show a narrow coil.) The voltages depend upon the position of the coil. For example, whenever the coil is at the 225° position, the voltage between its terminals is 7 V, and b is (+) with respect to a.

Suppose we build a special armature having *eight* coils identical to the coil in Fig. 11-32. The coils are uniformally distributed in a circle, at 45 degrees to each other (Fig. 11-37). They are numbered 1 to 8 so that we can readily identify them.

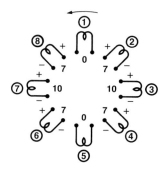

Figure 11-37
Snapshot of the voltages induced in an eight-coil armature.

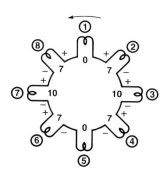

Figure 11-38
The coils of Fig. 11-37 can be connected in a closed loop without causing a current to flow.

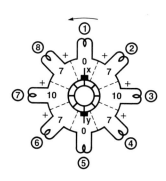

Figure 11-39
The coils are connected to a commutator and the brushes are placed to obtain the highest dc voltage.

Let us rotate this assembly of eight coils ccw at the same speed as before. Each coil will generate a voltage and polarity corresponding to its position. These voltages are the same as those induced in the single coil of Fig. 11-36. If we freeze the action when coil 1 is at zero degrees, the voltages in the coils will instantaneously be as shown in Fig. 11-37. A moment later, when the entire assembly has moved through 45°, the voltage in coil 1 will be +7 V, that in coil 2 will be zero, that in coil 3 will be −7 V, and so forth around the circle.

So far, the coils are isolated from one another, but we can connect them in series to form a closed loop, as shown in Fig 11-38. No current will flow in the loop because the algebraic sum of the voltages around the loop is zero. In other words, the voltage rises are equal to the voltage drops. The coil voltages in Fig. 11-38 are therefore the same as in Fig. 11-37.

The next step is to connect the coils to the commutator. These connections are shown by the dotted lines in Fig. 11-39. The voltages between the segments are obviously equal to the corresponding coil voltages.

Let us now place brushes x and y in the position indicated in Fig. 11-39. The voltage between the brushes is equal to the sum of the voltages between the segments which, at this moment, is $7 + 10 + 7 = 24$ V. Thus, $E_{xy} = +24$ V. When the armature rotates 45° from its original position, the induced voltages are the same except they are generated by a different set of coils. Consequently, the voltage between the brushes remains fixed at 24 V, and brush x is always positive with respect to brush y. E_{xy} is therefore a dc voltage. Note, however, that when the armature has turned 22.5°, there will be four coils between the brushes, and the sum of their voltages will be slightly different from 24 V. That is why the dc voltage fluctuates, giving rise to commutator ripple.

Brush x touches two commutator segments in Fig. 11-39, thus placing a momentary short-circuit on coil 1. In the same way, brush y short-circuits coil 5. But because the voltage in these coils is zero, no harm will result. However, if the brushes were shifted cw by 45°, they would short-circuit coils 2 and 6. The 7 V induced in these coils would produce a large current in the brushes, and sparking could result. When brushes spark, the commutation is said to be poor.

11.14 Commutation of current

Figure 11-40 shows the current in the armature coils when the load draws 100 A. The current flows through the armature in two parallel paths, with 50 A flowing in each coil.

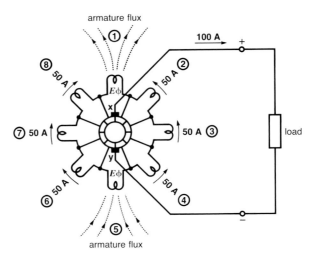

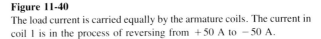

Figure 11-40
The load current is carried equally by the armature coils. The current in coil 1 is in the process of reversing from +50 A to −50 A.

However, the currents in coils 2, 3, and 4 flow in the opposite direction to those in coils 6, 7, and 8. This means that when the armature rotates, the current in each coil reverses periodically. This current reversal is called *commutation*.

The problem is that the current must change from +50 A to −50 A in a very short time. The time is equal to that needed for the commutator to move the width of one brush. It is therefore only a fraction of a millisecond long. The coils possess inductance, and such a rapid change of current produces a self-induced voltage E_L, given by $E_L = L\Delta I / \Delta t$. This voltage opposes the change in the current and the result is sparking at the brushes. In Figure 11-40, the current in coil 1 is the process of changing from +50 A. That in coil 5 is changing from −50 A to +50 A.

Example 11-3:

The individual coils in Fig. 11-40 have a self-inductance of 40 μH. The diameter of the commutator is 100 mm, and the width of the brushes is 8 mm. If the armature turns at 1800 r/min, calculate the average value of the induced voltage E_L.

Solution:

1. The commutator makes one turn in 60/1800 = 1/30 s

2. The circumference of the commutator is $\pi d = \pi \times 100 = 314$ mm

3. The commutator therefore moves at a surface speed of
   $$314 \times 30 = 9\ 420 \text{ mm/s}$$

4. The time to move a distance of 8 mm (the width of a brush) is
   $$\Delta t = 8/9420 = 8.49 \times 10^{-4} \text{ s}$$

5. The change in current is $\Delta I = 50 - (-50) = 100$ A

6. The induced voltage is
   $$
   \begin{aligned}
   E_L &= L\ \Delta I / \Delta t \\
   &= 40 \times 10^{-6} \times 100/(8.49 \times 10^{-4}) \\
   &= 4.7 \text{ V}
   \end{aligned}
   $$

11.15 Armature reaction and commutation

We recall that when current flows in the armature, an armature flux is created. This flux is strongest in the region of coils 1 and 5 (Fig. 11-40). A voltage $E\phi$ is therefore induced in these coils as they cut the armature flux. Because they are short-circuited by the brushes, a large current can result. This again contributes to poor commutation under load.

11.16 Commutating poles

We can improve the commutation by placing *commutating poles* between the main poles of the generator. They are connected in series with the armature in order to produce a mmf equal and opposite to the armature mmf. Consequently, they eliminate the armature flux that induced the undesired voltage $E\phi$ in coils 1 and 5 (Fig. 11-40).

In practice, the mmf of the commutating poles is made slightly greater than the armature mmf. This produces a small net flux whose direction is *opposite* to that shown in Fig. 11-40. The resulting induced voltage opposes the voltage E_L due to self-induction, and so the total voltage induced in coils 1 and 5 is practically zero. As a result, the commutation is much improved.

Figure 11-41 shows the commutating poles in a two-pole generator. The total mmf ($2M_C$) created by the two commutating poles is slightly greater than the mmf

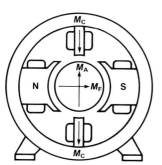

Figure 11-41
Commutating poles reduce sparking at the brushes.

(M_A) created by the armature. The mmf (M_F) of the shunt field is constant, but M_A and M_C vary in direct proportion to the load current. Figure 11-29 shows the commutating poles in a 4-pole generator.

11.17 Multipolar machines

Direct current machines rated below 1 kW usually have two poles. Larger machines have four, six, eight, or more poles. The main reason for increasing the number of poles is to reduce weight and cost, and to improve machine performance. Nevertheless, the basic properties of multipolar machines are identical to those we have discussed here.

11.18 Summary

This chapter enabled us to put to practical use several of the basic principles covered in previous chapters. We saw how the equation $E = BLv$ fits into the theory of dc generators. Then, we applied the phenomenon of saturation, covered in Chapter 5, to explain the shape of the no-load saturation curve. Finally, we were able to explain how the very rapid change in armature current produces sparking at the brushes, due to the phenomenon of self-induction.

The principles also apply to dc motors; consequently, much of the subject matter covered here will be useful in the next chapter.

TEST YOUR KNOWLEDGE

11-1   A commutator and its brushes together form a mechanical rectifier.

☐ true   ☐ false

11-2   If the armature core in Fig. 11-11 were made of plastic instead of iron, the armature voltage would

a. remain the same   b. diminish greatly
c. become zero

11-3   When the brushes of a two-pole dc generator are in the neutral position the angle between the armature mmf and the field mmf

a. is usually 90°   b. is always 90°
c. depends upon how the armature is
wound

11-4   In Fig. 11-25, if the battery terminals are reversed,

a. I_a will reverse   b. I_a will become zero
c. I_a will remain unchanged

11-5   In Fig. 11-39, if the armature turns 90°, the voltage in coil 8 will be

      a. zero   b. 7 V   c. 10 V

11-6   In Fig. 11-40, **the** instantaneous power delivered by coil 4 is

      a. 350 W   b. 2400 W   c. 350 J

11-7   In Fig. 11-41, the armature mmf is 6000 At. The interpoles should each develop a mmf of

      a. 6000 At   b. 3000 At   c. slightly more than 3000 At

QUESTIONS AND PROBLEMS

11-8   Why should we always place the brushes of a dc generator in the neutral position?

11-9   What effects does an increase in the field current have on a dc generator?

11-10   Why does the terminal voltage of a separately excited dc generator decrease with increasing load?

11-11   A separately excited dc generator develops an induced voltage of 127 V when the armature turns at 1400 r/min. (a) If the armature resistance is 2 Ω, what is the armature terminal voltage when the armature current is 12 A? (b) How much heat is lost in the armature?

11-12   A separately excited generator produces an open-circuit voltage of 115 V. Explain in detail what will happen if

      a. the speed is increased 20 percent?
      b. the direction of rotation is reversed?
      c. the field current is increased 20 percent?
      d. the polarity of the shunt field is reversed.

11-13   A flat-compound generator rated at 100 kW, 250 V has a shunt field of 2000 turns, and a series field of 7 turns. The shunt field has a resistance of 100 Ω, and is connected across the armature terminals in series with a rheostat of 25 Ω, as shown in Fig. 11-28. Calculate the mmf per pole when the machine operates (a) at no-load; and (b) at full-load.

11-14   The coils in Fig. 11-40 carry an ac current, but the output current is dc. Explain.

11-15   The voltage in the coils of Fig. 11-40 is ac, but the output voltage is dc. Explain.

CHAPTER **12**

Direct Current Motors

Introduction and Chapter Objectives

Direct current motors transform electrical energy into mechanical energy. They drive machines and devices such as hoists, fans, pumps, punch-presses, and cars. These devices may have a definite torque-speed characteristic (such as a pump or fan) or a highly variable one (such as a hoist or car). The torque-speed characteristic of the motor must be adapted to the type of load it will drive, and this requirement has given rise to three basic types of motors:

1. shunt motors,
2. series motors,
3. compound motors.

Direct current motors are seldom used in ordinary industrial applications because all electric utility systems furnish alternating current. However, for special applications such as in steel mills and electric trains, it is advantageous to transform the alternating current into direct current in order to use dc motors. The reason is that the torque-speed characteristics of dc motors can be varied over a wide range while retaining high efficiency.

DC motors and generators are built the same way. A dc generator can therefore be used as a dc motor, and vice versa. The problems of armature reaction, commutation, and temperature rise are also common to both machines.

12.1 Construction and torque-speed characteristics of dc motors

Torque and speed are the two most important properties of an electric motor. Note that it is the torque — and not the power — that determines the size of a motor. Thus, a motor that develops a torque of 160 N.m at a speed of 2000 r/min is about the same size as one developing 160 N.m at 500 r/min. But from a power standpoint, the first machine delivers four times as much power as the second.

We will briefly describe the basic features and torque-speed characteristics of shunt, series, and compound motors before explaining the corresponding theory.

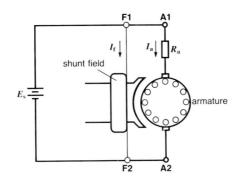

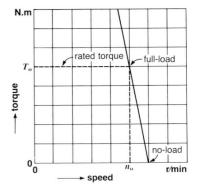

Figure 12-1
Shunt motor connected to the dc source.

Figure 12-2
Torque-speed characteristic of a shunt motor.

Figure 12-3
Four-pole stator of a 40 hp, 1750 r/min, 600 V dc stabilized-shunt motor, showing main poles and commutating poles. (*Courtesy Gould*)

The *shunt motor* has a shunt field and an armature, both of which are connected to the dc source E_s (Fig. 12-1). The figure shows only one pole, but it represents the two, four, six, or more poles that are on the actual machine. The motor usually has commutating poles, but they are considered to be part of the armature circuit and are not shown. Consequently, R_a is the sum of the armature resistance and the resistance of the commutating poles.

The torque-speed curve for this motor is shown in Fig. 12-2. The rated torque is T_o and the rated speed is n_o. The no-load speed is slightly higher than the rated speed.

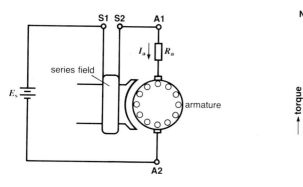

Figure 12-4
Series motor connected to the dc source

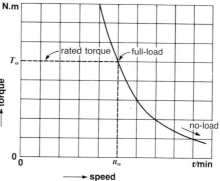

Figure 12-5
Torque-speed characteristic of a series motor.

The *series motor* has a series field and an armature, connected to the dc source E_s, as shown in Fig. 12-4. Because the series field carries the armature current, it is made of large wire. The field has relatively few turns because the large current does not require many turns to produce a substantial field mmf. The motor also has commutating poles, connected in series with the armature. Consequently, R_a represents the sum of the resistances of the armature, commutating poles, and series field.

The torque-speed curve of a series motor is shown in Fig. 12-5. The rated torque and speed are again T_o and n_o. The no-load speed is dangerously high, and so this motor should never be disconnected from its load. Compared with a shunt motor, the speed falls sharply with increasing torque.

The *compound motor* has both a series field and a shunt field, connected as shown in Fig. 12-6. The series mmf adds to the shunt mmf, and so the resultant

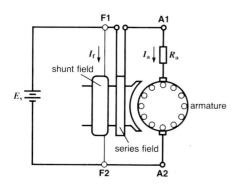

Figure 12-6
Compound motor connected to the dc source.

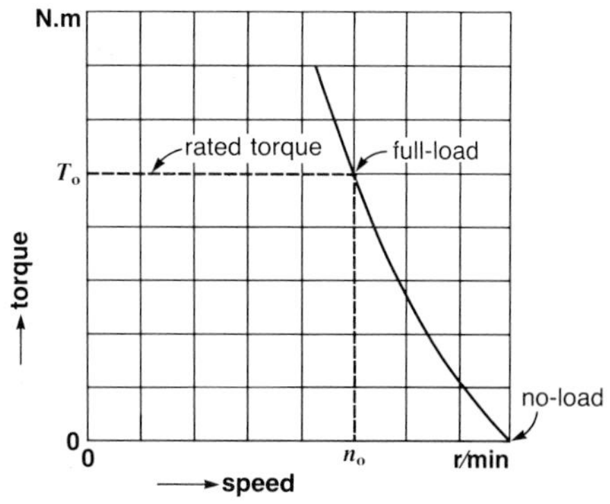

Figure 12-7
Torque-speed characteristic of a compound motor.

field flux increases with increasing load. This motor also has commutating poles, and so R_a is again the sum of armature, series field, and commutating pole resistances.

The torque-speed curve for this motor is shown in Fig. 12-7. It is similar to that of a shunt motor, but because of the series field, the speed falls more rapidly with increasing torque. However, it does not fall as steeply as the curve for a series motor. The no-load speed is typically 30 percent higher than the rated speed n_o.

In comparing these motors, it is important to remember that if they have the same rated torque, they will be the same size. In other words, the size of any dc motor — shunt, compound or series — depends on its rated torque.

12.2 Motor losses and efficiency

The dc motor converts electrical energy into mechanical energy, and it therefore has losses (see Section 1-8). Figure 12-8 schematically shows the power flow and losses in the machine.

The total electrical input P_e provides power P_f to the shunt field and power P_a to armature. Power P_f is dissipated as heat in the field coils and field rheostat. Power P_a supplies the I^2R losses of the series field, commutating poles, and armature. When these losses are subtracted from P_a, we obtain the electrical power that is converted into "internal" mechanical power P_{mi}. A part of this mechanical power is

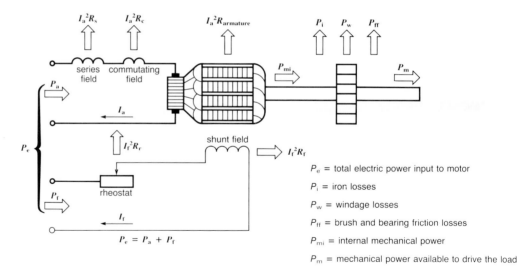

Figure 12-8
Input power and losses in a dc motor.

used up in brush friction, bearing friction, and windage losses. The windage losses are due to the fan and air turbulence around the armature. Another part of P_{mi} is used up in the hysteresis and eddy current losses in the armature. Although these iron losses are electrical in nature, they show up as a mechanical drag on the armature. The resulting mechanical power P_m available at the shaft is therefore less than the internal mechanical power P_{mi}. The total losses in the motor are equal to $P_e - P_m$, and they appear as heat.

Some of the losses in Fig. 12-8 depend upon the mechanical load while others do not. Thus, as the load increases, the armature current I_a increases. This causes the $I_a^2 R_a$ losses to increase. However, the other losses remain essentially fixed. For example, the brush friction loss decreases when the speed drops by a few percent, but the change is not significant. Thus, the graph of motor losses versus power output has the typical shape given in Fig. 12-9.

The losses cause the motor to heat up, and adequate cooling means are needed to keep its temperature from becoming too high (Fig. 12-10).

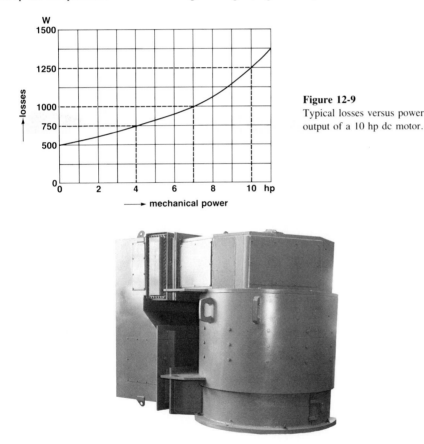

Figure 12-9
Typical losses versus power output of a 10 hp dc motor.

Figure 12-10
Totally-enclosed 2000 hp dc motor cooled with an air-to-water heat exchanger. Hot air from the motor circulates in the piping from the upper, commutator end, down through the heat exchanger on the left, and the cooled air reenters the motor at the lower fan end. This motor is used to position the platform over the hole in off-shore oil/gas well drilling. (*Courtesy General Electric Company, USA*)

Example 12-1:

A 10 hp motor has the loss curve given in Fig. 12-9. Calculate the efficiency of the motor when it delivers an output of 7 hp.

Solution:

1. 7 hp output corresponds to $P_m = 7 \times 746 = 5222$ W

2. The corresponding losses are 1000 W, taken from the graph

3. The power input to the motor is $P_e = 5222 + 1000 = 6222$ W

4. eff $= 100 \times \dfrac{5222}{6222} = 83.9\%$ Eq. 1-11

12.3 Starting a dc motor

When a dc motor is running, the armature is connected across the dc source. But if we attempt to start the motor this way, the line fuses will blow. The reason is that when the armature is not turning, the only opposition to current flow is the armature resistance. It is so low that the locked-rotor current may be as much as 50 times the rated current. (The term *locked-rotor* means that the armature is not turning.)

We must therefore place a resistor R_s in series with the armature during start-up.

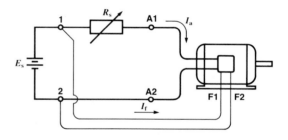

Figure 12-11
Circuit to start a dc motor. The shunt field is connected across the source.

The shunt field is fully excited during this time to produce maximum flux and maximum starting torque. A simple starting circuit is shown in Fig. 12-11. To get a particularly smooth start, resistor R_s should be gradually reduced to zero as the motor speed increases. However, we usually design the circuit so that R_s drops to zero in one or more steps. This may be done manually or automatically by using a dc starter.

Example 12-2:

A 15 hp, 225 V, 950 r/min dc shunt motor has a rated armature current of 60 A. The armature resistance is 0.15 Ω, and the shunt field resistance is 50 Ω. Calculate (1) the initial starting current if the motor is directly connected across the 225 V line and (2) the value of the starting resistance R_s needed to limit the locked-rotor current to 90 A.

Solution:

1. Without a current-limiting resistor, the locked-rotor current would be

$$I_a = E/R_a = 225/0.15 = 1500 \text{ A (25 times the rated current)}$$

2. a) To limit I_a to 90 A requires a total resistance in the armature circuit of
   $$R = E/I_a = 225/90 = 2.5 \ \Omega$$

   b) Because the armature already contributes 0.15 Ω, the external resistance
   $$R_s = 2.5 - 0.15 = 2.35 \ \Omega$$

12.4 Theory of the dc motor

Now that we have a general idea of the properties of dc motors, we can study the underlying theory. Figure 12-12 is the schematic diagram of a separately excited

Figure 12-11A
A hot-strip rolling mill compresses a steel slab to the desired thickness. The stand is driven by large dc motors, one of which is seen on the platform at the right. The white-hot strip is carried along on rollers driven by dozens of 3 hp dc motors mounted side by side. (*Courtesy General Electric*)

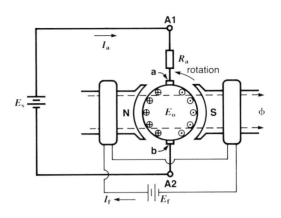

Figure 12-12
Circuit enabling the development of the basic equations of a dc motor.

shunt-wound motor that is connected to a dc source E_s. The motor draws a current I_a, and the armature is turning ccw at a speed of n turns per minute. The shunt field current I_f produces a flux ϕ, which is cut by the rotating armature conductors. As a result, a voltage E_o is induced in the armature. This voltage is *identical* to the voltage that would be induced if the machine were operating as a generator. Finally, the armature resistance R_a produces a small voltage drop I_aR_a. If follows that the induced voltage E_o is only slightly less than the applied voltage E_s. We now can develop some of the basic equations and relationships of a dc motor.

1. Armature current I_a must flow out of the (+) terminal of the source because the source is supplying power to the motor. The power absorbed by the armature is

$$P_a = E_s I_a \tag{12-1}$$

2. The armature resistance produces losses equal to $I_a^2 R_a$. The net power supplied to the armature is therefore

$$P_{mi} = E_s I_a - I_a^2 R_a \tag{12-2}$$

This electric power is entirely converted into the "internal" mechanical power mentioned previously.

Equation 12-2 can be simplified as follows:

$$P_{mi} = (E_s - I_a R_a) I_a$$

from which

$$\underline{P_{mi} = E_o I_a} \tag{12-3}$$

Thus, the internal mechanical power is equal to the product of the induced voltage E_o times the armature current I_a. This same result was obtained for a dc generator (see Section 11-11).

3. What is the polarity of E_o? Point a must be (+) with respect to b because the armature winding absorbs electric power. The induced voltage E_o is sometimes called the *counter emf* (cemf) of the motor. The reason is that its polarity tends to produce a current in the armature circuit opposite (counter) to the direction of I_a.

4. The current in the armature conductors flows in the direction given by the crosses and dots of Fig. 12-12. We come to this conclusion because the force ($F = BLI$) on the conductors next to the N pole must act downward in order to produce the ccw rotation. Similarly, the force on the conductors facing the S pole acts upward.

5. The current flowing in the armature conductors produces an armature mmf M_a that acts downward, as shown in Fig. 12-13. As a result, the flux crossing the armature is distorted, and pole tips c and d tend to saturate. Thus, as in the case of a generator, the armature reaction tends to reduce the flux ϕ, at the same time as it tends to cause sparking at the brushes.

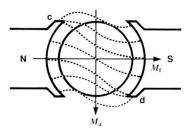

Figure 12-13
Armature reaction in a dc motor distorts the field flux, and saturates pole tips c and d.

12.5 Torque of a dc motor

Returning to Fig. 12-12, each armature conductor is subjected to a force given by F = BLI. Furthermore, the flux density B is directly proportional to the flux ϕ. Consequently, the torque produced by the motor depends on both ϕ and I_a. We can establish the exact relationship as follows.

According to Eq. 1-9, the torque T is given by

$$T = \frac{9.55\,P}{n} = \frac{9.55\,P_{mi}}{n} \qquad (12\text{-}4)$$

Substituting Eq. 12-3 in Eq. 12-4, we obtain

$$T = \frac{9.55\,E_o I_a}{n} \qquad (12\text{-}5)$$

But the induced voltage E_o is a generator voltage, which is given by Eq. 11-1:

$$E_o = Kn\phi \qquad \text{Eq. 11-1}$$

If we substitute Eq. 11-1 in Eq. 12-5, we obtain the torque equation

$$T = 9.55 \, K\phi I_a \tag{12-6}$$

where

T = torque, in newton meters (N.m)

ϕ = flux per pole, in webers (Wb)

I_a = armature current, in amperes (A)

K = a constant that depends upon the winding on the armature

9.55 = a constant to take care of units

Although we used the shunt-wound motor of Fig. 12-12 to arrive at this torque equation, it applies to any type of dc motor. For example, the flux ϕ may be produced by the combined effect of a series and shunt field. Equation 12-6 also tells us that the torque depends only upon the value of ϕ and I_a; the speed of the motor does not matter. However, the speed *does* affect the value of I_a and sometimes even the flux ϕ. Consequently, the speed has an indirect effect upon the torque, and this creates the various torque-speed curves we saw in Figs. 12-2, 12-5 and 12-7.

12.6 Speed of a dc motor

Referring again to Fig. 12-12, the induced voltage E_o is given by:

$$E_o = E_s - I_a R_a \tag{12-7}$$

When the motor runs between no-load and full-load, the $I_a R_a$ drop is always small compared with the source voltages E_s. Consequently, E_o is nearly equal to E_s. Using Eq. 11-1, we can therefore write

$$n = \frac{E_o}{K\phi} = \frac{E_s - I_a R_a}{K\phi} \quad \text{(exactly)} \tag{12-8}$$

from which we get

$$n = \frac{E_s}{K\phi} \quad \text{(approximately)} \tag{12-9}$$

where

n = speed of rotation, in revolutions per minute (r/min)

E_s = voltage across the armature, in volts (V)

ϕ = flux per pole, in webers (Wb)

K = constant that depends upon the winding on the armature

To interpret the meaning of Eq. 12-9, suppose that ϕ is held constant. The speed will then vary directly with the voltage E_s applied to the armature. This means that we can vary the motor speed from zero to maximum simply by varying E_s. This is one of the important features of a dc motor.

Next, suppose E_s is fixed and we vary ϕ. If the flux decreases, the denominator in Eq. 12-9 decreases, causing speed n to increase. Conversely, if ϕ increases, the speed will drop. This means that we can vary the speed by varying the field current I_f. This is another important feature of a dc motor.

12.7 Acceleration of a shunt motor

Consider Fig. 12-14 in which a shunt motor having an armature resistance R_a is about to be connected to a source E_s by means of switch S. An external resistor R_s is used to limit the starting current to about 1.5 times its rated value. The total resistance of the armature circuit is $R_t = R_a + R_s$. The shunt field is excited and produces flux ϕ. At the instant the switch closes, a current I_a will flow in the armature. Because the armature is not yet turning, the current is given by

$$I_a = E_s/R_t$$

The current will produce a strong starting torque, causing the motor to accelerate. As it picks up speed, the conductors cut across the flux lines, inducing a

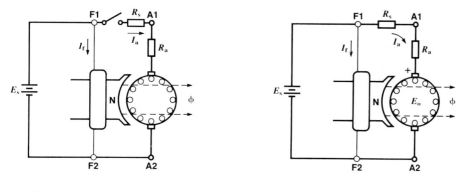

Figure 12-14

Shunt motor with the armature at rest.

Figure 12-15

A counter emf E_o is generated in the armature when the motor turns.

voltage E_o (Fig. 12-15). The greater the speed, the greater this cemf will become. The net voltage acting on the circuit is $E_s - E_o$. The resulting current I_a is therefore

$$I_a = (E_s - E_o)/R_t \tag{12-10}$$

Equation 12-10 shows that the current will decrease as E_o increases. This means that I_a will decrease with increasing speed. Is there a limit to the speed? Yes, for suppose the speed rises to a level at which $E_o = E_s$. Then, according to Eq. 12-10, current I_a will be zero. As a result, the torque will also be zero, and the motor will cease to accelerate.

In practice, the speed will not quite rise to the point at which $E_o = E_s$ because the motor has to develop *some* torque to keep running. A relatively small torque is needed to balance the opposing torque due to windage, friction, and iron losses.

Example 12-3:

A permanent-magnet dc generator (Fig. 12-16) produces 60 V when it is rotated at 500 r/min. The armature resistance is 2 Ω. If this machine operates as a motor connected to a 200 V source, calculate

1.   the locked-rotor current

2.   the induced voltage (counter-emf) when the motor turns at 1200 r/min

3.   the armature current at 1200 r/min

4.   the "internal" mechanical power at 1200 r/min

5.   the "internal" torque developed at 1200 r/min

6.   the approximate final speed of the motor

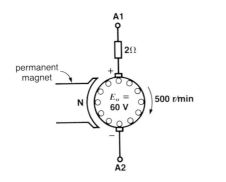

Figure 12-16
See Example 12-3. Machine operating as a generator.

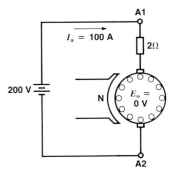

Figure 12-17
See Example 12-3. Machine operating as a motor, but not yet turning.

Solution:

1.   The motor circuit at the moment of start-up is shown in Fig. 12-17. The induced voltage is zero because the armature is not yet turning. Consequently

$$I_a = E_s/R_a = 200/2 = 100 \text{ A}$$

2.   When the motor turns at 1200 r/min (Fig. 12-18), the induced voltage, or cemf, is

$$E_o = \frac{1200}{500} \times 60 = 144 \text{ V}$$

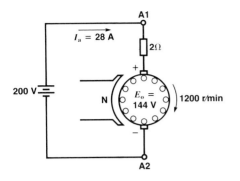

Figure 12-18

See Example 12-3. Machine operating as a motor, at 1200 r/min.

3. The voltage across R_a = $(200 - 144)$ = 56 V
    The armature current I_a = E/R_a = 56/2 = 28 A

4. The mechanical power P_{mi} = $E_o I_a$ = 144 × 28 = 4052 W
    which, in horsepower = 4032/746 = 5.4 hp

5. T = $9.55 \, P/n$ Eq. 1-9
       = $9.55 \times 4032/1200$ = 32.09 N.m

6. If the friction losses were negligible, the motor speed would rise until
    E_o = 200 V. Because we know that E_o = 60 V corresponds to a generator
    speed of 500 r/min, it follows that E_o = 200 V corresponds to
    500 × (200/60) = 1667 r/min. Thus, the motor will turn at a final speed
    slightly below 1667 r/min.

12.8 Torque and speed characteristics of a shunt motor*

Equations 12-1 to 12-10 enable us to determine the torque and speed of a dc motor
when we know its basic properties. We thus can determine the torque and speed of a
shunt motor between no-load and full-load. Using a numerical example, we assume
the motor has the following ratings:

$$E_s = 250 \text{ V} \qquad\qquad \text{rated } I_a = 500 \text{ A}$$

$$R_a = 0.02 \text{ }\Omega \qquad K = 3 \qquad \text{rated } n = 800 \text{ r/min}$$

From Eq. 12-8, the flux ϕ is

$$\phi = \frac{E_o}{Kn} = \frac{250 - 500 \times 0.02}{3 \times 800} = \frac{240}{2400} = 0.1 \text{ Wb}$$

From Eq. 12-6, the full-load torque is

$$T = 9.55 \, K\phi \, I_a = 9.55 \times 3 \times 0.1 \times 500 = 1432.5 \text{ N.m}$$

In a shunt motor, the flux ϕ remains constant. It follows from Eq. 12-6 that the

---

*   Sections 12-8, 12-9, and 12-10 are somewhat more advanced and may be deferred for later study.

torque T is directly proportional to I_a. This straight-line relationship is shown in Fig. 12-19. We have also plotted the value of E_o; it slopes downward because of the I_aR_a drop. Knowing E_o and using Eq. 12-8, we can find the speed n for any load I_a: For example, at no-load, $E_o = E_s = 250$ V and $I_a = 0$ which gives us

$$n = \frac{E_o}{K\phi} = \frac{250}{3 \times 0.1} = 833.3 \text{ r/min} \qquad \text{Eq. 12-8}$$

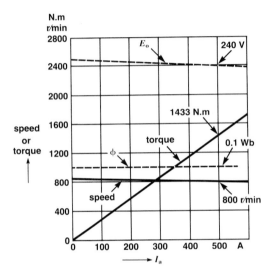

Figure 12-19
Torque and speed versus armature current of a shunt motor.

At full-load, $E_o = E_s - I_aR_a = 250 - 500 \times 0.02 = 240$ V

Therefore $n = \dfrac{E_o}{K\phi} = \dfrac{240}{3 \times 0.1} = 800 \text{ r/min}$

The speed curve is a straight line ranging from n = 833 r/min at $I_a = 0$ to n = 800 r/min when $I_a = 500$ A.

The torque and speed can be replotted as a torque-speed curve, and this will yield the general result shown in Fig. 12-2.

In Section 12-5 we mentioned that due to armature reaction, the field flux decreases when the motor is under load. This becomes a serious problem in large shunt motors because it causes them to run faster as the load increases. To obtain a drooping speed characteristic, a weak series field is added, and such machines are called *stabilized shunt* motors (see Fig. 12-3).

12.9 Torque and speed characteristics of a series motor

We shall now determine the torque and speed of a series motor from no-load to full-load. We assume the motor has the same full-load rating as the shunt motor studied in the previous section. Thus, the full-load flux is again 0.1 Wb. However, in a series motor, the flux decreases as the current decreases, and the assumed relationship between the two is shown in Fig. 12-20. It is not a straight line because we assume the iron begins to saturate as the current in the series winding increases.

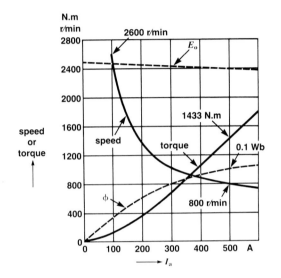

Figure 12-20
Torque and speed versus armature current of a series motor.

Figure 12-21
This metro train is driven by two 153 hp, 360 V series motors having a speed range from zero to 2160 r/min. Each motor weighs 625 kg, and is insulated with 155°C insulation.

Using the same technique as for a shunt motor, we can calculate the torque and speed of the series motor for any load I_a. For example, when $I_a = 200$ A, $\phi = 0.06$ Wb (read off from the graph), and so

$$n = \frac{E_o}{K\phi} = \frac{250 - 200 \times 0.02}{3 \times 0.06} = \frac{246}{0.18} = 1367 \text{ r/min}$$

and $T = 9.55\, K\phi I = 9.55 \times 3 \times 0.06 \times 200 = 344$ N.m

The solid curves in Fig. 12-20 show the torque and speed.

It is obvious that the no-load speed of a series motor can rise to dangerous levels.

If we replot the torque and speed on a torque-speed curve, we obtain the general result shown in Fig. 12-5.

Series motors are used to drive large exhaust fans on ships, and electric trains and buses (Fig. 12-21).

12.10 Torque and speed characteristics of a compound motor

Figure 12-22 shows the torque and speed of a compound motor. We assume this machine has the same full-load rating as the shunt motor studied in Section 12-8. However, the flux is assumed to increase from 0.06 Wb at no-load to 0.1 Wb at full-load.

The torque and speed at different load currents are calculated the same way as before.

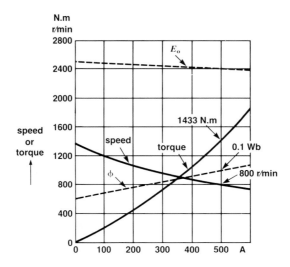

Figure 12-22
Torque and speed versus armature current of a compound motor.

12.11 Speed control of a dc motor

The speed of a shunt motor can be varied over a wide range by changing the field current, with a rheostat. When the speed is varied this way, the armature current and armature voltage can retain their rated values. It follows that the motor can deliver its rated horsepower for all values of I_f.

On the other hand, if the field current is kept at its rated value, the same motor can operate below rated (base) speed if we apply a lower-than-rated voltage E_s to the armature. However, the armature current can retain its rated value. Consequently, the motor can develop its full rated torque for any value of armature voltage.

Thus, as far as rating is concerned, a shunt motor is a constant-torque machine when it operates below base speed and a constant-horsepower machine when it operates above base speed.

In some cases, the load has to be accelerated and decelerated very quickly. Field control is usually not feasible because the field current cannot change quickly enough because of the field inductance. Thus, if we want fast response we have to vary the armature voltage. Acceleration is obtained by increasing E_s. We also can obtain *forced* deceleration by decreasing E_s. For example, suppose the motor in

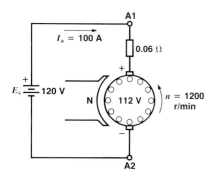

Figure 12-23
DC motor running normally at a speed of
1200 r/min with a supply voltage at 120 V.

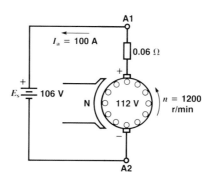

Figure 12-24
Reducing the supply voltage to 106 V causes
the armature current and torque to reverse.

Fig. 12-23 is running at constant speed and exerting rated torque. It is connected to a source E_s of 120 V, and its cemf is 112 V. The net voltage acting on the armature circuit is $(E_s - E_o) = (120 - 112) = 6$ V, and so $I_a = E/R_a = 6/0.06 = 100$ A. Current flows in the direction shown because E_s exceeds E_o.

If we suddenly reduce E_s to 106 V, we obtain the condition shown in Fig. 12-24. E_o is now greater than E_s and so the current I_a reverses. As a result, a strong *braking* torque is exerted on the motor, causing its speed to drop rapidly. In Fig. 12-24, the motor is actually operating as a generator, delivering power to the 106 V "source" which is now acting as a load.

Such fast speed control is obtained by the so-called Ward-Leonard system (Fig. 12-25). A dc generator G produces a variable voltage E_s that is applied to the

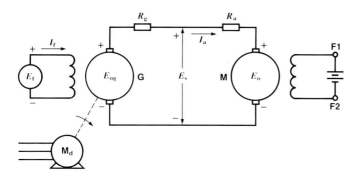

Figure 12-25
Ward-Leonard speed control system.

armature of the motor we want to control. The motor field is fixed. The generator is driven by a motor M_d connected to the ac power line in the plant. By varying the field excitation I_f of the generator, voltage E_s varies, which in turn controls the motor speed. The speed can be reversed by reversing the polarity of E_f.

Rapid speed changes and reversals are needed in the operation of mine hoists, and rolling mills in the steel industry. Figure 12-26 shows two 7000 hp motors that operates in tandem to drive a slabbing mill.

Figure 12-26
Two 7000 hp, 30 r/min dc motors operate together to drive a metal rolling mill. The motor on the left is complete with its 24-pole stator. The partly-assembled motor on the right shows the armature with its man-size commutator. (*Courtesy General Electric Company, USA*)

12.12 Dynamic braking and plugging

Whenever a motor has to be brought to a rapid stop, a method called *dynamic braking* may be used. It consists of connecting a resistor R_d across the armature terminals immediately after the armature has been disconnected from the line (Fig. 12-27). The motor, now acting as a generator, dissipates its kinetic energy in the resistor until it comes to rest.

If emergency stopping is needed, dynamic braking may not be quick enough. In such cases, we use a method called *plugging*. It consists of reversing the polarity of the supply voltage E_S across the terminals of the armature (Fig. 12-28). The net voltage acting in the armature circuit suddenly changes from $(E_s - E_o)$ to $(-E_s - E_o)$. To limit the very high braking current I_a that would otherwise result, a resistor R_p is connected in series with the armature while the motor is stopping.

When the motor has come to rest and is about to reverse, a switch S opens automatically, disconnecting the armature from the line.

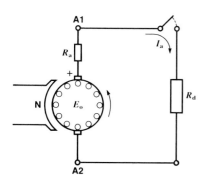

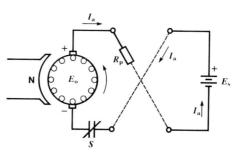

Figure 12-27
A motor can be stopped quickly by
using dynamic braking.

Figure 12-28
A dc motor can be stopped even
faster by plugging.

12.13 Starters for dc motors

Many types of starters are available to bring dc motors up to speed. We will limit
our study to one type of manual starter and then cover the general principles of
automatic starters. Figure 12-29 shows the schematic diagram of a manual face-
plate starter for a shunt motor. Bare copper contacts are connected to resistors R_1,
R_2, R_3, and R_4. Conducting arm 1 sweeps across the contacts when it is pulled to the
right by means of insulated handle 2. In the position shown, the arm touches dead
copper contact M and the motor circuit is open. As we draw the handle to the right,
the conducting arm first touches fixed contact N.

The supply voltage E_s immediately causes full field current I_f to flow, but the
armature current I is limited by the four resistors in the starter box. The motor
begins to turn and, as the cemf builds up, the armature current gradually falls. When
the motor speed ceases to rise any more, we pull the arm to the next contact, thereby
removing resistor R_1 from the armature circuit. The current immediately jumps to a

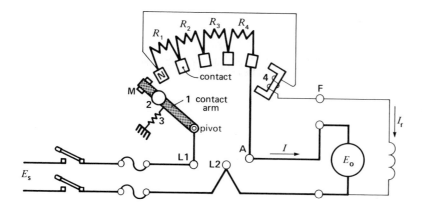

Figure 12-29
Manual face-plate starter for a dc motor.

higher value and the motor quickly accelerates to the next higher speed. When the speed is again stable, we move to the next contact until the arm finally touches the last contact. The arm is magnetically held in this position by a small electromagnet 4, which is in series with the shunt field.

If the supply voltage is suddenly interrupted or if the field excitation should accidentally be cut, the electromagnet releases the arm, allowing it to return to its dead position, under the pull of spring 3. This safety feature prevents the motor from starting by itself when the supply voltage is re-established.

Automatic starters perform the same function as manual starters. The only difference is that the current-limiting resistors are short-circuited, in succession, by a set of magnetic contactors. The timing is effected by slow-action pneumatic or electronic relays whose contacts close several seconds after their holding coils are energized.

Automatic starters will be more fully discussed in Chapter 22 because they involve symbols and principles better covered in a separate chapter.

12.14 Summary

In this chapter, we were able to use many of the principles explained in the previous chapter on dc generators. In particular, we learned that a dc motor generates a counter-emf E_o that is exactly the same as the voltage induced in a generator. Furthermore, the electromechanical energy conversion process again takes place in the armature by virtue of the expression $P_{mi} = E_o I_a$, where I_a is the armature current.

The same fundamental equation $E_o = Kn\phi$ then led us to the conclusion that the speed of a dc motor can be varied either by varying the field flux ϕ or the voltage applied to the armature. We were also able to determine the actual torque and speed characteristics of a shunt, series, and compound motor by applying the fundamental equations to a numerical example. The link between a dc motor and generator was again noted when we covered dynamic braking and the Ward Leonard system of speed control. In both cases, rapid deceleration of the motor is obtained by having it operate briefly as a generator.

Finally, we learned that current-limiting resistors must be used in series with the armature whenever we start a dc motor.

TEST YOUR KNOWLEDGE

12-1   The shunt field of a motor is connected

   a. in parallel with the source   b. in series with the armature

12-2   Referring to Fig. 12-5, the rated torque (T_o) is 250 N.m and speed (n_o) is 1500 r/min. The rated power of the dc motor is:

   a. 375 hp   b. 52.6 hp   c. 502.7 hp

12-3   Figure 12-1 represents a four-pole shunt motor. The resistance of each field coil is 60 Ω, and E_s = 300 V. The field current I_f is

        a. 1.25 A   b. 20 A   c. 5 A

12-4   Figure 12-6 represents a four-pole compound motor. The armature resistance is 0.2 Ω, and each series field coil has a resistance of 0.05 Ω. If each interpole has a resistance of 0.02 Ω, the value of R_a is

        a. 480 mΩ b. 0.27 Ω c. 480 mΩ

12-5   The iron losses in a dc motor are due to

        a. the heat released by the field coils
        b. hysteresis and eddy currents in the armature
        c. hysteresis and eddy currents in the armature and frame

12-6   A 450 hp dc motor has total losses of 26 kW. The efficiency is

        a. 91.6 %   b. 94.5 %   c. 92.8 %

12-7   In Fig. 12-12, armature reaction is due to

        a. I_a b. I_f c. I_a and I_f

12-8   A compound motor running at 600 r/min has a field flux of 150 mWb. If the flux is reduced to 120 mWb, the speed will

        a. rise to 750 r/min   b. rise to 937.5 r/min   c. fall to 480 r/min

12-9   A four-pole compound-wound motor connected to a 250 V source draws an armature current of 600 A. The induced voltage E_o is 240 V. The total mechanical power developed is

        a. 144 kW   b. 150 kW   c. 201 hp

12-10   When a shunt-wound motor suddenly acts as a generator,

        a. the armature current reverses   b. it develops a greater torque
        c. its speed may run away

QUESTIONS AND PROBLEMS

12-11   Name three types of motors, and make a sketch of the armature and field-coil connections.

12-12   What determines the magnitude of the counter-emf of a given dc motor?

12-13   What determines the polarity of the cemf of a dc motor?

12-14   The cemf of a dc motor is always slightly less than the applied armature voltage. Explain.

12-15   Name two ways to vary the speed of a dc motor.

12-16   Why is it dangerous to start a dc motor without a starting resistance?

12-17   In Fig. 12-11, if the shunt field is connected to terminals A1 and A2 during the starting period, the motor will probably not be able to bring the load up to speed. Explain.

12-18   In Fig. 12-11, if F1 and F2 are interchanged, what will happen?

12-19   In Fig. 12-11, if A1 and A2 are interchanged, what will happen?

12-20   A 500 V shunt-wound motor has a rated armature current of 1200 A. If the total resistance of the armature circuit is 0.008 Ω, calculate
    a.   the cemf at full-load

    b.   the total "internal" mechanical power developed by the motor, in horsepower

    c.   the armature circuit heat loss

12-21   In Problem 12-20, calculate the locked-rotor current if the armature were directly connected to the line.

12-22   The motor of Fig. 12-6 has 1200 turns on the shunt winding and 4 turns on the series winding, per pole. The shunt field has a total resistance of 500 Ω, and the rated armature current I_a is 50 A. If the motor is fed by a 250 V line, calculate
    a.   the mmf per pole at no-load

    b.   the mmf per pole at full-load

CHAPTER 13

Simple Alternating Current Circuits

Introduction and Chapter Objectives

The voltages and currents in an electric circuit are often determined by the presence of three basic components: resistors, inductors, and capacitors. In this chapter, we will show how these components behave in an ac circuit. You will encounter several new terms, such as reactance, active power, and reactive power. These terms, and the basic theory introduced here will be used throughout the rest of the book.

   Before you read this chapter, we recommend a brief review of Chapter 7, on ac voltages and currents, and Chapters 9 to 10, on inductors and capacitors.

13.1 Resistive circuit

Let us consider a 20 Ω resistor connected across a 60 Hz source E_{ab} (Fig. 13-1). The voltage is sinusoidal and has a peak value of 200 V. One cycle corresponds to

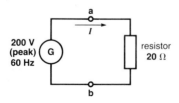

Figure 13-1
Resistor connected to a 60 Hz ac source that generates a peak voltage of 200 V.

1/60 s and, as we saw in Chapter 7, this period may be represented by an angle of 360° (Fig. 13-2). Because the voltage varies sinusoidally, we can calculate its

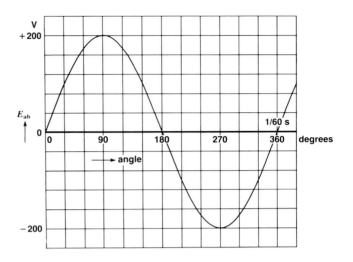

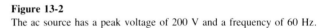

Figure 13-2
The ac source has a peak voltage of 200 V and a frequency of 60 Hz.

instantaneous value at any angle. We can therefore calculate the corresponding value of current I using Ohm's law. Thus:

At 30°, voltage E_{ab} = 200 sin 30 = 200 × 0.5 = 100 V;

   current $I = E/R$ = 100/20 = 5A

At 60°, E_{ab} = 200 sin 60 = 200 × 0.866 = 173 V;

   $I = E/R$ = 173/20 = 8.66 A

At 90°, E_{ab} = 200 sin 90 = 200 × 1.0 = 200 V

   $I = E/R$ = 200/20 = 10 A

At another angle such as 210°, we find

   E_{ab} = 200 sin 210 = 200 × (−0.5) = − 100 V

Consequently, $I = E/R$ = − 100/20 = − 5A

If we plot the value of I, we obtain a sine wave having a peak value of 10 A (Fig. 13-3). Note that the peak positive current occurs at the same instant as the peak positive voltage. The voltage and current are therefore in phase. If we represent the two quantities by phasors, we obtain the diagram of Fig. 13-4.

13.2 Power dissipated in a resistive circuit

The instantaneous power in a circuit is always given by the product of voltage times current. Thus, in Fig. 13-3, if we multiply the instantaneous values of voltage and

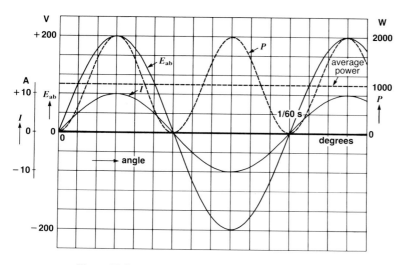

Figure 13-3
Instantaneous current and power in the circuit of Fig. 13-1.

Figure 13-4
This phasor diagram represents the voltage and current in Figs. 13-1 and 13-3.

current, we obtain the power from one instant to the next. For example:

At 0°, $E_{ab} = 0$ V, $I = 0$ A: $P = EI = 0 \times 0 = 0$ W

At 30°, $E_{ab} = 100$ V, $I = 5$ A: $P = 100 \times 5 = 500$ W

At 60°, $E_{ab} = 173$ V, $I = 8.66$ A: $P = 173 \times 8.66 = 1500$ W

At 90°, $E_{ab} = 200$ V, $I = 10$ A: $P = 200 \times 10 = 2000$ W

When the voltage and current are both negative, the power is still positive because the product of two negative numbers is positive. For example, at 210°, $E_{ab} = -100$ V, $I = -5$A : $P = (-100) \times (-5) = +500$ W.

If we plot these instantaneous values of power (Fig. 13-3), we obtain a curve that starts at zero, rises to a maximum of 2000 W, falls to zero, rises again to 2000 W, and so forth. The power dissipated in the resistance varies from instant to instant, but the average power is 1000 W, a value that is found by analyzing the power curve. Consequently, the resistor will reach a temperature corresponding to a continuous heat dissipation of 1000 W.

13.3 Effective value of voltage and current

So far, we have specified the magnitude of an ac voltage by its peak value. An ac current has been similarly specified. Another way of specifying the value of an ac current is by the *heating effect* it produces in a resistor. In order to establish a meaningful relationship, it is convenient to compare the heating effect of an ac current with that of an equivalent dc current. For example, what value of dc current will produce the same heating effect in a 20 Ω resistor as current I does in

Fig. 13-1? We know that the average power dissipated is 1000 W, and so the equivalent dc current can be found using Eq. 2-5 :

$$P = I^2R$$
$$1000 = I^2 \times 20$$

therefore $I^2 = 50$, and so $I = \sqrt{50} = 7.07$ A

Thus, a dc current of 7.07 A produces the same heating effect as an ac current whose peak value is 10 A. We say that the *effective value* of the ac current is 7.07 A.

By definition, the effective value of an ac current is equal to the value of the dc current that produces the same amount of heat in the same resistor.

When an ac current varies sinusoidally, the effective value is always equal to 0.707 of its peak value. More precisely, it is $1/\sqrt{2}$ of its peak value, where $1/\sqrt{2} = 0.707106781....$.

In the same way, when an ac voltage varies sinusoidally, its effective value is 0.707 times its peak value.

In practice, when we speak of the value of an ac voltage or current, we always mean the effective value. Most voltmeters and ammeters are calibrated to display the effective value of the voltage and current — not the peak value. Thus, if an ac voltmeter is connected into the circuit of Fig. 13-1 it will indicate $200 \times 0.707 = 141.4$ V. Similarly, an ac ammeter will indicate a current of $10 \times 0.707 = 7.07$ A. (Fig. 13-5).

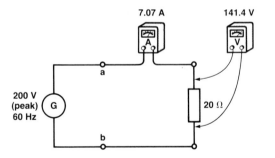
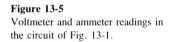

Figure 13-5
Voltmeter and ammeter readings in the circuit of Fig. 13-1.

Example 13-1:

An ac voltmeter connected to the outlet in a home shows a reading of 120 V. Calculate the peak value of the voltage.

Solution:

The peak voltage is $E = 120/0.707 = 169.7$ V
The peak value is also given by $120\sqrt{2} = 120 \times 1.414 = 169.7$ V

When effective values of voltage and current are used, we can apply Ohm's law, as well as the power equations $P = EI$, $P = I^2R$, and $P = E^2/R$, just as if we were working with dc. This means that the dc theory covered in Chapter 3 can also be applied to ac resistive circuits. That is one of the advantages of using effective values.

13.4 Inductive circuit

In Section 9-10, we learned that when a sinusoidal voltage is applied to an inductor, the resulting current is sinusoidal. Furthermore, the current lags 90° behind the voltage. Knowing this, let us determine the relationship between voltage, current, and power when a peak current of 10 A flows in a coil. The coil has an inductance of 53 mH, and the frequency of the current is 60 Hz (Fig. 13-6). We assume an odd value like 53 mH because we want to relate our findings to the previous case in which we used a 20 Ω resistor.

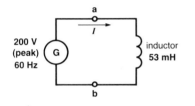

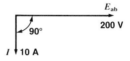

Figure 13-6
Inductor connected to a 60 Hz ac source that generates a peak voltage of 200 V.

Figure 13-7
Phasor diagram of the voltage and current in the inductive circuit of Fig. 13-6.

You will recall that the voltage E_{ab} induced in a coil is given by

E_{ab} = L × rate of change of current   Eq. 9-4

Furthermore, we know from Section 7-9 that the maximum rate of change of current is given by

$(\Delta I / \Delta t)_{max} = 2\pi f I_m = 6.28 \times 60 \times 10 = 3768$ A/s   Eq. 7-2

The corresponding peak value of E_{ab} is

E_{ab}(peak) $= L$ × maximum rate of change of current   Eq. 9-4
            $= 53 \times 10^{-3} \times 3768 = 200$ V

Thus, the peak value of the sinusoidal ac voltage is 200 V.

The voltage and current are therefore the same in Fig. 13-6 as they were in the resistive circuit of Fig. 13-1. The only difference is that the current lags 90° behind the voltage. The corresponding phasor diagram is given in Fig. 13-7.

If a voltmeter and ammeter are connected into the circuit of Fig. 13-6, the voltmeter will read 141.4 V and the ammeter 7.07 A (Fig. 13-8). These readings

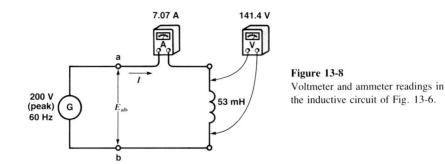

Figure 13-8
Voltmeter and ammeter readings in the inductive circuit of Fig. 13-6.

are the same as those in the resistive circuit of Fig. 13-5. It appears therefore that the power in the inductive circuit is

$$P = EI = 141.4 \times 7.07 = 1000 \text{ watts}$$

But we discover that the coil is not warm at all. There is no evidence that it is dissipating 1000 W. What, then, is happening in the circuit?

13.5 Reactive power in an inductive circuit, the var

Figure 13-9 shows the waveshape of the voltage and current in the inductor, with the current lagging 90° behind the voltage. According to the explanations given in Section 7-14, the respective equations are:

$$E_{ab} = 200 \sin \theta$$
$$I = 10 \sin (\theta - 90°)$$

Let us multiply the instantaneous values of voltage and current, as we did in the case of the resistor.

First, the power is zero at 0°, 90°, 180°, and so forth because either the voltage or current is zero at these moments. But midway between these zero points, the power reaches a maximum. Thus, it is maximum at 45°, 135°, 225°, and so forth.

At 135°, $E_{ab} = 200 \sin 135° = 141.4$ V and $I = \sin(135 - 90) = 10 \sin 45° = 7.07$ A. Therefore, the maximum power is $P = EI = 141.4 \times 7.07 = 1000$ W. Note that both E_{ab} and I are positive at this moment, and so P is also positive. Power is flowing into the coil.

The next power maximum occurs at 225°. The current is positive but E_{ab} is negative. Consequently, the power is negative. Power is therefore flowing out the coil. Its value is again 1000 W. The resulting positive and negative power swings are shown in Fig. 13-9.

These cyclic power surges go on indefinitely. The coil absorbs energy during a quarter cycle and then returns it to the source during the next quarter cycle. The average power absorbed by the coil is therefore zero, and that is why the coil does not get hot. In effect, magnetic energy is stored up in the coil for a quarter cycle and is then released during the next quarter cycle.

In an inductor, the product of effective voltage times effective current is called

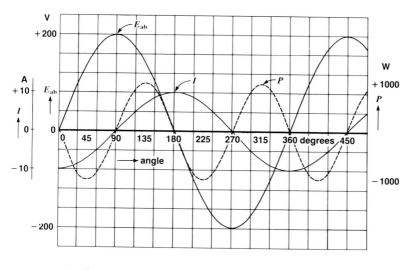

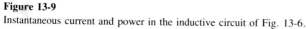

Figure 13-9
Instantaneous current and power in the inductive circuit of Fig. 13-6.

reactive power. It is designated by the symbol Q. This power is not expressed in watts, but in *reactive voltamperes.* The unit of reactive voltamperes is the var. The power absorbed by the coil is therefore 1000 var, and not 1000 W.

For reasons given later, engineers assume that an inductor *absorbs* reactive power. It is convenient to think that an inductor absorbs reactive power in order to create its alternating magnetic field.

13.6 Inductive reactance (X_L)

The voltmeter and ammeter readings in Fig. 13-8 (E = 141.4 V, I = 7.07 A) would lead us to believe that the "resistance" of the coil is E/I = 141.4 V/7.07 A = 20 Ω. But a coil does not have the properties of a resistor, and so the ratio E/I is called the *inductive reactance* of the coil. Inductive reactance (symbol X_L) is expressed in ohms. Unlike a resistor, the inductive reactance is not constant, but varies with the frequency. Its value may be calculated by the equation:

$$X_L = 2\pi fL \qquad (13\text{-}1)$$

where

X_L = inductive reactance of the coil, in ohms (Ω)

f = frequency of the source, in hertz (Hz)

L = inductance of the coil, in henries (H)

For example, the inductive reactance of a 53 mH coil at a frequency of 60 Hz is

Figure 13-10
This 100 Mvar, 425 kV, 600 Hz reactor is used to prevent
overvoltages on long transmission lines. The reactor is immersed in
mineral oil. and the high insulating bushing enables the reactor to be
connected to the EHV line. (*Courtesy Hydro-Québec*)

$$X_L \ = \ 2\pi fL \ = \ 6.28 \ \times \ 60 \ \times \ 0.053 \ = \ 20 \ \Omega$$

This calculated value agrees with the value we found previously.

Equation 13-1 shows that the reactance of a coil increases with the inductance
and the frequency of the applied voltage.

Example 13-2:

A coil having an inductance of 2 H is connected to a 4 kV, 50 Hz source. Calculate:

1. the inductive reactance X_L of the coil

2. the effective current I flowing in the coil

3. the reactive power Q absorbed by the coil

Solution:

1. $X_L = 2\pi fL = 6.28 \times 50 \times 2 = 628\ \Omega$ Eq. 13-1

2. Because the 4 kV is not stated to be peak or effective, it is *understood* to be the effective value
$I = E/X_L = 4000/628 = 6.37$ A

3. $Q = EI = 4000 \times 6.37 = 25\ 480$ var $= 25.48$ kvar

Figure 13-9 shows a very large reactor used on extra high voltage (EHV) transmission lines.

13.7 Capacitive circuit, capacitive reactance (X_C)

In Section 10-15, we saw that when a sinusoidal voltage is applied to a capacitor, a sinusoidal current is produced. Furthermore, the current leads the voltage by 90°. We also learned that the current is given by the equation:

current = capacitance × rate of change of voltage   Eq. 10-8

Knowing this, let us determine the current and power when a 60 Hz sinusoidal voltage is applied to a capacitor of 133 μF (Fig. 13-11). The peak value of the voltage is 200 V.

First, we calculate the maximum rate of change of voltage:

$(\Delta E\ /\ \Delta t)_{max} = 2\pi fE_m$ Eq. 7-1
$= 6.28 \times 60 \times 200 = 75\ 360$ volts per second

Then, using Eq. 10-8, we calculate the corresponding peak current:

I (peak) $= C \times$ maximum rate of change of voltage
$= 133 \times 10^{-6} \times 75\ 360 = 10$ A

Figure 13-11
Capacitor connected to a 60 Hz ac source having a peak voltage of 200 V.

Figure 13-12
Phasor diagram of the voltage and current in the circuit of Fig. 13-11.

Thus, the voltage and current have the same magnitude as in the case of the 20 Ω resistor. The only difference is that the current leads the voltage by 90°. The phasor diagram is given in Fig. 13-12.

If a voltmeter and ammeter were connected into the circuit, the voltmeter would

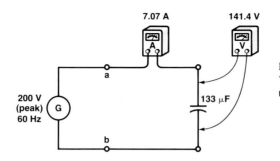

7.07 A   141.4 V

200 V
(peak) G
60 Hz

a

133 µF

b

Figure 13-13
Voltmeter and ammeter readings in
the capacitive circuit of Fig. 13-11.

read 141.4 V and the ammeter 7.07 A (Fig. 13-13). The capacitor appears therefore to have a resistance of $E/I = 141.4 / 7.07 = 20\ \Omega$. However, because a capacitor is so much different from a resistor, this opposition to current flow is named *capacitive reactance* (symbol X_C).

13.8 Reactive power in a capacitive circuit

As in the case of the inductor, we can calculate the instantaneous power in the capacitor by multiplying the voltage by the current. This yields the power curve shown in Fig. 13-14. The power reaches successive positive and negative peaks of 1000 W. The net power absorbed by the capacitor is therefore zero. Consequently, an ideal capacitor does not heat up.

The positive values of power correspond to those intervals when the capacitor is charging, and storing electric energy. The negative values occur during the periods when the capacitor is discharging and returning its energy to the source.

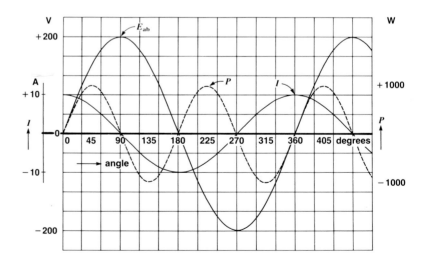

Figure 13-14
Instantaneous current and power in the capacitive circuit of Fig. 13-11.

Returning to Fig. 13-13, it appears that the power absorbed by the capacitor is

$$P = EI = 141.4 \times 7.07 = 1000 \text{ W}$$

But the capacitor does not heat up, and so this capacitive "power" cannot be expressed in watts. As in the case of an inductor, it is called *reactive power* (symbol Q). The unit of capacitive reactive power is also the var. The reactive power in Fig. 13-13 is therefore 1000 var.

It is convenient to assume that a capacitor *generates* reactive power. On the other hand, as mentioned previously, an inductor is assumed to absorb reactive power.

13.9 Calculation of capacitive reactance

We can calculate the capacitive reactance of a capacitor by the equation:

$$X_C = \frac{1}{2\pi f C} \qquad \text{(13-2)}$$

where

X_C = capacitive reactance, in ohms (Ω)

f = frequency of the source, in hertz (Hz)

C = capacitance, in farads (F)

2π = constant (approximate value 6.28)

This equation shows that the reactance decreases as the capacitance becomes larger. Furthermore, X_C decreases as the frequency increases.

Applying this equation, we find that the capacitive reactance of the 133 μF capacitor at a frequency of 60 Hz is

$$X_C = \frac{1}{2\pi f C} = \frac{1}{6.28 \times 60 \times 133 \times 10^{-6}} = 20 \ \Omega$$

This agrees with the value we found previously.

Example 13-3:

A 10 μF capacitor is connected to a 120 V, 400 Hz source. Calculate:

1. the capacitive reactance X_C of the capacitor

2. the peak current I_m in the capacitor

3. the reactive power Q generated by the capacitor

Solution:

1. $X_C = \dfrac{1}{2\pi f C} = \dfrac{1}{6.28 \times 400 \times 10 \times 10^{-6}} = 40 \ \Omega$

2. the effective value of current $I = E/X_C = 120/40 = 3$ A
   the peak current $I_m = 3/0.707 = 4.24$ A

3. the reactive power produced by the capacitor is
   $Q = EI = 120 \times 3 = 360$ var

Figure 13-15 shows a large capacitor bank.

Figure 13-15
Capacitor bank rated 360 kvar, 600 V, 3-phase, 60 Hz composed of six 60 kvar, 3-phase units. The capacitors are contained in a control cubicle that connects and disconnects the 60 kvar units in order to maintain the plant power factor at 90 percent. Each 60 kvar unit is 45 cm high, 35 cm wide and 15 cm deep, and dissipates only 30 W at rated voltage and frequency. (*Courtesy Electro-Mecanik*)

13.10 Active and reactive power

If we compare the power flow in a resistor with that in an inductor or capacitor, we note some distinctive features. Power in a resistor is always positive (Fig. 13-3). It continually flows from the ac source to the load. On the other hand, power in an inductor surges back and forth between the inductor and the source (Fig. 13-9). The same remarks apply to a capacitor.

In ac circuits we therefore recognize two types of power: active power and reactive power.

Power that does real work or produces heat is called *active power (P)*. Active

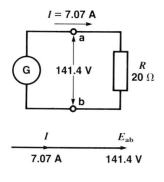

$I = 7.07$ A

141.4 V

R
20 Ω

I	E_{ab}
7.07 A	**141.4 V**

$P = EI = 141.4 \times 7.07$

$P = 1000$ W

resistor absorbs

active power P

R

Figure 13-16
Phasor diagram and power in
a resistive circuit.

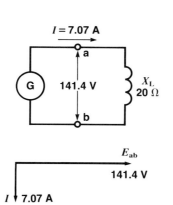

$I = 7.07$ A

141.4 V

X_L
20 Ω

E_{ab}

141.4 V

I **7.07 A**

$Q = EI = 141.4 \times 7.07$

$Q = (+)\ 1000$ var

inductor absorbs

reactive power Q

$X_L = 2\pi f L$

Figure 13-17
Phasor diagram and power in
an inductive circuit.

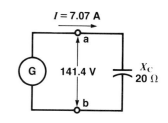

$I = 7.07$ A

141.4 V

X_C
20 Ω

I **7.07 A**

E_{ab}

141.4 V

$Q = EI = 141.4 \times 7.07$

$Q = (-)\ 1000$ var

capacitor generates

reactive power Q

$X_C = 1/2\pi f C$

Figure 13-18
Phasor diagram and power in
a capacitive circuit.

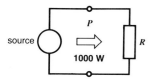

P
1000 W

source

R

Figure 13-19
A resistor absorbs active
power.

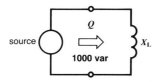

Q
1000 var

source

X_L

Figure 13-20
A reactor absorbs reactive
power.

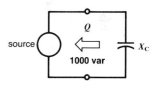

Q
1000 var

source

X_C

Figure 13-21
A capacitor produces reactive
power.

power is always expressed in watts. By definition, active power is the product EI of
the effective voltage E times the effective current I when E and I are in phase.

Power that oscillates back and forth, such as in an inductor or capacitor, is called
reactive power. Reactive power is expressed in vars. By definition, reactive is the
product EI of the effective voltage E times the effective current I when E and I are
90° out of phase.

A capacitor is assumed to produce, or generate reactive power. An inductor is
assumed to absorb reactive power. This is somewhat analogous to saying that an ac
generator produces active power and a resistor absorbs it.

Because we know that the instantaneous power in an inductor or capacitor merely surges back and forth, it is somewhat surprising to state that an inductor *absorbs* reactive power, and a capacitor *generates* reactive power. However, this conventional interpretation of reactive power has been approved by the Standards Committee of the Institute of Electrical and Electronics Engineers, by the American National Standards Institute, and internationally by the Electrotechnical Commission. The reason for assuming that capacitors generate reactive power, and inductors absorb it is that it greatly simplifies the treatment of power in ac circuits.

One final important feature of active and reactive power is that one cannot be converted into the other. Each exists in a separate world, so to speak.

13.11 Phasor diagrams and effective values

So far, the phasor diagrams have always shown the peak values of voltage and current. However, because effective values are usually given, we can change the phasor diagrams accordingly. Thus, from now on, and unless indicated otherwise, we show effective values in the phasor diagrams.

13.12 Recapitulation of *R, L, C* circuits

The relationships found in the *R, L, C* circuits are summarized in Figs. 13-16 to 13-18. You should examine and compare them to ensure that everything is clear. Note that effective values of voltage and current are now used.

Next, to clarify the notion of active and reactive power flow, Figs. 13-19 to 13-21 show how powers *P* and *Q* are assumed to flow over the line. The resistor absorbs active power, and so the active power *P* flows from the source to the load. The inductor absorbs reactive power, and so the reactive power *Q* also flows from the source to the load. On the other hand, the capacitor *delivers* reactive power *Q* to the source to which it is connected.

13.13 Summary

Electrical energy is almost entirely supplied in the form of alternating current and most industrial and domestic equipment is designed to operate on ac. Many devices contain coils in order to set up their magnetic fields. In this chapter, we learned that it takes reactive power to create an ac magnetic field. As a result, motors, relays and other magnetic devices absorb reactive power. The reactive power can be supplied by the generators of the electrical utility company, but it can also be supplied by capacitors. Indeed, one of the most important uses of capacitors is to supply the reactive power absorbed by industry, and by the distribution system.

We learned that electric power in ac circuits can be grouped into active power *P* and reactive power *Q*. In the next chapter we will see how these powers are combined when different devices are connected to the same line.

TEST YOUR KNOWLEDGE

13-1   An ac ammeter shows a reading of 64 A. The peak value of current is

   a. 45.2 A   b. 90.5 A   c. 128 A

13-2   A 5 Ω resistor dissipates 46 kW. The peak ac current in the resistor is

   a. 9200 A   b. 95.9 A   c. 135.6 A

13-3   A 60 Hz sinusoidal voltage has an effective value of 240 kV. The maximum rate of change of this voltage occurs when its value is

   a. 240 kV   b. 339.4 kV   c. zero

13-4   In Problem 13-3, the maximum rate of change of voltage is

   a. 90 432 000 V/s   b. 127 890 000 V/s   c. 127 890 V/s

13-5   An inductor has a reactance of 200 Ω at a frequency of 60 Hz. The reactance at 120 Hz is

   a. 400 Ω   b. 100 Ω   c. 200 Ω

13-6   An inductor has an inductance of 3 H when used in a 60 Hz line. If it is placed in a circuit in which the frequency is 300 Hz, the inductance is

   a. 15 H   b. 600 mH   c. 3 H

13-7   The unit of reactive power is

   a. the Var   b. the VAR   c. the var

13-8   The idea that a capacitor generates reactive power is based on a technical convention

   ☐ true   ☐ false

13-9   A capacitor connected to a 120 V, 60 Hz source draws a current of 10 A. The capacitance is

   a. 12 Ω   b. 31 400 μF   c. 0.000221 F

13-10   A 60 W, 120 V incandescent lamp draws a peak power of

   a. 60 W   b. 120 W   c. 84.85 W

QUESTIONS AND PROBLEMS

13-11   a.   Draw the waveshape of a sinusoidal voltage E_{12} having a peak value of 200 V and a frequency of 50 Hz. (The voltage is zero and increasing positively at $t = $ O.)

       b.   What is the voltage at $t = $ 5 ms, and what is the polarity of terminal 2 with respect to terminal 1?

13-12   A sinusoidal voltage of 120 V is applied to a 10 Ω resistor. Calculate:

       a.   the effective current in the resistor

       b.   the peak voltage across the resistor

13-13   A capacitor has a rating of 500 kvar, 440 V, 60 Hz. Calculate:

       a.   the rated effective current

       b.   the capacitive reactance

       c.   the capacitance

       d.   the peak voltage across the capacitor terminals

       e.   the peak energy stored in the capacitor, in joules

13-14   The large reactor* shown in Fig. 13-10 has a rating of 110 Mvar, 425 kV, 60 Hz. Calculate:

       a.   the rated current of the reactor

       b.   the inductive reactance of the reactor

       c.   the inductance of the reactor

       d.   the peak current in the reactor

       e.   the peak energy stored in the magnetic field, in joules

---

*   An inductor that operates on ac is often called a reactor

CHAPTER 14

Solving Single-Phase AC Circuits

Introduction and Chapter Objectives

In Chapter 13 we learned about the three basic elements of an ac circuit: resistors, inductors and capacitors. It is a remarkable fact that most ac power circuits can be expressed in terms of one or more of these elements. Even an ac motor can be represented by a circuit containing only R and X_L elements. Of course, this circuit is only a *model* of the motor, but it enables us to predict its behavior without testing the machine in a laboratory. Circuits, therefore, are useful in analyzing electrical apparatus and systems.

This chapter is concerned with single-phase circuits, meaning those that are connected to a one-voltage source. Three-phase circuits are covered in Chapter 15.

In solving an ac circuit, we are usually interested in knowing what the voltages and currents are. This is the standard circuit approach. But when we are dealing with electrical machinery, transmission lines, and industrial loads, we are often more interested in power, and power flow. We then use a power-circuit approach to calculate such quantities as reactive power, power factor, and apparent power. In this chapter, we will introduce both methods of solving ac circuits. You will also become familiar with the meaning of impedance and how phasors are used to solve ac circuits.

14.1 Voltages and currents in an RL circuit

Consider the circuit of Fig. 14-1, in which a 40 Ω resistor and a 30 Ω reactor are connected across a source E having a peak voltage of 120 V. We want to determine the voltage, current, and power in this circuit.

Because the voltage is common to both elements, current I_r is 3A (peak value)

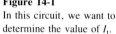

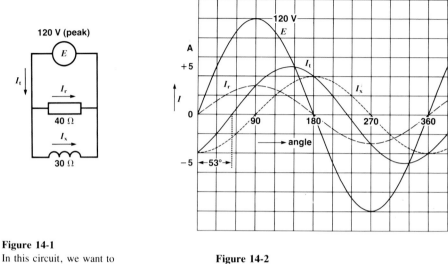

Figure 14-1
In this circuit, we want to
determine the value of I_t.

Figure 14-2
I_t is found by adding instantaneous currents.

and I_x is 4 A. If we use the rules for solving dc circuits, we would believe that the
total current I_t is equal to $(3 + 4) = 7$ A (peak value). But current I_x lags 90°
behind the voltage, while I_r is in phase with it. Therefore, to find the true peak value
of I_t we must add the *instantaneous* values of I_r and I_x. This can be done by first
drawing the waveshapes of E, I_r, and I_x, as shown in Fig. 14-2. Then, by adding
the instantaneous values of I_r and I_x we obtain the waveshape of I_t. We note that it is
sinusoidal, and its peak value is 5 A. Furthermore, by scaling off the angles, we
discover that I_t lags 53° behind the voltage. These results are summarized in
Fig. 14-3.

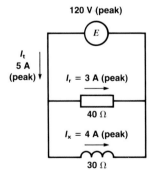

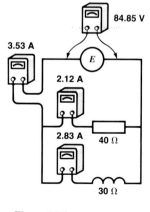

Figure 14-3
I_t is not equal to the arithmetic sum of I_x and I_r.

Figure 14-4
Effective voltage and currents
in the circuit of Fig. 14-1.

If voltmeters and ammeters were introduced into the circuit, we would obtain the effective values shown in Fig. 14-4. (The effective values are 0.707 times the peak values).

The method we have just described enables us to calculate the total current in a circuit, and its phase angle with respect to the applied voltage. But it is laborious, and it would be useful if a faster method could be found. The following method is much easier because it involves the addition of phasors, rather than the addition of instantaneous currents and voltages.

14.2 Addition of phasors

Suppose we want to find the sum of two ac voltages E_1 and E_2 that have the same frequency. They are represented by the phasor diagram of Fig. 14-5. Referring to

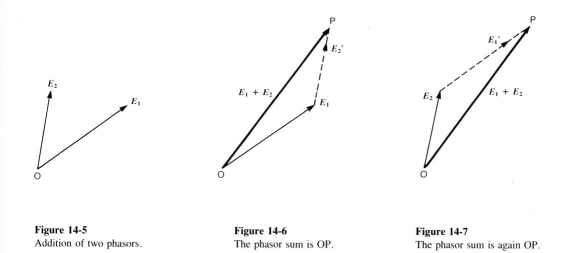

Figure 14-5
Addition of two phasors.

Figure 14-6
The phasor sum is OP.

Figure 14-7
The phasor sum is again OP.

Fig. 14-6, we proceed as follows:
1. We choose any one of the two phasors, say E_1, as a "starting" phasor.

2. From the arrowhead on phasor E_1, we draw a phasor E_2' having the same length and direction as phasor E_2.

3. Phasor OP, which starts at the origin 0 and ends at the arrowhead of E_2', is equal to the sum of ($E_1 + E_2$). This is referred to as the phasor sum of the two phasors.

We could also start with phasor E_2 and then add phasor E_1 to obtain the phasor sum. This does not change the final result (Fig. 14-7).

The same method can be used to find the phasor sum of any number of phasors. Thus, in Fig. 14-8, the phasor sum of $E_1 + E_2 + E_3$ yields phasor OP. In this graphical construction, we first found the sum of $E_2 + E_3$ and then added phasor E_1 (Fig. 14-9).

The phasor sum can be found using either the peak or effective values of voltage and current.

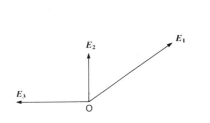

Figure 14-8
Addition of three phasors.

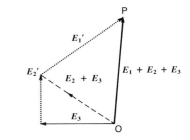

Figure 14-9
The phasor sum is OP.

Example 14-1:

Determine the phasor sum of the peak currents I_r and I_x in Fig. 14-3.

Solution:

The two currents have peak values of 3 A and 4 A and they are out of phase by 90°. Drawing the phasors to scale and then finding their sum, we discover that I_t has a length corresponding to 5 A (Fig. 14-10). Furthermore, the phase angle between I_t

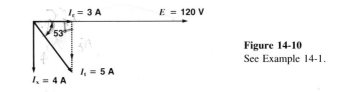

Figure 14-10
See Example 14-1.

and E is found to be 53°, as measured with a protractor. Clearly, this method of finding the magnitude and phase angle of I_t is much easier and faster than the method used in Fig. 14-2.

In the next example, we use *effective* values of voltage and current to determine the phasor sum. Most phasor diagrams show effective values.

Example 14-2:

In Fig. 14-11, a source having an effective voltage of 360 V is connected to a parallel R, L, C circuit, as shown. Draw the phasor diagram showing the *effective* voltage and effective current in each element. Determine the value of the line current I_t and its phase angle with respect to the line voltage.

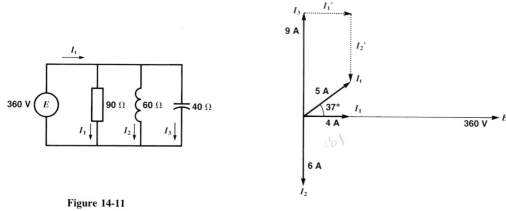

Figure 14-11
See Example 14-2.

Figure 14-12
See Example 14-2.

Solution:

The phasor for 360 V is drawn horizontally using an appropriate scale, such as 1 mm = 6 V. The voltage is used as a reference phasor because it is common to the three parallel elements. We then draw the phasors for I_1, I_2, and I_3.

$I_1 = E/R = 360/90 = 4$ A. I_1 is in phase with E because the element is a resistor. (We use a scale of 3 mm = 1 A.)

$I_2 = E/X_L = 360/60 = 6$ A. I_2 lags 90° behind E because the element is an inductor.

$I_3 = E/X_C = 360/40 = 9$ A. I_3 is 90° ahead of E because the element is a capacitor.

I_t is equal to the phasor sum of $I_1 + I_2 + I_3$, as shown in Fig. 14-12. Note that I_t is *not* equal to 4 + 6 + 9 = 19 A.

The length of I_t is 15 mm and so its value is 15/3 = 5 A. By using a protractor, we find I_t is 37° ahead of E. Because I_t leads E, the capacitance predominates over the inductance.

14.3 X and Y components of a phasor

In the previous examples, we used a graphical method to determine the magnitude and phase angle of a current. However, the availability of hand calculators makes it much easier to solve such problems by trigonometry. The procedure in making these calculations is very simple, as the following explanation shows. We begin by explaining what is meant by the X and Y component of a phasor.

Figure 14-13 shows a phasor having a magnitude V that is inclined at an angle θ to the X axis. It is designated by the symbol V $\lfloor$θ. The angle may have any value between zero and 360°. Starting from the arrowhead, perpendicular lines are

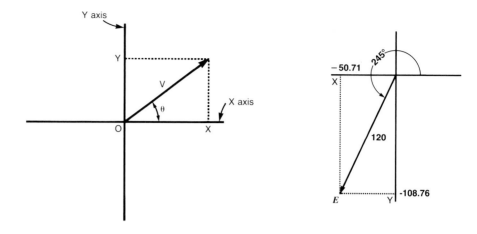

Figure 14-13
X and Y components of a phasor.

Figure 14-14
See Example 14-3.

dropped on the X and Y axes. By definition, the length OX is called the X component of the phasor, and OY is the Y component. We designate these components by the symbols **X** and **Y**. With these definitions, it is clear that

a)   X component of $V \lfloor \theta = \mathbf{X} = V \cos \theta$ (14-1)

b)   Y component of $V \lfloor \theta = \mathbf{Y} = V \sin \theta$ (14-2)

c) $\mathbf{Y/X} = \tan \theta$ (14-3)

d) $\theta = \arctan \mathbf{Y/X}$ (14-4)

Example 14-3:

Determine the X and Y components of the voltage phasor $E = 120 \lfloor 245°$.

Solution:

$\mathbf{X} = 120 \cos 245 = 120 \times (-0.4226) = -50.71$ V   Eq. 14-1

$\mathbf{Y} = 120 \sin 245 = 120 \times (-0.9063) = -108.76$ V   Eq. 14-2

The phasor and its components are shown in Fig. 14-14.

14.4 Constructing a phasor from its X and Y components

In making phasor calculations, we sometimes know the X and Y components of a phasor. How can we determine the magnitude of the phasor and its phase angle? Let us suppose that the components are $\mathbf{X}_s$ and $\mathbf{Y}_s$ (Fig. 14-15). It is clear from the figure that

1.   angle of phasor $\mathbf{V}_s$ is $\theta_s = \arctan \mathbf{Y}_s/\mathbf{X}_s$ (14-5)

2.   magnitude of phasor $\mathbf{V}_s$ is $V_s = \mathbf{X}_s/\cos \theta_s$ (14-6)

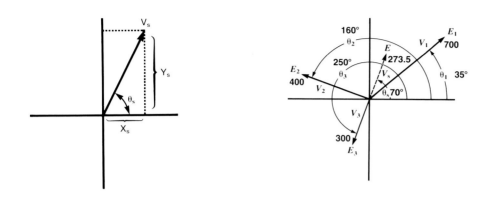

Figure 14-15
Constructing a phasor from its X and
Y components.

Figure 14-16
Determining the phasor sum by trigonometry. See Example 14-4.

14.5 Calculating the phasor sum by trigonometry

Suppose we want to calculate the phasor sum V_s of several phasors. We cannot obtain V_s directly, but we *are* able to determine the magnitude of its X and Y components. Knowing the components, it is easy to determine the magnitude and phase angle of the phasor sum by the method explained in the previous section.

It can be proved that the X component of V_s is given by:

X_s = sum of the X components of the phasors to be added   (14-7)

Similarly, the Y component of V_s is:

Y_s = sum of the Y components of the phasors to be added   (14-8)

For example, suppose we want to find the phasor sum (V_s) of the phasors E_1, E_2, and E_3 shown in Fig. 14-16. The X_s component of the V_s is given by

$$X_s = X_1 + X_2 + X_3$$ Eq. 14-7

$$X_s = V_1 \cos \theta_1 + V_2 \cos \theta_2 + V_3 \cos \theta_3$$ Eq. 14-1

The Y_s component of the V_s is given by:

$$Y_s = Y_1 + Y_2 + Y_3$$ Eq. 14-8

$$Y_s = V_1 \sin \theta_1 + V_2 \sin \theta_2 + V_3 \sin \theta_3$$ Eq. 14-2

Example 14-4:

The phasors in Fig. 14-16 have the following values:

$E_1 = 700 \underline{|35°}$; $E_2 = 400 \underline{|160°}$; $E_3 = 300 \underline{|250°}$

Calculate their phasor sum, in magnitude and phase angle.

Solution:

$$\mathbf{X_s} = \mathbf{X_1} + \mathbf{X_2} + \mathbf{X_3} \qquad\qquad\qquad\text{Eq. 14-7}$$

$$= 700 \cos 35 + 400 \cos 160 + 300 \cos 250$$

$$= 573.4 - 375.9 - 102.6 = +94.9$$

$$\mathbf{Y_s} = \mathbf{Y_1} + \mathbf{Y_2} + \mathbf{Y_3}$$

$$= 700 \sin 35 + 400 \sin 160 + 300 \sin 250$$

$$= 401.5 + 136.8 - 281.9$$

$$= +256.4$$

$$\theta_s = \arctan \mathbf{Y_s}/\mathbf{X_s} \qquad\qquad\qquad\text{Eq. 14-5}$$

$$= \arctan 256.4/94.9 = \arctan 2.70$$

$$= 69.7°$$

$$V_s = \mathbf{X_s}/\cos \theta_s \qquad\qquad\qquad\text{Eq. 14-6}$$

$$= 94.9 \,/\, \cos 69.7°$$

$$= 273.5$$

The phasor sum is therefore $\mathbf{E} = 273.5 \,\underline{|69.7°}$

14.6 Impedance of a circuit

Consider an electric circuit composed of resistors, inductors, and capacitors enclosed in a box. The components are interconnected in any way, but the final circuit has only two terminals 1 and 2 (Fig. 14-17). Suppose that an ac voltage E applied to the terminals produces a corresponding ac current I.

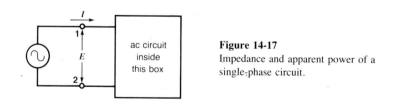

Figure 14-17
Impedance and apparent power of a single-phase circuit.

The ratio E/I is called the *impedance* of the circuit, and it is measured in ohms. The impedance (symbol Z) represents the "opposition" the circuit offers to the flow of ac current. For example, if the ac voltage across a circuit is 120 V and the current is 4 A, the impedance of the circuit is $Z = E/I = 120 \text{ V}/4\text{A} = 30 \ \Omega$.

14.7 Apparent power of a circuit, the voltampere (VA)

The apparent power of an ac circuit is equal to the product of the effective voltage E across its terminals times the effective current I it absorbs. Referring to Fig. 14-17, the apparent power (symbol S) is given by $S = EI$.

Apparent power is expressed in voltamperes (VA), kilovoltamperes (kVA), or megavoltamperes (MVA).

Because the circuit in Fig. 14-17 may contain anything, the apparent power includes every form of electric power, including active power (watts), reactive power (vars), or any combination of the two. In summary, we can therefore write the equations:

$$\overline{Z = E/I} \qquad (14\text{-}9)$$

and

$$\overline{S = EI} \qquad (14\text{-}10)$$

in which

$Z =$ impedance of the circuit, in ohms (Ω)

$E =$ effective sinusoidal voltage across the circuit, in volts (V)

$I =$ effective sinusoidal current flowing in the circuit, in amperes (A)

$S =$ apparent power, in voltamperes (VA)

14.8 Solution of a parallel R, C circuit

Example 14-5:

The circuit of Fig. 14-18 is composed of a 30 Ω resistor and a 16 Ω capacitive reactance connected in parallel across a 240 V ac source. We want to calculate:

1. the line current I and its phase angle with respect to line voltage E

2. the impedance of the circuit

3. the active, reactive, and apparent power of the circuit

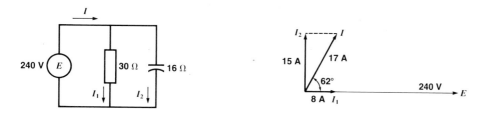

Figure 14-18
See Example 14-5

Figure 14-19
See Example 14-5

Solution:

1. To determine the line current I, we must draw a phasor diagram

    a) Because this is a parallel circuit, the voltage is common to both elements; we therefore adopt it as the reference phasor. Let us draw it horizontally, along the X axis (Fig. 14-19).

    b) The current in the resistor is
        $I_1 = E/R = 240/30 = 8$ A. We draw this current phasor in phase with E because the element is resistive.

    c) The current in the capacitor is
        $I_2 = E/X_C = 240/16 = 15$ A. Because the element is capacitive, phasor I_2 is drawn 90° ahead of E.

    d) Line current I is the phasor sum of $I_1 + I_2$. Using the method described in Section 14-4, we find:

        $\theta = \arctan I_2/I_1 = \arctan 15/8 = 61.9°$

        $I = I_1/\cos \theta = 8/\cos 61.9 = 17$ A

2. The impedance of the circuit is $Z = E/I = 240/17 = 14.1 \ \Omega$. Note that the impedance is a property of the circuit. It is independent of the value of the applied voltage. For example, if the voltage E is reduced by half (to 120 V), all the currents will be reduced by half, but the value of Z will remain the same. Furthermore, all the phase angles remain the same. (The impedance and the phase angles would change if the frequency of the source were changed because the value of X_C depends on frequency.)

3. The active, reactive, and apparent powers are found as follows:

    a) The resistor consumes an active power P
        $P = EI_1 = 240 \times 8 = 1920$ watts $= 1920$ W

    b) The capacitor generates a reactive power Q
        $Q = EI_2 = 240 \times 15 = 3600$ vars $= 3600$ var

    c) The apparent power of the circuit is
        $S = EI = 240 \times 17 = 4080$ voltamperes $= 4080$ VA

Note that the apparent power is *not* equal to the arithmetic sum of the active and reactive power in the circuit. In other words, 4080 VA is not equal to the sum (1920 + 3600) = 5520. The reason for this will be explained in Section 14-14.

14.9 Solution of a series R, L circuit

Example 14-6:

The circuit of Fig. 14-20 is composed of a 12 Ω resistor in series with a 5 Ω inductive reactance. The effective current is 10 A and we want to calculate:

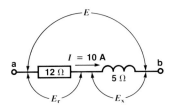

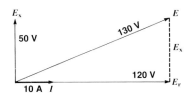

Figure 14-20
See Example 14-6

Figure 14-21
See Example 14-6

1. the voltage between terminals a and b, and its phase angle with respect to I

2. the impedance of the circuit

3. the apparent power of the circuit

Solution:

1. Because this is a series circuit, the current is common to both elements, and so we use I as the reference phasor. We draw it horizontal along the X axis (Fig. 14-21). However, we could have selected any other direction.

   a) The voltage across the resistor is:
   $E_r = IR = 10 \times 12 = 120$ V. E_r is in phase with I because the element is resistive

   b) The voltage across the reactor is:
   $E_x = IX_L = 10 \times 5 = 50$ V. E_x is drawn 90° ahead of I because in an inductive reactance the current lags 90° behind the voltage.

   c) The voltage across the terminals is equal to the phasor sum of $E_r + E_x$. Using the method described previously, we find:

   $\theta = \arctan E_x/E_r = \arctan 50/120 = 22.6°$

   $E = E_r/\cos \theta = 120/\cos 22.6° = 130$ V

2. The impedance of the circuit is
   $Z = E/I = 130/10 = 13$ Ω

3. The apparent power of the circuit is
   $S = EI = 130 \times 10 = 1300$ voltamperes $= 1.3$ kVA

14.10 Impedance of two elements in series

We often have to calculate the impedance of two elements that are connected in series. If they are of the same type, we simply add the impedances arithmetically. For example, a 3 Ω reactor and a 5 Ω reactor connected in series gives a total impedance of 8 Ω. But if the impedances are not of the same kind, we have to use phasor methods to determine their total impedance. This could be done by employing the techniques we have just studied. However, it saves time to use standard formulas to calculate the series impedance directly. These formulas are given in Table 14A.

TABLE 14A IMPEDANCE OF ELEMENTS IN SERIES

Circuit	Elements in series	Resulting impedance	
	R, X_L	$Z_s = \sqrt{R^2 + X_L{}^2}$	(14-11)
	R, X_C	$Z_s = \sqrt{R^2 + X_C{}^2}$	(14-12)
	X_L, X_C	$Z_s = X_L - X_C$	(14-13)
	R, X_L, X_C	$Z_s = \sqrt{R^2 + (X_L - X_C)^2}$	(14-14)

14.11 Impedance of two elements in parallel

The impedance of two impedances Z_1 and Z_2 connected in parallel is given by the equation:

$$Z_p = \frac{Z_1 Z_2}{Z_1 + Z_2} = \frac{Z_1 Z_2}{Z_s} \qquad (14\text{-}15)$$

where

Z_p = value of the parallel impedance, in ohms (Ω)
Z_1 = impedance of Z_1, in ohms (Ω)
Z_2 = impedance of Z_2, in ohms (Ω)
Z_s = impedance that would be obtained if Z_1 and Z_2 were connected in series, in ohms (Ω)

Note that the value of Z_s in Eq. 14-15 is calculated from the equations given in Table 14A.

Table 14B lists formulas for some common parallel circuits.

TABLE 14 B   IMPEDANCE OF ELEMENTS IN PARALLEL

Circuit	Elements in parallel		Resulting impedance	
	Z_1	Z_2	Z_p	
	R	X_L	$Z_p = \dfrac{RX_L}{\sqrt{R^2 + X_L^2}}$	(14-15)
	R	X_C	$Z_p = \dfrac{RX_C}{\sqrt{R^2 + X_L^2}}$	(14-16)
	X_L	X_C	$Z_p = \dfrac{X_L X_C}{X_L - X_C}$	(14-17)
	R, X_L in series	X_C	$Z_p = \dfrac{X_C \sqrt{R^2 + X_L^2}}{\sqrt{R^2 + (X_L - X_C)^2}}$	(14-18)

Example 14-7:

Calculate the impedance of the circuit of Fig. 14-20 and determine the value of E.

Solution:

1. Using Eq. 14-11

$$Z_s = \sqrt{R^2 + X_L^2} = \sqrt{12^2 + 5^2} = \sqrt{144 + 25} = \sqrt{169} = 13 \ \Omega$$

2. $E = IZ = 10 \times 13 = 130$ V

These values correspond to those found in Section 14-9.

Example 14-8:

What is the impedance of a circuit composed of a resistance R in parallel with a reactance X_L?

Solution:

1. The impedance of $Z_1 = R$ ohms

2. The impedance of $Z_2 = X_L$ ohms

3. If Z_1 and Z_2 were connected in series, the impedance would be:

$$Z_s = \sqrt{R^2 + X_L^2} \qquad\qquad \text{Eq. 14-11}$$

4. The impedance in parallel is therefore

$$Z_p = \frac{Z_1 Z_2}{Z_s} = \frac{RX_L}{\sqrt{R^2 + X_L^2}} \qquad\qquad \text{Eq. 14-15}$$

Example 14-9:

Calculate the impedance of the circuit in Fig. 14-18.

Solution:

$$
\begin{aligned}
Z_p &= RX_C/\sqrt{R^2 + X_C^2} = 30 \times 16/\sqrt{30^2 + 16^2} \qquad\qquad \text{Eq. 14-16}\\
Z_p &= 480/\sqrt{900 + 256} = 480/\sqrt{1156}\\
&= 480/34 = 14.1 \ \Omega
\end{aligned}
$$

This agrees with the value previously found in Section 14-8

Example 14-10:

A coil has a resistance of 2 Ω and an inductance of 8 mH. It is connected in parallel with a capacitor of 80 μF. The entire circuit is then connected to a 200 V, 200 Hz source (Fig. 14-22). Calculate

1. the current in the coil

2. the current in the capacitor

3. the impedance of the parallel circuit

4. the current drawn from the 200 V source

Solution:

1. a) The inductive reactance of the coil is
   $$X_L = 2\,\pi f L = 6.28 \times 200 \times 8 \times 10^{-3} = 10 \ \Omega \qquad \text{Eq. 13-1}$$

   b) The impedance of the coil is $Z_1 = \sqrt{R^2 + X_L^2} = \sqrt{2^2 + 10^2} = 10.2 \ \Omega$

   c) Current in the coil $I_1 = E/Z_1 = 200/10.2 = 19.6$ A

2. a) The capacitive reactance of the capacitor is

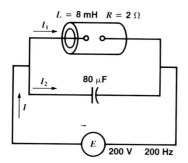

Figure 14-22
See Example 14-10

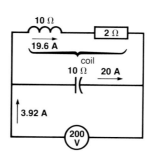

Figure 14-23
See Example 14-10

$$X_C = \frac{1}{2\pi f C} = \frac{1}{6.28 \times 200 \times 80 \times 10^{-6}} = 10 \ \Omega$$

b) current in the capacitor $I_2 = E/X_C = 200/10 = 20$ A

3. a) The impedance of the entire circuit is

$$Z_p = \frac{X_C \sqrt{R^2 + X_L^2}}{\sqrt{R^2 + (X_L - X_C)^2}} \qquad \text{Eq. 14-18}$$

$$= \frac{10 \sqrt{2^2 + 10^2}}{\sqrt{2^2 + (10 - 10)^2}}$$

$$= 10 \times 10.2/2 = 51 \ \Omega$$

Note that the impedance of the circuit is greater than the impedance of its separate components.

4. $I = E/Z = 200/51 = 3.92$ A

The impedances and currents are shown in Fig. 14-23.

This circuit illustrates the phenomenon of *parallel resonance*. It occurs when $X_C = X_L$. When there is parallel resonance, the currents in the coil and capacitor are much larger than the line current (compare 20 A with 3.92 A).

14.12 Series resonance

A circuit composed of an inductor in series with a capacitor can also produce *series resonance*. Series resonance, just like parallel resonance, occurs when $X_L = X_C$. Very high voltages can then appear across both the inductor and capacitor, compared with the voltage of the source.

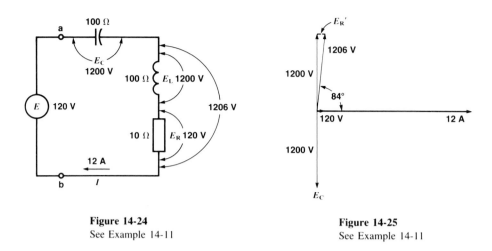

Figure 14-24
See Example 14-11

Figure 14-25
See Example 14-11

Example 14-11:

A coil having a resistance of 10 Ω and a reactance of 100 Ω is connected in series with a capacitor whose capacitive reactance is 100 Ω (Fig. 14-24). Calculate the voltage across the coil and the capacitor when the circuit is connected to a 120 V source.

Solution:

1. The impedance of the circuit is

$$Z_{ab} = \sqrt{R^2 + (X_L - X_C)^2} = \sqrt{10^2 + 100 - 100^2} = 10\ \Omega \quad \text{Eq. 14-14}$$

2. $I = E/Z = 120/10 = 12$ A. The current is in phase with the voltage because the circuit is effectively resistive.

3. The voltage E_c across the capacitor lags 90° behind I

4. $E_c = IX_C = 12 \times 100 = 1200$ V

5. The impedance of the coil is $Z_c = \sqrt{R^2 + X_L^2} = \sqrt{10^2 + 100^2} = 100.5\ \Omega$

6. Voltage across the coil is $E = IZ_c = 12 \times 100.5 = 1206$ V

The phasor diagram is shown in Fig. 14-25.

POWER IN AC CIRCUITS

14.13 Measurement of active and reactive power

Special instruments are available that can measure the active and reactive power in a circuit. They are respectively called wattmeters and varmeters.

A *wattmeter* has four terminals, two for the line current and two for the line voltage. In portable instruments, two of the terminals always bear a ± sign. For our purposes, the current terminals are marked ±A and A, while the voltage terminals are marked ±V and V. The instrument is connected into an ac line exactly as shown in Fig. 14-26.

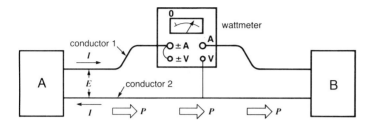

Figure 14-26
Conventional method of connecting a wattmeter. An upscale reading means that active power is flowing from left to right.

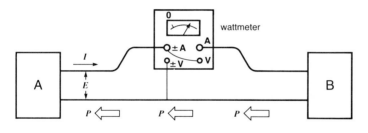

Figure 14-27
Reverse method of connecting a wattmeter. An upscale reading means that active power is flowing from right to left.

Suppose the line connects two boxes, A and B, one of which contains a source, the other a load. We have no way of telling on which side the source and the load happen to be. The wattmeter, however, will indicate in which direction the active power is flowing. Thus, if the pointer moves upscale, giving a positive reading, power is flowing over the line from left to right. Consequently, the source is in box A.

Conversely, if the wattmeter shows a negative reading, we know that power is flowing from right to left. The source is therefore in box B.

Most wattmeters have the zero on the left-hand end of the scale, and so they cannot register negative readings directly. In order to measure negative power, we must reverse the wires connected to the ±V and V terminals (Fig. 14-27). When connected this way, we know that a positive, upscale reading means that power is flowing from right to left.

A wattmeter is a useful instrument because it tells us both the magnitude of the active power and the direction in which it is flowing. The phase angle between current *I* and voltage *E* can have any value. In effect, the wattmeter is designed to read only the active power flowing over the line.

It is important to make a distinction between power flow and current flow. Referring back to Fig. 14-26, it is clear that if current *I* in conductor 1 flows momentarily in the direction shown, an equal current *I* will flow in the opposite direction in conductor 2. Power flow is quite different: it does not flow down one conductor and return by the other. Power flows in the same direction along *both* conductors. Like the banks of a stream, the two conductors act as a channel along which power can flow.

Figure 14-28 shows a portable wattmeter.

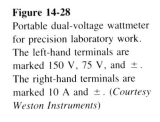

Figure 14-28
Portable dual-voltage wattmeter for precision laboratory work. The left-hand terminals are marked 150 V, 75 V, and ±. The right-hand terminals are marked 10 A and ±. (*Courtesy Weston Instruments*)

A *varmeter* is similar to a wattmeter because it also has two voltage and two current terminals. In portable instruments, two of the terminals always bear a ± sign. It is connected as shown in Fig. 14-28. Most varmeters have the zero at

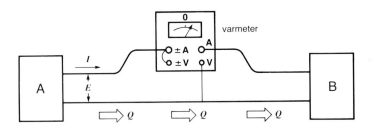

Figure 14-29
Conventional method of connecting a varmeter. A positive reading means that reactive power is flowing from left to right.

the center of the scale, so we can read positive and negative vars without having to change connections.

With the varmeter connected as shown, a positive reading means that reactive power is flowing from left to right over the two wires. The source of reactive power is therefore in box A. Conversely, if the pointer gives a negative reading, reactive power is flowing from right to left. The reactive *load* is then in box A.

A varmeter is specially designed to indicate only the reactive power flowing over the line. The phase angle between E and I may have any value, and the varmeter will always show the correct value of the reactive power. Figure 14-30 shows a varmeter used in educational equipment.

Figure 14-30
Varmeter with center-scale zero, and fully protected against accidental short-circuits and overvoltages. The term OUT means that reactive power is flowing left to right, from terminals 1, 2, 3 to terminals 4, 5, 6. (*Courtesy Lab-Volt Ltd.*)

Suppose that a wattmeter and varmeter are connected into an ac line *exactly* as shown in Fig. 14-31. In this case, we happen to know that a reactor and resistor are on the right-hand side of the instruments and that the 600 V electrical outlet is on the left. The reactor draws 6 A, and the resistor draws 10 A. As we would expect from the previous discussion, the wattmeter gives a positive reading of $P = EI = 600 \times 10 = 6000$ W.

The varmeter also gives a positive reading of $Q = EI = 600 \times 6 = 3600$ var. The reading is positive because a reactor absorbs reactive power. The direction of power flow for P and Q is shown by the heavy arrows.

If we replace the reactor by a capacitor that draws 7 A, the wattmeter will still register $+6000$ W, but the varmeter will show a *negative* reactive power of $Q = EI = 600 \times 7 = 4200$ var (Fig. 14-32). The reading is negative because a capacitor generates reactive power, as mentioned in Section 13-8. The direction of active and reactive power flow is shown by the heavy arrows. It is surprising to see two powers flowing in opposite directions over the same line. But we must

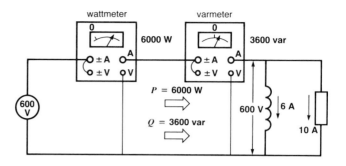

Figure 14-31
Wattmeter and varmeter readings obtained with an inductive and resistive load.

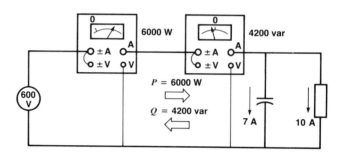

Figure 14-32
Wattmeter and varmeter readings obtained with a capacitive and resistive load.

remember that active power is not the same as reactive power, and they flow independently of each other.

14.14 Relationship between *P*, *Q*, and *S*

Figure 14-33 again shows an ac circuit enclosed in a box. The circuit is connected to an ac source *E* and draws a line current *I*. The circuit is composed of various devices, such as resistors, inductors, capacitors, motors, transformers, generators, and so forth. Each device either absorbs or delivers a certain amount of active power. When all these individual active powers are added up, let us assume that the circuit as a whole absorbs *P* watts.

Similarly, each device either absorbs or generates a certain amount of reactive power. When all these individual reactive powers are added up, let us assume that the circuit as a whole *absorbs Q* vars.

If a wattmeter and varmeter are connected into the line, they will respectively indicate *P* watts and *Q* vars. Furthermore, both instruments will give a positive reading.

We can make a model of this complex circuit by assuming it is composed of a

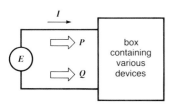

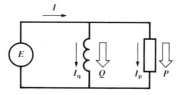

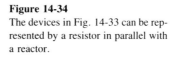

Figure 14-33
The devices in the box absorb a total
of P watts, and a total of Q vars.

Figure 14-34
The devices in Fig. 14-33 can be represented by a resistor in parallel with
a reactor.

resistor in parallel with a reactor. The resistor consumes all the active power P.
Similarly, the reactor consumes all the reactive power Q. The result is the simple
parallel circuit of Fig. 14-34. In such an inductive circuit, the current I lags by some
angle θ behind the voltage E.

Referring to Fig. 14-34, we can write

$P = EI_p$ therefore $I_p = P/E$, and I_p is in phase with E

$Q = EI_q$ therefore $I_q = Q/E$, and I_q lags 90° behind E

$S = EI$ therefore $I = S/E$, and I lags θ degrees behind E

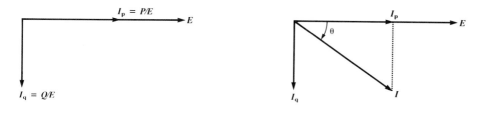

Figure 14-35
The magnitudes of I_p and I_q are
determined.

Figure 14-36
The magnitude of the line current I is
determined.

The phasors for I_p and I_q are drawn in Fig. 14-35. Current I is the phasor sum of
I_p and I_q, and so the position of I is as shown as in Fig. 14-36. According to the
relationship between the sides of a right-angle triangle, we have

$$I^2 = I_p^2 + I_q^2$$ Eq. 1-13

therefore

$$(S/E)^2 = (P/E)^2 + (Q/E)^2$$

from which

$$S^2 = P^2 + Q^2$$ (14-19)

Equation 14-19 can be put into the following three forms:

$$S = \sqrt{P^2 + Q^2} \qquad P = \sqrt{S^2 - Q^2} \qquad Q = \sqrt{S^2 - P^2} \qquad (14\text{-}20)$$

The relationship between P, Q, and S is sometimes easier to remember by referring to the so-called power triangle of Fig. 14-37.

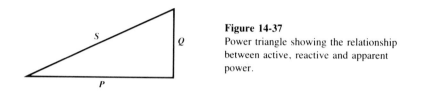

Figure 14-37
Power triangle showing the relationship between active, reactive and apparent power.

From the way Eq. 14-19 was derived, it is obvious that we don't have to know how the circuit elements are connected inside the box. As long as we know the active and reactive power that is absorbed (or generated) we can apply Eqs. 14-19 and 14-20.

14.15 Power factor

In power circuits, we are mainly interested in the active (or real) power P absorbed by the load. This is the power that does work or produces heat. However, the apparent power S drawn by the load is often greater than P. By definition, the ratio of active power to apparent power is called the *power factor* of the load.

$$\text{power factor} = P/S \qquad (14\text{-}21)$$

The power factor can never be greater than 1. It may be expressed as a decimal, such as 0.83, or as a percentage, such as 83%.

Referring to Fig. 14-36, we recall that $P = EI_p$ and $S = EI$. Consequently,

$$\text{power factor} = P/S = \frac{EI_p}{EI} = \frac{I_p}{I} = \cos\theta \qquad (14\text{-}22)$$

and

$$P = S\cos\theta = EI\cos\theta \qquad (14\text{-}23)$$

The power factor of a circuit is therefore equal to the cosine of the angle between the line voltage E and the line current I. This important result holds true for any single-phase ac circuit.

The power factor is said to be *lagging* when the current lags behind the voltage. The power factor is said to be *leading* when the current leads the voltage.

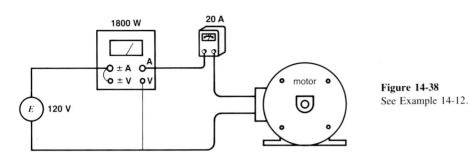

Figure 14-38
See Example 14-12.

Example 14-12:

A motor connected to a 120 V line draws a current of 20 A. A wattmeter indicates 1800 W (Fig. 14-38). Calculate

1. the apparent power S consumed by the motor

2. the power factor of the motor

3. the phase angle between the voltage and current

Solution:

1. $S = EI = 120 \times 20 = 2400$ VA

2. power factor $= P/S = 1800/2400 = 0.75$, or 75%   Eq. 14-21

3. $\cos \theta = 0.75$; therefore, $\theta = $ arcos $0.75 = 41.4°$. The current lags 41.4° behind the voltage because the motor absorbs reactive power.

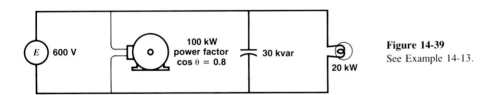

Figure 14-39
See Example 14-13.

Example 14-13:

An industrial load (Fig. 14-39) consists of several motors, a capacitor, and some incandescent lighting. The motors draw a total of 100 kW at a power factor of 80%. The capacitor has a rating of 30 kvar, and the lighting load is 20 kW. The incoming feeder voltage is 600 V. Calculate:

1. the active, reactive, and apparent power absorbed by the plant

2. the overall power factor of the plant

3. the current in the feeder line

Solution:

1. In all such problems we have to find (1) the total active power, and (2) the total reactive power absorbed by the load. The units kVA, kW, and kvar tell us whether the stated power is apparent, active or reactive, and so we must watch them carefully.

    a) Apparent power drawn by the motors is

    $$S = P/\cos \theta = 100/0.8 = 125 \text{ kVA}$$

    b) Reactive power absorbed by the motors is   Eq. 14-21

    $$Q = \sqrt{S^2 - P^2} = \sqrt{125^2 - 100^2} = 75 \text{ kvar}$$

    c) Total active power absorbed by the plant

    $$P = P_{\text{motor}} + P_{\text{capacitor}} + P_{\text{lighting}}$$

    $$P = 100 + 0 + 20 = 120 \text{ kW}$$

    d) Total reactive power absorbed by the plant

    $$Q = Q_{\text{motor}} + Q_{\text{capacitor}} + Q_{\text{lighting}}$$

    $$Q = 75 - 30 + 0 = 45 \text{ kvar}$$

    e) The apparent power drawn by the plant is

    $$S = \sqrt{P^2 + Q^2} = \sqrt{120^2 + 45^2} = \sqrt{16425} = 128.2 \text{ kVA}$$

2. The overall power factor is
    power factor $= P/S = 120/128.2 = 0.936$ or 93.6 %

3. The current in the feeder line is
    $I = S/E = 128200 / 600 = 213.7 \text{ A}$

14.16 Summary

This chapter concludes our study of basic ac circuit theory. We learned how to find the phasor sum of voltages and currents by graphical methods, and by trigonometry. This, in turn, enabled us to solve series and parallel ac circuits. The general term impedance was also introduced to signify any opposition to ac current flow.

The phenomenon of series and parallel resonance showed that by combining a capacitor with an inductor, it is possible to obtain some unexpected results. Thus, in a parallel resonant circuit the impedance is much higher than when the inductor or capacitor is acting alone. Conversely, in a series resonant circuit, the impedance is much lower. Resonance will again appear in the study of long transmission lines.

We were also able to tie together the relationship between active, reactive, and apparent power. The equation $S^2 = P^2 + Q^2$ will frequently be used in the study of ac machines and power distribution. Of equal importance is the fact that all the reactive powers in a circuit can be added to obtain the total reactive power without concern for the actual circuit connections. The same is true when adding up the active powers.

We will be applying this basic circuit theory to many different circuits and machines. The principles remain the same in each application, and, as a result, our knowledge of circuits will be broadened and strengthened.

TEST YOUR KNOWLEDGE

14-1   A resistor and capacitor are connected in series to an ac source. The currents in the two elements are

      a. different   b. in phase   c. 90° out of phase

14-2   An inductor and a capacitor are connected in parallel to an ac source. The respective currents are 30 A and 40 A. The total current drawn from the source is

      a. 70 A   b. 50 A   c. 10 A

14-3   The X component of a voltage E is 30 V and the Y component is 60 V. The magnitude and phase angle of E is

      a. 90 V, 63.4°   b.67.08 V, 63.4°   c. 67.08 V, 26.5°

14-4   The unit of apparent power is

      a. the VA   b. the va   c. the watt

14-5   A load draws 60 kW and 120 kvar from a line. The power factor of the load is

      a. 50 %   b. 44.7 %   c. 33.3 %

14-6   A coil has a resistance of 20 Ω and an inductance of 15 H. If it is connected to a 12 V battery, the final current is

      a. 0.6 A   b. 0.48 A   c. zero

14-7   An ac line carries simultaneously 120 kW and 70 kvar. If a varmeter is connected into the line, it will read

      a. 50 kvar   b. 70 kvar   c. 138.9 kvar

14-8   A motor draws 7 kW from a 240 V, 60 Hz line. If the line current is 36 A, the reactive power absorbed by the motor is

      a. 1640 var   b. 5.06 kvar   c. 11.1 kvar

14-9   A capacitor connected to an ac line has a power factor that is

      a. leading and zero   b. lagging and zero   c. 100 percent

14-10 A baseboard heater in a home has a power factor of

a. 1   b. zero   c. 0.707

QUESTIONS AND PROBLEMS

14-11 The current in a single-phase motor lags 50° behind the voltage. What is the power factor of the motor?

14-12 A large motor absorbs 600 kW at a power factor of 88 percent. Calculate (a) the apparent power and (b) the reactive power absorbed by the machine.

14-13 A 200 μF capacitor is connected to a 240 V, 60 Hz source. Calculate the reactive power it generates.

14-14 In Fig. 14-11, what value of capacitive reactance must be connected in series with the reactance of 40 Ω for the circuit to become resonant?

14-15 In Fig. 14-40, calculate:

a. the current in each element

b. the current delivered by the source

c. the power factor of the circuit

d. the impedance of the circuit

e. draw the complete phasor diagram

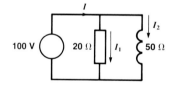

Figure 14-40
See Problem 14-15.

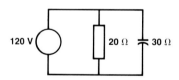

Figure 14-41
See Problem 14-16.

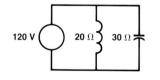

Figure 14-42
See Problem 14-17.

14-16   Repeat Problem 14-15 for the circuit of Fig. 14-41

14-17   Repeat Problem 14-15 for the circuit of Fig. 14-42

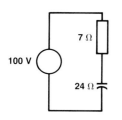

Figure 14-43
See Problem 14-18.

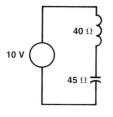

Figure 14-44
See Problem 14-19.

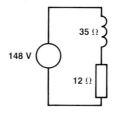

Figure 14-45
See Problem 14-20.

14-18   In Fig. 14-43, calculate:

    a.   the impedance of the circuit

    b.   the voltage across each element

    c.   the active power supplied by the source

    d.   draw the complete phasor diagram

14-19   Repeat Problem 14-18 for the circuit of Fig. 14-44

14-20   Repeat Problem 14-18 for the circuit of Fig. 14-45

14-21   In Fig. 14-46, calculate in the following order:

    a.   the current in the reactor

    b.   the current in the 30 Ω resistor

    c.   the current in the 40 Ω resistor

    d.   the total active power absorbed by the circuit

    e.   the total reactive power absorbed by the circuit

    f.   the apparent power absorbed by the circuit

    g.   the current supplied to the circuit

    h.   the voltage of the source

    i.   the power factor of the circuit

    j.   the phase angle between E_1 and I

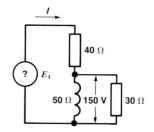

Figure 14-46
See Problem 14-21.

14-22   In Fig. 14-47, calculate in the following order:

a.   the impedance of the circuit

b.   the voltage across the inductor

c.   the voltage across the capacitor

d.   the supply voltage E_2

e.   the current in the resistor

f.   the active power consumed in the circuit

g.   the reactive power absorbed by the inductor

h.   the reactive power generated by the capacitor

i.   the reactive power delivered to the source E_2

j.   the apparent power of the source

k.   the power factor of the source

l.   the phase angle between E_2 and I

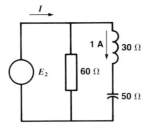

Figure 14-47
See Problem 14-22.

14-23   In Fig. 14-47, calculate the current flowing in the capacitor if the frequency of the source is doubled, while keeping E_2 constant.

CHAPTER **15**

Three-Phase Circuits

Introduction and Chapter Objectives

Electric power is generated, transmitted, and distributed in the form of three-phase power. Homes and small establishments are wired for single-phase power, but this represents only a portion of the basic three-phase system. Three-phase power is preferred to single-phase power for several important reasons:

a) 3-phase motors, generators, and transformers are simpler, cheaper and more efficient;

b) three-phase transmission lines can deliver more power for a given weight and cost;

c) the voltage regulation of three-phase transmission lines is inherently better.

A knowledge of three-phase power and three-phase circuits is essential to an understanding of power technology. Fortunately, the techniques we used to solve single-phase circuits in the previous chapter, can be directly applied to three-phase circuits.

In this chapter, we will investigate the properties of wye and delta connections with regard to voltage, current, and power. We conclude with a discussion on the measurement of three-phase power, and the meaning of phase sequence.

15.1 Polyphase systems

We can gain a preliminary understanding of polyphase systems by referring to the common gasoline engine. A single-cylinder engine having one piston is comparable to a single-phase machine. A two-cylinder engine is comparable to a two-phase machine, and the more common six-cylinder engine could be called a six-phase machine. In a six-cylinder engine, identical pistons move up and down inside identical cylinders, but they do not move in unison. They are staggered in such a way as to deliver power to the shaft in successive pulses rather than at the same time. This produces a smoother-running engine and a smoother output torque.

Similarly, in a three-phase electrical system, the three phases are identical, and deliver power at different times. As a result, the total instantaneous power flow from the three phases is constant. Furthermore, because the phases are identical, one phase* may be used to represent the behavior of all three.

Although we must be careful not to carry analogies too far, the above description reveals that a three-phase system is basically composed of three single-phase systems which operate in sequence. Once this is understood, any mystery surrounding three-phase systems disappears.

15.2 Generating a three-phase voltage

We have already seen that a single coil revolving between the N and S poles of a permanent magnet produces an ac voltage (see Section 8-6). If we place three identical coils on the armature, symmetrically spaced at 120° to each other (Fig. 15-1), the voltage generated by each coil will have the same effective value. However, the peak positive values will not occur at the same time, but will be displaced at 120° to each other. If the terminals of the coils are labeled a-n, b-n, and c-n, the corresponding voltages can be represented by the sine waves of Fig. 15-2. The peak value of each voltage is E_m, but E_{bn} lags 120° behind E_{an}, while E_{cn} lags 240° behind E_{an}. The three voltages can be represented more simply by the phasors of Fig. 15-3.

Each winding generates a single-phase voltage between its terminals. Thus, voltages E_{an}, E_{bn}, and E_{cn} are single-phase voltages, but together they make up a three-phase system. The coil ends n are usually connected to create a neutral

---

*   The term "phase" is used to designate different things, consequently, it has to be read in context to be understood. The following examples show some of the ways in which "phase" is used.

1.   the current is out of phase with the voltage (refers to phasor diagram);

2.   the three phases of a transmission line (meaning the three conductors of the line);

3.   the phase-to-phase voltage (meaning line voltage);

4.   the phase sequence (the order in which the phasors follow each other);

5.   the burned-out phase (the burned-out winding of a 3-phase machine);

6.   the three-phase voltage (the line voltage of a 3-phase system);

7.   the 3-phase currents are unbalanced (the currents in a 3-phase line or machine are unequal and not displaced at 120°);

8.   phase-shift transformer (a device which can change the phase angle of the output voltage with respect to the input voltage);

9.   phase-to-phase fault (a short circuit between two line conductors);

10.   phase-to-ground short-circuit (short circuit between one line conductor and ground);

11.   the phases are unbalanced (the line voltages, or the line currents are unequal or not displaced at 120° to each other).

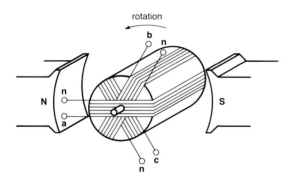

Figure 15-1
A simple three-phase generator.

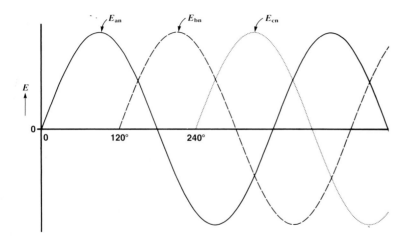

Figure 15-2
The generated voltages are displaced 120° from each other.

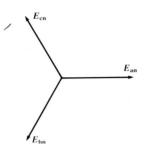

Figure 15-3
Phasor diagram representing the generated voltages.

terminal **n**. Consequently, voltages will appear between terminals a, b, and c as well as between terminals an, bn, and cn (Fig. 15-4).

The line voltages E_{ab}, E_{bc}, and E_{ca} between terminals a, b, and c are also equal and displaced at 120° to each other. Their magnitude and phase angle can be determined from phasors E_{an}, E_{bn}, and E_{cn} and the circuit diagram of Fig. 15-4.

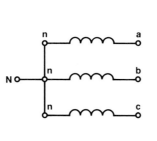

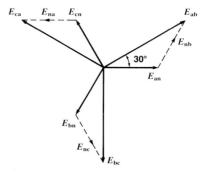

Figure 15-4
The generator windings are connected to form a common neutral.

Figure 15-5
The voltages between the terminals are found by phasor addition.

For example, by looking at Fig. 15-4, we see that $E_{ab} = E_{an} + E_{nb}$. The phasor sum E_{ab} is constructed as shown in Fig. 15-5. (Phasor E_{nb} is equal and opposite to phasor E_{bn}.) It shows that E_{ab} is larger than E_{an} and that it is 30° ahead of E_{an}. Phasors E_{bc} and E_{ca} are similarly found from the phasor sums $E_{bc} = E_{bn} + E_{nc}$, and $E_{ca} = E_{cn} + E_{na}$.

There is a definite relationship between the magnitude of the line voltages (such as E_{ab}) and the line-to-neutral voltages (such as E_{an}). Referring to Fig. 15-5 and from the discussion in Section 1-20, we have $E_{ab} = \sqrt{3}\ E_{an}$. We can therefore write:

$$E = \sqrt{3}\ E_{LN} \qquad (15\text{-}1)$$

where

E = line voltage, in volts (V)

E_{LN} = line-to-neutral voltage, in volts (V)

$\sqrt{3}$ = 1.732...

In order to appreciate the meaning of these three-phase relationships, Fig. 15-6 shows the voltages indicated by voltmeters connected across the terminals of a 3-phase, 440 V generator. Note that the line-to-neutral voltages are indeed $440/\sqrt{3}$ = 254 V. Furthermore, the three line voltages have the same values, as do the three line-to-neutral voltages.

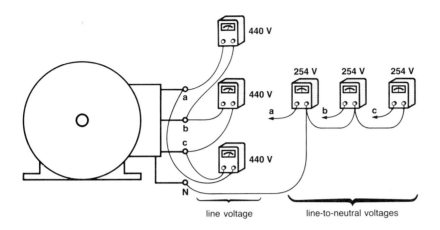

440 V

254 V   254 V   254 V

440 V

a

b

c

440 V

N

line voltage   line-to-neutral voltages

Figure 15-6
Voltages generated by a 3-phase, 440 V ac generator.

15.3 Three-phase currents

When a three-phase source is connected to a three-phase load, only the line terminals a, b, and c need to be connected. The neutral terminal is left open (Fig. 15-7). The currents I_a, I_b, and I_c in the three lines will have the same values, but they, too, are out of phase with each other by 120° (Fig. 15-8). If the effective line current is 106 A and ammeters are connected in each line, they will all indicate a current of 106 A. The peak current in each line is therefore 106/0.707 = 150 A.

In order to understand the nature of this three-phase current flow, Fig. 15-9 shows the instantaneous currents i_a, i_b, and i_c at successive intervals of 30°. On a 60

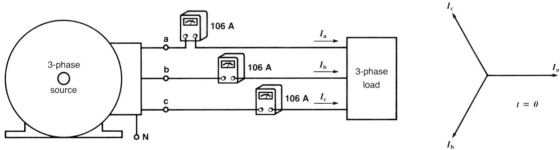

106 A

a

I_a

3-phase
source

b

106 A

I_b

3-phase
load

c

106 A I_c

N

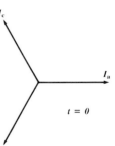

I_c

I_a

$t = 0$

I_b

Figure 15-8
Phasor diagram of the three-phase currents in Fig. 15-7. Each phasor corresponds to a peak current of 150 A.

Figure 15-7
A balanced three-phase load causes three-phase currents to flow in the generator, line, and load.

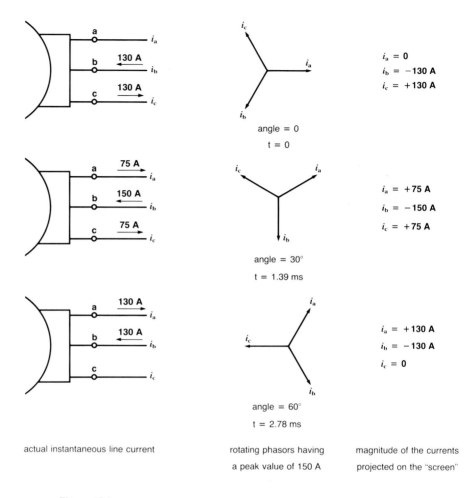

actual instantaneous line current   rotating phasors having   magnitude of the currents
                                         a peak value of 150 A   projected on the "screen"

Figure 15-9
The magnitude and direction of the currents in the three lines change continually. The magnitudes of the instantaneous currents can be visualized by assuming that the rotating phasors are projected on a screen at the right. The magnitudes are given at three instants, separated by intervals of 30°.

Hz line, 30° corresponds to an interval of 1.39 ms. The instantaneous line currents at three successive moments are shown. Note that the revolving phasors enable us to visualize the relative magnitudes and signs of the currents. This is a direct application of the concept discussed in Section 1-22.

However, we can also calculate the instantaneous currents from the equations:

$$i_a = 150 \sin \theta$$

$$i_b = 150 \sin (\theta - 120)$$

$$i_c = 150 \sin (\theta - 240)$$

These equations are obtained from the sine waves represented by the three phasors of Fig. 15-8, and using the technique explained in Section 7-14.

When the line currents are equal and displaced at 120°, and the line voltages are equal and displaced at 120°, the three-phase system is said to be *balanced*. We always try to maintain a balanced condition.

15.4 Wye and delta connection

A balanced three-phase load is actually composed of three identical single-phase loads. The loads are connected symmetrically, and there are two basic ways of accomplishing this. The single-phase loads are connected either in *wye* or in *delta* (Figs. 15-10 and 15-11). These schematic drawings try to capture the phase relationship between the line and line-to-neutral voltages by drawing the loads at 120° angles. However, the circuit diagrams can also be drawn as shown in Figs. 15-12 and 15-13. This, of course, does not alter the voltages or currents.

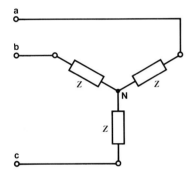

Figure 15-10
Three single-phase loads connected in wye.

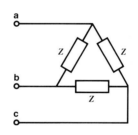

Figure 15-11
Three single-phase loads connected in delta.

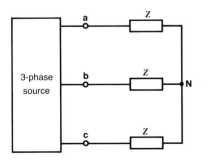

Figure 15-12
Alternative method of showing a
wye connection.

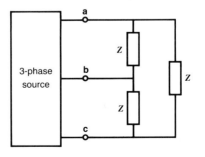

Figure 15-13
Alternative method of showing a
delta connection.

15.5 Currents and voltages in a wye connection

Figure 15-14 shows the line voltages and line-to-neutral voltages when a wye-connected load is powered by a 208 V, 3-phase source. The voltages across the loads are equal in magnitude, and so we show only one voltmeter reading. The line-to-neutral voltage (120 V) is less than the line voltage (208 V). The ratio between the two is again $\sqrt{3}$.

The current I_L in each load is obviously equal to the line current I, which is assumed to have a value of 45 A.

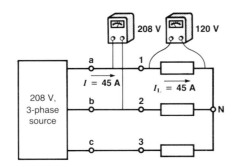

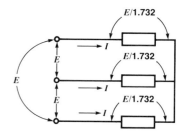

Figure 15-14
Currents and voltages in a 208 V
3-phase line and a wye-connected
load.

Figure 15-15
General current and voltage relationships in a balanced wye-connected
load.

These results are generalized in Fig. 15-15. Two simple rules apply to a wye connection: if the line voltage is E and the line current is I,

1. the current in each load is equal to I

2. the voltage across each load is equal to $E/1.732$

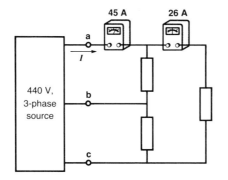

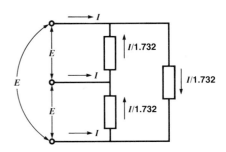

Figure 15-16
Currents and voltages in a 440 V,
3-phase line and delta-connected
load.

Figure 15-17
General current and voltage relationships in a balanced delta-connected
load.

15.6 Currents and voltages in a delta connection

Figure 15-16 shows the line currents and load currents when a delta-connected load is fed by a 440 V, 3-phase source. The currents in the loads are all equal and so we show only one ammeter reading. The load current (26 A) is less than the line current (45 A). The ratio between the two is $\sqrt{3}$ or 1.732.

The voltage across each load is obviously equal to the line voltage.

These results are generalized in Fig. 15-17. Two simple rules apply to a delta connection: if the line voltage is E and the line current is I,

1. the voltage across each load is equal to E

2. the current in each load is equal to $I/1.732$.

15.7 Power and power factor in a three-phase line

In discussing wye and delta connections, we did not specify the nature of the single-phase loads. We only said they were identical. Consequently, they may be resistors, inductors, capacitors, motors, or any other devices. However, the active power P_L absorbed by each of the three single-phase loads is the same. The apparent powers S_L are also the same. The power factor ($\cos \theta$) of each single-phase load is therefore $\cos \theta = P_L/S_L$.

The 3-phase line is delivering $P = 3P_L$ watts to the three loads. It is also supplying $S = 3S_L$ voltamperes. It follows that the power factor of the three-phase line is $3P_L/3S_L = P_L/S_L$, which is the same as the power factor of the individual single-phase loads. Thus, the power factor of a balanced three-phase line is given by

$$\cos \theta = \frac{\text{active power of the line}}{\text{apparent power of the line}} = \frac{P}{S} \qquad (15\text{-}2)$$

What is the apparent power S supplied by a three-phase line? To answer the question, let us return to Fig. 15-17. It shows a three-phase load connected to a line in which the line current is I and the line voltage is E. The current in each single-phase branch is $I/\sqrt{3}$ and the voltage across it is E. The apparent power of each branch is therefore $EI/\sqrt{3}$ voltamperes. For the three loads, the apparent power is three times as great, so we have:

$$S = 3\,EI/\sqrt{3} = EI\sqrt{3}$$

and so

$$S = 1.732\,EI \qquad (15\text{-}3)$$

where

$S =$ apparent power of a three-phase line or load, in voltamperes (VA)

$E =$ line voltage, in volts (V)

$I =$ line current, in amperes (A)

Equations 15-2 and 15-3 can be applied to any three-phase balanced load or line. The load may be connected in any way as long as it is balanced.

15.8 Solving three-phase circuits

The secret to solving a three-phase circuit is to consider only one phase. Because we know how to solve single-phase circuits, it is advantageous to reduce a three-phase circuit to a single-phase equivalent. Another important technique is to draw a sketch of the three-phase circuit, listing the important values. The following examples illustrate the procedure.

Example 15-1:

Three resistors are connected in wye across a 440 V, 3-phase line. If they dissipate a total of 6000 W, calculate:

1. the current in each line

2. the value of each resistor

Solution:

1. First, we make a sketch of the circuit (Fig. 15-18)

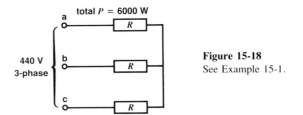

Figure 15-18
See Example 15-1.

2. Next, we make a list of everything we know about the single-phase load:

    a) the power dissipated by one resistor is $P = 6000/3 = 2000$ W

    b) the voltage across the resistor is $E = 440/1.732 = 254$ V

    c) the current in the resistor is $I = P/E = 2000/254 = 7.87$ A

    d) the line current $= 7.87$ A

    e) the resistance of the resistor is $R = E/I = 254/7.87 = 32.3$ Ω

Example 15-2:

A 4 kV, 3-phase line carries a line current of 90 A. The power factor is 0.85 lagging, and the load is connected in delta. Calculate:

1. the impedance of the load, per phase

2. the reactive power, per phase

Solution:

1. We first make a sketch of the delta-connected load (Fig. 15-19).

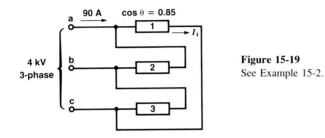

Figure 15-19
See Example 15-2.

2. Next, we make a list of everything we know about the single-phase loads, taking load 1 as a sample.

a) the current in load 1 is I_1 = 90/1.732 = 52 A

b) the voltage across load 1 is E = 4000 V

c) Impedance of load 1 is Z = E/I = 4000/52 = 76.9 Ω

d) Apparent power of load 1 is

$$S = EI = 4000 \times 52 = 208\ 000 = 208\ \text{kVA}$$

e) Active power of load 1 is

$$P = S \cos \theta = 208 \times 0.85 = 176.8\ \text{kW}$$

f) Reactive power is

$$Q = \sqrt{S^2 - P^2} = \sqrt{208^2 - 176.8^2} = 109.6\ \text{kvar} \qquad \text{Eq. 14-20}$$

15.9 Measuring three-phase power

The active power P flowing in a 3-phase, 3-wire line can be measured by using two single-phase wattmeters. The instruments are connected as shown in Fig. 15-20. The total active power consumed by the three phases is equal to the sum of the two wattmeter readings. When the power factor is less than 50 percent, one of the instruments will give a negative reading. To measure it, we must reverse the leads connected to the voltage terminals of the instrument. The total power is then equal to the difference of the two readings.

Note that this so-called 2-wattmeter method of measuring power yields the correct value even if the three-phase system is unbalanced.

In some three-phase installations, the neutral wire is also used. For example, in a 208 V/120 V system, motor loads are connected between lines, while lighting loads and service outlets are connected between the lines and neutral. This is called a 3-phase, 4-wire system. We need three wattmeters to measure the active power in

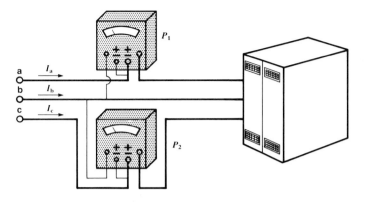

Figure 15-20
Measuring the active power in a 3-phase, 3-wire line.

such a system. They are connected as shown in Fig. 15-21. The total power is equal to the sum of the three wattmeter readings. This method also gives correct results even if the loads are unbalanced.

Reactive power can be measured by means of varmeters. The varmeter connections for 3-wire and 4-wire systems are the same as for wattmeters. However, varmeters give meaningful readings only when the loads are balanced.

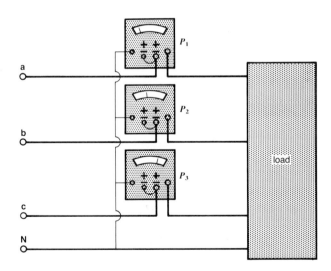

Figure 15-21
Measuring the active power in a 3-phase, 4-wire line.

Example 15-3:

A test on a three-phase motor running at light load on a 440 V line gives wattmeter readings $P_1 = +2300$ W and $P_2 = -500$ W. We assume the motor is connected in wye. If the line current is 10 A, calculate the power factor of the motor and its line-to-neutral impedance.

Solution:

1. We first draw a diagram of the problem, showing all the given information (Fig. 15-22).

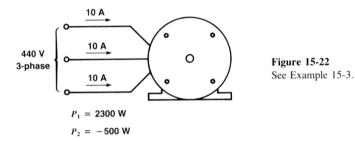

Figure 15-22
See Example 15-3.

2. Next, we make a list of everything we know about one phase of the motor:

a) voltage from line-to-neutral is $E = 440/1.732 = 254$ V (the neutral connection is made inside the motor).

b) current in one phase is $I = 10$ A because of the wye connection

c) impedance per phase $Z = E/I = 254/10 = 25.4 \, \Omega$

d) apparent power per phase $S = EI = 254 \times 10 = 2540$ VA

e) total 3-phase active power is
   $P = P_1 + P_2 = 2300 - 500 = 1800$ W

f) active power in one phase is $P = 1800/3 = 600$ W

g) $\cos \theta$ of one phase $= P/S = 600/2540 = 0.236$ or 23.6 percent

h) power factor of the three-phase motor is 23.6 percent

In many three-phase loads, the three individual phases are placed in a single enclosure, and so it is impossible to tell whether they are connected in wye or in delta. In such cases we *assume* a wye connection because it is a slightly simpler arrangement than the delta connection.

15.10 Phase sequence

An important property of a three-phase system is its *phase sequence*. It is the order in which the line voltages become positive. Suppose that the voltages between the lines a, b, and c of Fig. 15-23 can be represented by the phasor diagram shown. Based upon our knowledge of revolving phasors (Section 7-12), the peak positive values of voltage occur in the sequence E_{ab}, E_{ca}, E_{bc}. The phase sequence is then said to be a - c - b - a - c - b . . ., or simply acb.

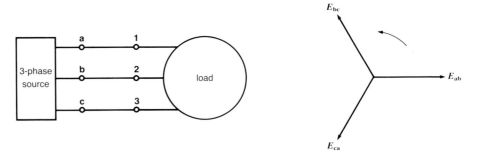

Figure 15-23
Load terminals 1, 2, 3 have the same phase sequence as line terminals a, b, c.

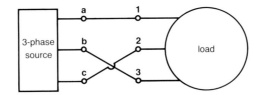

Figure 15-24
Load terminals 1, 2, 3 have the reverse phase sequence of line terminals a, b, c.

When a load having terminals 1, 2, and 3 is connected to lines a, b, and c as shown, the phase sequence for the load is obviously in the order 1 - 3 - 2. However, as far as the load is concerned, we can change the phase sequence by interchanging any two leads, as shown in Fig. 15-24. The line phase sequence a - c- b now produces a load phase sequence 1 - 2 - 3. Thus, we can change the phase sequence of a three-phase line by interchanging any two lines.

Phase sequence is important because it determines the direction of rotation of three-phase motors. It is also important whenever three-phase generators or transformers have to be connected in parallel.

15.11 Summary

This chapter provided another opportunity to use phasors and phasor diagrams. We learned that a balanced three-phase load is essentially composed of three identical single-phase loads. The loads may be connected in wye or in delta. Power in a 3-phase, 3-wire line can be measured with two wattmeters. Power in a 3-phase, 4-wire line can be measured with three wattmeters.

The basic principles covered here will be used in the study of electrical machinery and transmission lines.

TEST YOUR KNOWLEDGE

15-1   The rotor in Fig. 15-1 turns at 1500 r/min. The time for E_{an} to complete one cycle is

      a. 25 s   b. 40 ms   c. 251.2 ms

15-2   The frequency of the voltages in Fig. 15-2 is 400 Hz. The time interval between the positive peaks of E_{an} and E_{bn} is

      a. 7.5 ms   b. 2.5 ms   c. 833.3 μs

15-3   In Fig. 15-2, E_{bn} lags behind E_{an} but leads E_{cn} by 120°.

      ☐   true   ☐   false

15-4   In Fig. 15-4, if $E_{ab} = 380$ V, then E_{bn} is

      a. 219 V   b. 657 V   c. 190 V

15-5   In Fig. 15-6, the peak voltage between terminals b and n is

      a. 440 V   b. 359 V   c. 254 V

15-6   In Fig. 15-9, when t = 5.56 ms, the value of i_a is

      a. 150 A   b. -130 A   c. $+130$ A

15-7   A 69 kV, 3-phase transmission line carries 350 A per conductor. The apparent power carried by the line is

      a. 24.15 MVA   b. 41.8 MW   c. 41.8 MVA

15-8   The load in Fig. 15-16 has a power factor of 82 percent, lagging. It consumes an active power of

      a. 28.1 kW   b. 28.1 kVA   c. 16.2 kW

15-9   The phase sequence of the currents in Fig. 15-8 is I_a - I_c - I_b.

      ☐   true   ☐   false

QUESTIONS AND PROBLEMS

15-10   The line voltage in Fig. 15-10 is 620 V, and the load is resistive.

   a.   What is the voltage across each resistor?

   b.   If each resistor has a value of 15 Ω, what is the current in each line?

   c.   Calculate the power supplied by the source

15-11   Three resistors are connected in delta. If the line voltage is 13.2 kV and the line current is 1202 A, calculate:

   a.   the current in each resistor

   b.   the voltage across each resistor

   c.   the power supplied to each resistor

   d.   the power factor of the load

15-12   Three 120 V, 60 W lamps are connected in wye across a 160 V, 3-phase line. Will the lamps be dim or bright?

15-13   A resistance of 18 Ω is measured between any two terminals of the three-phase resistor bank shown in Fig. 15-12. If the load is connected to a 460 V, 3-phase line, calculate the total power that is dissipated.

15-14   The windings of a 3-phase motor are connected in delta. If the resistance between any two terminals is 0.6 Ω, what is the resistance of each winding? (Hint: use the principles of Section 3-5)

15-15   In Fig. 15-2, the peak voltage is 340 V. Express the three voltages in trigonometric form, using the method explained in Section 7-14.

15-16   A three-phase motor connected to a 440 V line draws a current of 136 A. Two wattmeters connected into the line as shown in Fig. 15-20 give readings of 59.5 kW and 23.8 kW. Calculate

   a.   the apparent power absorbed by the motor

   b.   the power factor of the motor

   c.   the reactive power absorbed by the motor

CHAPTER 16

Transformers

Introduction and Chapter Objectives

The transformer is one of the simplest electrical devices and yet one of the most useful. It can change the voltage or current in a circuit, can isolate one circuit from another, and can make it possible to transmit large amounts of power over distances as great as 1000 miles.

This chapter covers the theory, construction and application of transformers. In addition to standard power transformers, we cover instrument transformers, autotransformers, and transformer connections in three-phase systems.

16.1 The transformer and power transmission

You will recall that the power delivered by a transmission line is equal to the product of the line voltage times the line current: $P = EI$. This means that for a given amount of power, the line current becomes smaller as the line voltage is increased. A smaller current reduces the required conductor size, as well as the I^2R losses.

For example, suppose we need to transmit 1000 kW over a distance of 20 miles. If we decide to use a line voltage of 1000 V, the line current is 1000 A, and this would require a large conductor. On the other hand, if we use 10 000 V, the current will be only 100 A. We can therefore use a much smaller conductor, producing considerable savings. That is why voltages as high as 14 400 V, 345 000 V and even 765 000 V are used to carry large blocks of energy over great distances.

Figure 16-1 shows a simple transmission system composed of a generating station, a transmission line, and a power consumer. At the generating station, a transformer T_1 raises the voltage from 4160 V to 69 kV to reduce the transmission line current and corresponding line losses. At the other end of the line a second transformer T_2 reduces the voltage from 69 kV to a more usable, practical value, such as 480 V.

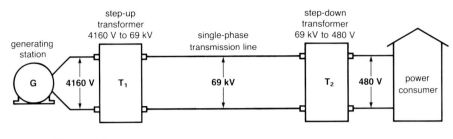

Figure 16-1
Transformers enable the economical transmission of power over a large distance.

16.2 Simple transformer

A transformer in its simplest form is composed of two coils that are mounted on a laminated iron core (Fig. 16-2). The coils are carefully insulated from each other and from the core.

One of the coils is connected to the ac source; it is called the *primary winding*, or simply primary. The other coil, connected to the load, is called the *secondary winding*, or simply secondary. Power therefore flows from the primary to the secondary winding.

Note that the power flow through a transformer is reversible: the primary can become a secondary, and vice versa. Thus, in Fig. 16-2, if terminals 1 and 2 are connected to a source, the secondary winding automatically becomes a primary winding. Furthermore, because terminals a and b are now connected to a load, the primary winding becomes the secondary winding.

Figure 16-3 shows how the HV coils of a large transformer are pre-formed on a slowly-rotating drum.

16.3 No-load operation

At no-load the secondary is on open-circuit and so the primary behaves like a simple inductor as far as the source is concerned (Fig. 16-2). Therefore, the current I_m drawn from the source lags 90° behind voltage E_p. This so-called *magnetizing* current produces an ac flux ϕ in the core. The flux, in turn, induces a small ac voltage v in each turn of the primary.

If the primary has N_p turns, the voltage induced across its terminals is equal to the sum of the voltages induced in all the turns, namely $v \times N_p$ volts. We therefore can write

$$E_p = v \times N_p \text{ volts} \tag{16-1}$$

The secondary links the same ac flux, therefore the voltage v induced in each turn must be the same as that induced in a primary turn. If the secondary has N_s turns, the voltage E_s induced across its terminals is given by

$$E_s = v \times N_s \tag{16-2}$$

From these results we obtain an important ratio, called the *transformer ratio*.

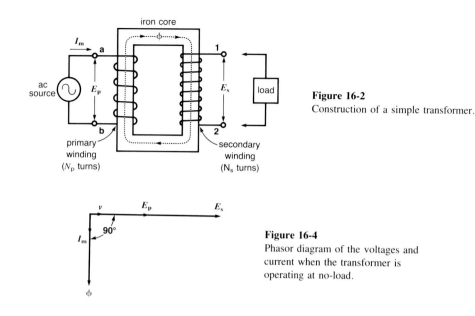

Figure 16-2
Construction of a simple transformer.

Figure 16-4
Phasor diagram of the voltages and
current when the transformer is
operating at no-load.

$$\frac{E_p}{E_s} = \frac{v \times N_p}{v \times N_s} = \frac{N_p}{N_s}$$

thus

$$\boxed{\frac{E_p}{E_s} = \frac{N_p}{N_s}}$$ (16-3)

where

E_p = voltage applied to the primary winding, in volts (V)

E_s = voltage induced in the secondary winding, in volts (V)

N_p = number of primary turns

N_s = number of secondary turns

   Equation 16-3 means that the ratio of the primary voltage to secondary voltage is
the same as the ratio of the number of turns. Furthermore, because the primary and
secondary voltages are produced by the same ac flux, they must be in phase. Thus,
when the primary voltage is zero, the secondary voltage is zero. Similarly, when E_p
is maximum positive, the secondary voltage E_s is maximum positive.
   The relationship between E_p, E_s, and I_m is shown in the phasor diagram of
Fig. 16-4. Phasor E_s is longer than phasor E_p because the secondary winding is
assumed to have more turns than the primary. Note that the flux ϕ in the core is nec-
essarily in phase with the magnetizing current I_m.

Figure 16-3
Winding the HV coil of a large transformer. (*Courtesy ASEA*)

Example 16-1:

A source of 4160 V is connected to a step-down transformer whose primary has 2600 turns, while the secondary has 300. Calculate:

1. the voltage E_s induced in the secondary

2. the volts per turn

3. the transformer ratio

Solution:

1. $$E_s = \frac{N_s}{N_p} \times E_p = \frac{300}{2600} \times 4160 = 480 \text{ V}$$ Eq. 16-3

2. The volts per turn is $v = E_p/N_p = 4160/2600 = 1.6$ V

3. The transformer ratio can be expressed either as 4160 V/480 V = 8.666 or 480 V/4160 V = 0.1154. We recommend using the ratio that gives a number greater than 1.

16.4 Transformer under load

If a *resistive* load is connected across the secondary winding of a transformer (Fig. 16-5), the secondary load current I_s will be in phase with E_s. The power delivered to the load is E_sI_s watts. Because this power must come from the ac source, the source must deliver a current I_p' so that $E_p \times I_p' = E_sI_s$ watts. This expresses the fact that the power received by the transformer must be equal to the power it delivers. Current I_p' must also be in phase with E_p because the load is resistive (Fig. 16-6).

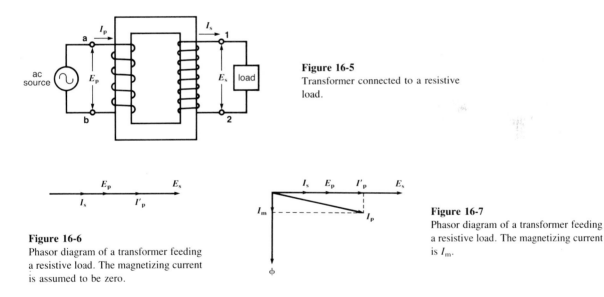

Figure 16-5
Transformer connected to a resistive load.

Figure 16-6
Phasor diagram of a transformer feeding a resistive load. The magnetizing current is assumed to be zero.

Figure 16-7
Phasor diagram of a transformer feeding a resistive load. The magnetizing current is I_m.

But we have just learned that the source must supply a current I_m to produce the flux in the core. This current lags 90° behind E_p. The total current I_p supplied by the source is therefore equal to the phasor sum of I_p' and I_m (Fig. 16-7).

Fortunately, the value of I_m is negligible compared with the rated value of I_p.

Consequently, at full-load I_p is very nearly equal to I_p', and so we can write the approximate equation:

$$E_p \times I_p = E_s \times I_s$$

or
$$\frac{I_p}{I_s} = \frac{E_s}{E_p} \qquad (16\text{-}4)$$

Equation 16-4 states that the ratio of currents is the inverse of the ratio of voltages. The current in a coil is large when its voltage is small and vice versa. In effect, what we gain in voltage, we lose in current.

Knowing that $E_p/E_s = N_p/N_s$, we obtain another useful equation

$$\frac{I_p}{I_s} = \frac{N_s}{N_p} \qquad (16\text{-}5)$$

or
$$I_p N_p = I_s N_s$$

This equation states that the ratio of primary to secondary current is inversely proportional to the ratio of the number of turns. It also states that the mmf of the primary is equal to the mmf of the secondary.

16.5 Losses, efficiency and rating of a transformer

Like any electrical machine, a transformer has losses. They are composed of

1. I^2R losses in the windings (commonly called the copper losses)

2. hysteresis and eddy current losses in the core (commonly called the iron losses)

The losses appear in the form of heat and produce an increase in temperature and a drop in efficiency. Under normal operating conditions, the efficiency of transformers is very high, usually above 98 percent.

The heat produced by the iron losses depends upon the flux in the core, which, in turn, depends upon the applied voltage. On the other hand, the heat dissipated in the windings depends upon the current they carry. Consequently, to keep the temperature of the transformer at an acceptable level, we must set limits to both the applied voltage and the current drawn by the load. These two limits determine the *rated* voltage and *rated* current of the transformer.

The power rating of a transformer is equal to the product of the rated voltage times rated current. The product is the same for either the primary or the secondary side. However, the power rating is not expressed in watts, as we would expect, because the phase angle between the voltage and current may have any value, depending upon the nature of the load. Consequently, the power-handling capacity of a transformer is expressed in terms of apparent power, with units in voltamperes

(VA). The rating of large transformers is given in kilovoltamperes (kVA), or in megavoltamperes (MVA), depending upon the size of the transformer.

The temperature rise of a transformer is directly related to the apparent power that flows through it. This means that a 500 kVA transformer will get just as hot feeding a 500 kvar inductive load as a 500 kW resistive load.

The rated kVA, frequency, and voltages are indicated on the nameplate. In large transformers, the corresponding rated currents are also shown. Figure 16-8 shows the construction of a large transformer.

Figure 16-8a

Three-phase transformer for an electric arc furnace rated 36 MVA, 13.8 kV/160 V to 330 V, 60 Hz. The primary and secondary coils are mounted on a three-legged core. The secondary voltage is adjustable from 160 V to 320 V by means of 32 taps on each primary winding. The taps from the three phases can be seen in the foreground. The secondary consists of three large busbars, each of which can deliver 65 000 A. The busbars can be seen protruding from behind the transformer. (*Courtesy Ferranti-Packard*)

Figure 16-8b
This photograph shows the same transformer ready for delivery. It is mounted on a depressed flatcar in order to obtain maximum overpass clearance. (*Courtesy Ferranti-Packard*)

Example 16-2:

The nameplate of a distribution transformer shows the rating to be 250 kVA, 60 Hz, primary 4160 V, secondary 480 V.

1. Calculate the rated value of the primary and secondary currents.

2. If we apply 2000 V to the 4160 V primary, can we still draw 250 kVA from the transformer?

Solution:

1. Rated current of the 4160 V winding is:

$$I_\mathrm{p} = \frac{\text{rated apparent power } S}{\text{rated primary voltage } E_\mathrm{p}} = \frac{250 \times 1000}{4160} = 60 \text{ A}$$

Rated current of the 480 V winding is:

$$I_\mathrm{p} = \frac{\text{rated apparent power } S}{\text{rated secondary voltage } E_\mathrm{s}} = \frac{250 \times 1000}{480} = 521 \text{ A}$$

2. By applying 2000 V to the primary (instead of 4160 V), the flux and the iron losses will be lower than normal, and the core will be cooler. However, the load current should not exceed its rated value, otherwise the windings will overheat. Consequently, the maximum power output using this lower voltage is:

$$S = 2000 \text{ V} \times 60 \text{ A} = 120 \text{ kVA}$$

16.6 Rated voltage of a transformer

Most transformers are designed so that the peak flux density in the iron core is slightly above the "knee" of the saturation curve when rated voltage is applied to the primary winding.

If we apply a voltage below the rated value, the core becomes less saturated, and so the iron losses and the magnetizing current become smaller.

On the other hand, if the applied voltage exceeds its rated value, the flux density increases, and the core becomes more saturated. This abnormal situation increases the iron losses slightly, but produces a very large increase in the magnetizing current. For example, if we apply twice the rated voltage to a winding, the magnetizing current can easily exceed the rated full-load current. As a general rule, we should not apply a voltage greater than 110 percent of the rated value.

16.7 Polarity of a transformer

Referring to Fig. 16-9, we recall that the primary and secondary voltages E_p and E_s attain their peak values at the same instant. Suppose that during one of these peak

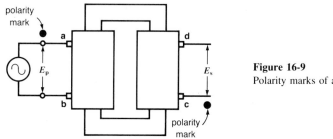

Figure 16-9
Polarity marks of a transformer.

moments primary terminal a is positive with respect to primary terminal b, and secondary terminal c is positive with respect to secondary terminal d. Terminals a and c are then said to possess the same polarity because they are "positive" at the same instant. This "sameness" can be shown by placing a dot beside primary terminal a and secondary terminal c. The dots are called *polarity marks*.

The polarity marks in Fig. 16-9 could equally well be placed next to terminals b and d because, as the voltage alternates, they, too, become simultaneously positive. Thus, the polarity marks could be placed beside terminals a and c *or* beside terminals b and d.

This dot-type marking is used on instrument transformers. On power transformers, however, the terminals are designated by the symbols H_1 and H_2 for the high-voltage (HV) winding, and by X_1 and X_2 for the low-voltage (LV) winding. By convention, H_1 and X_1 have the same polarity.

Although the polarity is known when the symbols H_1, H_2, X_1 and X_2 are given, it is common practice to mount the four terminals in a standard way so that the transformer has either *additive* or *subtractive* polarity. A transformer is said to have additive polarity when terminal H_1 is diagonally opposite to terminal X_1. A transformer has subtractive polarity when terminal H_1 is adjacent to terminal X_1 (Fig. 16-10).

Subtractive polarity is standard for all single-phase transformers above 200 kVA, provided the high-voltage winding is rated above 8660 V. All other transformers have additive polarity.

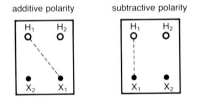

Figure 16-10
Definition of additive and subtractive polarity.

16.8 Polarity tests

Suppose the terminals of a transformer are not marked in any way. To determine whether it has additive or subtractive polarity, we connect the high-voltage winding to an ac source E_g (Fig. 16-11). A jumper J is connected between any two adjacent HV and LV terminals, and a voltmeter E_x is connected across the remaining HV and LV terminals. Another voltmeter E_p is connected across the HV winding.

If E_x gives a higher reading than E_p, the polarity is additive. This means that the terminals diagonally opposite to each other must be marked H_1 and X_1. On the

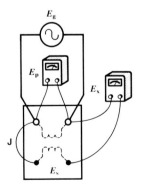

Figure 16-11
Determining the polarity of a transformer.

other hand, if E_x gives a lower reading then E_p, the polarity is subtractive and terminals H_1 and X_1 are adjacent.

In this polarity test, jumper J effectively connects the secondary voltage E_s in series with the primary voltage E_p. Consequently, E_s either adds to, or subtracts from E_p. In other words, $E_x = E_p + E_s$ or $E_x = E_p - E_s$, depending on the polarity. That is how the terms additive and subtractive originated.

In making the polarity test, an ordinary 120 V, 60 Hz source can be connected to the HV winding, even if its voltage rating is several hundred kilovolts.

Example 16-3:

During a polarity test on a 150 kVA, 69 kV/600 V transformer (Fig. 16-11), the following readings are obtained: $E_p = 118$ V, $E_x = 119$ V. Determine the polarity markings of the terminals.

Solution:

The polarity is additive because E_x is greater than E_p. Consequently, the HV and LV terminals connected by the jumper must respectively be labeled H_1 and X_2. (However, they could also be labeled H_2 and X_1.)

16.9 Transformer taps

It often happens that the voltage at an electric utility load center is below normal. For example, a utility transformer having a ratio of 2400 V to 120 V may be connected to a transmission line whose voltage is never higher than 2000 V. Under these conditions, the voltage across the secondary is considerably less than 120 V. Incandescent lamps glow dimly, electric stoves take longer to cook food, and motors may stall as soon as they are slightly overloaded.

To correct this problem, taps are provided on the primary windings of distribution transformers (Fig. 16-12). Taps enable us to change the transformer ratio so as to raise the secondary voltage. We can therefore maintain a satisfactory secondary voltage, even though the primary voltage may be below normal. For example, sup-

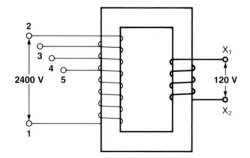

high-voltage terminals	percent tap	primary voltage	secondary voltage
1 - 2	0	2400 V	120 V
1 - 3	$4^1/_2$	2292 V	120 V
1 - 4	9	2184 V	120 V
1 - 5	$13^1/_2$	2076 V	120 V

Figure 16-12
A transformer with taps permits various primary to secondary voltage ratios.

pose the 2400 V/120 V transformer of Fig. 16-12 is connected to a line whose voltage is only 2076 V. We can still obtain 120 V on the secondary side, if the HV line is connected to taps 1 and 5.

Some transformers are designed to change taps automatically whenever the secondary voltage is above or below a preset level. Such tap-changing transformers can maintain the secondary voltage within ±2 percent of its rated value, even though the primary voltage may fluctuate by as much as ±10.

Figure 16-13 shows the high-voltage taps of a three-phase transformer.

Figure 16-13
Three-phase transformer prior to being placed in its protective oil tank. The coils are mounted on a three-legged core with the HV winding on top of the LV windings. Each HV winding is equipped with five taps. The HV terminals of the transformer are seen at the top left, while the busbars lead to the LV terminals at the top right. Single-phase transformers are built the same way, except that two sets of HV/LV coils are wound on a two-legged core.

16.10 Transformers in parallel

When a growing load eventually exceeds the power rating of an installed transformer, we sometimes connect a second transformer in parallel with it. The voltage rating of the two transformers must be the same. Furthermore, only those terminals having the same polarity marks must be connected (Fig. 16-14).

Polarity marks are then of crucial importance because a wrong connection produces the same effect as a short-circuited winding on both transformers.

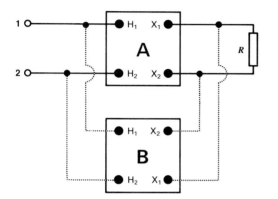

Figure 16-14
Two transformers connected in parallel.

16.11 Cooling methods

To prevent rapid deterioration of the insulation inside a transformer, the windings and core must be adequately cooled.

Low-power transformers below 50 kVA can be cooled by the natural circulation of air. The metal housing is fitted with ventilating louvers so that convection currents can flow over the windings and around the core (Fig. 16-15). Larger transformers are built the same way, but forced circulation of clean air must be used. Such dry-type transformers are used inside buildings, where the air is clean.

Figure 16-15
Single-phase transformer rated 15 kVA, for indoor use. (*Courtesy Hammond Manufacturing Company*)

Figure 16-16
Two single-phase oil-filled transformers rated 75 KVA, 14.4 kV/240 V, 60 Hz. The radiators at the side increase the effective cooling area.

Transformers above 200 kVA are usually immersed in mineral oil, and enclosed in a steel tank. Oil carries the heat from the transformer to the tank, where it is dissipated by radiation and convection to the outside air. Oil is a much better insulator than air is; consequently, it is always used in high-voltage transformers.

As the power rating increases, we add external radiators to increase the cooling surface of the oil-filled tank (Fig. 16-16). For still higher ratings, cooling fans blow air over the radiators.

Figure 16-17
Three-phase transformer rated 36/48/60 MVA, depending upon the type of cooling used.

Some large transformers are designed to have a triple power rating, depending on the method of cooling used. Thus, a transformer may have a rating of 18/24/32 MVA depending on whether it is cooled by

a) the natural circulation of air (18 MVA),

b) forced-air cooling using fans (24 MVA), or

c) the forced circulation of oil accompanied by forced-air cooling (32 MVA).

These elaborate cooling systems are used because they permit a bigger output from a transformer of given size (Fig. 16-17).

16.12 Transformer impedance

Figure 16-18 shows an *ideal* transformer having a rated secondary voltage E_s and delivering rated current I_s to a load Z_L. The rated load impedance Z_L is defined as

$$Z_L = E_s/I_s \qquad (16\text{-}6)$$

In an ideal transformer, E_s remains constant from no-load to full-load, as long as E_p is fixed.

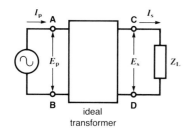

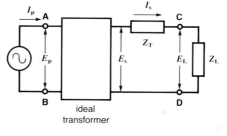

Figure 16-18
In an ideal transformer, the secondary voltage remains the same from no-load to full-load.

Figure 16-19
In a practical transformer, the secondary voltage varies slightly from no-load to full-load because of the internal impedance Z_T.

Unfortunately, transformers are not perfect, and so the secondary voltage at full-load is different from what it is at no-load. The reason is that the windings have a certain amount of resistance and reactance. The resulting transformer impedance Z_T produces an internal voltage drop when the transformer is under load. A practical transformer can therefore be represented by the equivalent circuit shown in Fig. 16-19. It consists of an ideal transformer and an internal impedance Z_T. The internal impedance is due to both the primary and secondary windings. However, it may be considered to be concentrated in the secondary winding alone. That is why it is shown in series with the secondary in Fig. 16-19. The actual terminals of the transformer are labeled A, B, C, and D.

The voltage E_L across the load is different from the ideal secondary voltage E_s. However, because Z_T is small compared with the rated load impedance Z_L, voltage E_L is nearly equal to E_s.

The relative magnitude of Z_T may be expressed as a percentage of Z_L. This percentage value is called the *percent impedance* of the transformer. By definition, the percent impedance is given by the equation:

$$\text{percent } Z_T = \frac{Z_T}{Z_L} \times 100 \qquad (16\text{-}7)$$

where

$$\text{percent } Z_T = \text{impedance of the transformer, in percent (\%)}$$
$$Z_T = \text{impedance of the transformer, in ohms } (\Omega)$$
$$Z_L = \text{rated load impedance, in ohms } (\Omega)$$

We can measure Z_T by short-circuiting the primary winding, and applying a low voltage e_s to the secondary (Fig. 16-20). Voltage e_s must be low enough so that the resulting current i_s does not exceed the rated value of the secondary current. Z_T is then given by

$$Z_T = e_s/i_s \qquad (16\text{-}8)$$

Knowing Z_T, we can calculate % Z_T by using Eqs. 16-6 and 16-7.

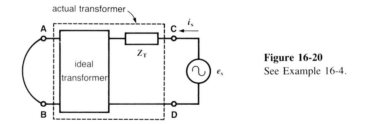

Figure 16-20
See Example 16-4.

Example 16-4:

An impedance test is made on a transformer rated at 500 kVA, 69 kV/4160 V, 60 Hz. The primary terminals are short-circuited, and the secondary is connected to a source e_s, as shown in Fig. 16-20. When $e_s = 75$ V, the resulting current $i_s = 35$ A. Calculate

1.   The rated impedance of the load, referred to the secondary

2.   The transformer impedance Z_T, referred to the secondary

3.   The percent impedance of the transformer

Solution:

1.   a.   The rated secondary current is

$$I_s = \frac{S}{E_s} = \frac{500\,000}{4160} = 120.2 \text{ A}$$

b.   The rated load impedance referred to the secondary is

$$Z_L = E_s/I_s = 4160/120.2 = 34.6 \ \Omega$$

2.   According to the test, $Z_T = e_s/i_s = 75/35 = 2.14 \ \Omega$

3.   The percent impedance is

$$\text{percent } Z_T = \frac{Z_T}{Z_L} \times 100 = \frac{2.14}{34.6} \times 100 = 6.18 \% \qquad \text{Eq. 16-6}$$

SPECIAL TRANSFORMERS

16.13 Distribution transformers

Transformers that supply electric power to residential areas generally have two secondary windings, each rated at 120 V. When the two windings are connected in series, the voltage between the lines is 240 V and that between the lines and the middle conductor is 120 V (Fig. 16-21). The middle conductor, called neutral, is always connected to ground. The reasons for using such a dual-voltage 120 V/240 V system will be explained in Chapter 22.

These so-called distribution transformers are often mounted on the poles of the electric utility company (Fig. 16-22). They may supply power to as many as twenty consumers.

The load on distribution transformers varies greatly throughout the day, depending upon customer demand. In residential districts, a peak occurs in the morning and another in the late afternoon. The power peaks seldom last for more than one or two hours, and so the transformers operate far below their rating most of the time. Because thousands of such units are connected to the electric utility system, they are specially designed to keep the no-load losses small. Towards this end, core steels are used that have particularly low hysteresis and eddy-current losses.

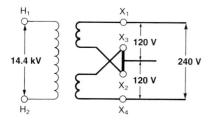

Figure 16-21
Schematic diagram of a distribution transformer.

Figure 16-22
Single-phase pole-mounted distribution transformer rated 25 kVA, 14.4 kV/120-240 V, 60 Hz. It is connected between one line and neutral of a 24 940/14 400 V, 60 Hz 3-phase transmission line

16.14 Voltage transformers

Voltage transformers (formerly called potential transformers) are high-precision transformers that are used to measure or monitor the voltage on transmission lines, and to isolate the metering equipment from these lines (Fig. 16-23). In these transformers, the ratio of primary voltage to secondary voltage is a known constant, which changes very little with load.* The rated secondary voltage is usually 115 V, which permits standard instruments and relays to be used.

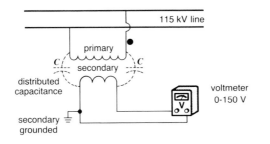

Figure 16-23
Voltage transformer installed on a 115 kV line. Note the distributed capacitance between the windings.

One terminal of the secondary winding is always connected to ground to reduce the danger of shock when touching one of the secondary leads. Although the secondary *appears* to be isolated from the primary, the distributed capacitance between the two windings makes an invisible connection which can produce a dangerously high voltage between the secondary winding and ground (Fig. 16-23). By grounding one of the secondary terminals, the highest voltage between the secondary lines and ground is limited to 115 V.

---

*   In the case of voltage transformers and current transformers, the load is called "burden".

Figure 16-24a
Voltage transformer installed on a 230 kV line in a substation.

Figure 16-24b
Voltage transformer rated at 7000 VA, 80.5 kV/115 V/115 V, 50/60 Hz, accuracy 0.3 %. The primary terminal at the top of the porcelain bushing is connected to the HV line, and the other primary terminal is connected to ground. The transformer has a height of 2565 mm and weighs 740 kg. (*Courtesy Ferranti-Packard*)

The construction of voltage transformers (VTs) is similar to that of conventional transformers. However, their power rating is usually less than 500 VA. Figure 16-24 shows a 7000 VA, 80.5 kV/115 V/115 V transformer. It has a tall ceramic bushing to isolate the HV line from the grounded case. The case houses the actual transformer. The insulation between the primary and secondary windings must be sufficient to withstand the full line voltage, as well as lightning and switching surges.

16.15 Current transformers

Current transformers are high-precision transformers in which the ratio of primary to secondary current is a known constant that changes very little with the burden.

Current transformers are used to measure or monitor the current in a line, while isolating the metering and relay equipment from it. The primary winding is connected in series with the line, as shown in Fig. 16-25. The rated secondary current is usually 5 A, irrespective of the primary current rating.

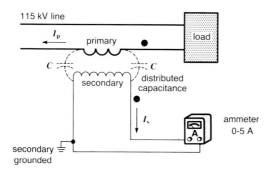

Figure 16-25
Current transformer installed on a 115 kV line. Note the polarity marks indicated by the dots.

Figure 16-26a
Current transformer rated at 50 VA,
100 A/5 A, 50/60 Hz for use on a 34.5 kV
line. (*Courtesy General Electric*)

Figure 16-26b
Current transformer in series with one phase
of a 230 kV line inside a substation.

Because current transformers (CTs) are used only for measurement and protection, their power rating is small — generally between 15 and 200 VA. As in the case of conventional transformers, the current ratio is inversely proportional to the number of turns on the primary and secondary windings. A current transformer having a ratio of 150 A/5 A has therefore 30 times more turns on the secondary than on the primary.

For safety reasons, current transformers must be used when measuring currents in HV transmission lines. The insulation between the primary and secondary windings must be sufficient to withstand the full line-to-neutral voltage. The maximum voltage that the CT can withstand is stamped on the nameplate.

As in the case of voltage transformers (and for the same safety reasons), one of the secondary terminals is always connected to ground. Another important precaution is to **never** open the secondary circuit of a current transformer. The reason is that a dangerous high voltage will then appear across the secondary, and the precision of the transformer may also be affected.

Figure 16-26 shows a 500 VA, 1000 A/5 A current transformer designed for a 230 kV line. The large bushing permits the line current to flow into and out of the primary winding. The CT is housed in the steel case at the lower end of the bushing.

At lower voltages and when the line current exceeds 100 A, we can sometimes use a toroidal current transformer. It consists of a ring-shaped core which carries the secondary winding. The primary is the line conductor that simply passes through the ring (Fig. 16-27). The conductor produces the same effect as a single primary turn. If the secondary possesses N turns, the ratio of transformation is N. Thus, a toroidal CT having a ratio of 300 A/5 A must have 60 turns on the secondary winding.

Toroidal CTs are simple and inexpensive and are widely used in LV (low voltage) and MV (medium voltage) indoor installations. They are also used internally in circuit breakers to monitor the current flowing through them.

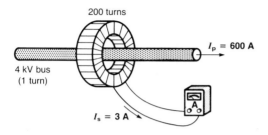

200 turns

I_p = 600 A

4 kV bus
(1 turn)

I_s = 3 A

Figure 16-27
Toroidal CT having a ratio of 1000 A/5 A, connected to
indicate the current in a 4 kV line.

Example 16-5:

A voltage transformer rated 14 400 V / 115 V, and a current transformer of 75/5 A are used to measure the voltage and current in a transmission line. If the voltmeter indicates 112 V and the ammeter reads 3.5 A, what are the voltage and current in the line ?

Solution:

1. Voltage on the line is:

    $$E = 112 \times (14\ 400/115) = 14\ 024\ V$$

2. Current in the line is:

    $$I = 3.5 \times (75/5) = 52.5\ A$$

16.16 Autotransformer

An autotransformer consists basically of a single winding mounted on an iron core. The "high" voltage is obtained across the entire winding and the "low" voltage appears between one end of the winding and an intermediate tap. For a given power output, an autotransformer is smaller and cheaper than a standard two-winding transformer. The cost and weight advantage is particularly important when the transformer ratio is close to 1. However, the absence of electrical isolation between primary and secondary is sometimes a serious drawback.

Autotransformers are used in reduced-voltage motor starters, and to increase or decrease transmission-line voltages by fixed amounts.

Consider an autotransformer (Fig. 16-28) having a tap T located at 80 percent of

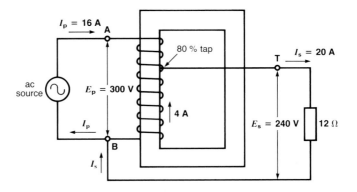

Figure 16-28
Construction and method of connecting an autotransformer to the source and load.

the winding (commonly called an 80 percent tap). If we apply a voltage $E_p = 300$ V to terminals A and B, the voltage E_s between terminals T and B will be 80 percent of 300 V, or 240 V. The transformer ratio is therefore 300/240 = 1.25.

If a 12 Ω resistive load is connected across the secondary terminals, it will draw a current of $I_s = E/R = 240/12 = 20$ A. The corresponding power is

$$P = E_s I_s = 240 \times 20 = 4800 \text{ W.}$$

But this load power must come from the 300 V source, and so the source must deliver a current I_p given by

$$I_p = P/E_p = 4800/300 = 16 \text{ A.}$$

Applying Kirchhoff's current law to point B, we discover that the current flowing in winding TB is equal to the *difference* between I_p and I_s. Its value is (20 − 16) = 4 A. This current is one fourth of that flowing in portion TA of the winding. Consequently, the conductor in portion TB need be only one quarter the size of that in portion TA.

16.17 Connecting a standard transformer as an autotransformer

A conventional two-winding transformer can be made into an autotransformer by connecting the windings in series. Depending upon the connection used, the secondary voltage will either add to, or subtract from, the primary voltage.

Consider, for example, the standard transformer of Fig. 16-29, which has a ratio of 600 V to 120 V. If terminals having the same polarity marks (H1 and X1, or H2 and X2) are connected, the low voltage subtracts from the high voltage. The resulting secondary voltage is therefore 600 V − 120 V = 480 V (Fig. 16-30). On the other hand, if terminals having opposite polarity marks (H1 and X2, or H2 and X1) are connected, the two voltages add. The resulting secondary voltage is 600 V + 120 V = 720 V (Fig. 16-31).

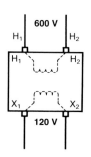

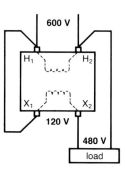

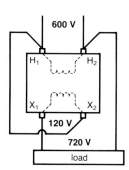

Figure 16-29
Conventional transformer having a ratio of 600 V/120 V.

Figure 16-30
Conventional transformer connected as an autotransformer so that the secondary voltage subtracts from the primary voltage.

Figure 16-31
Conventional transformer connected as on autotransformer so that the secondary voltage adds to the primary voltage.

Example 16-6:

A standard single-phase transformer has a rating of 30 kVA, 600 V/120 V, 60 Hz. We want to reconnect it as an autotransformer to give a ratio of 600 V/480 V. Calculate the maximum load it can carry.

Solution:

1. Rated current of the 600 V winding is

   $$I_1 = S/E_1 = 30\ 000/600 = 50 \text{ A}$$

2. Rated current of the 120 V winding is

   $$I_2 = S/E_2 = 30\ 000/120 = 250 \text{ A}$$

3. To obtain 480 V, the low voltage (120 V) must subtract from the high voltage (600 V). Consequently, terminals having the same polarity are connected, as shown in Fig. 16-30. The corresponding schematic diagram is shown in Fig. 16-32.

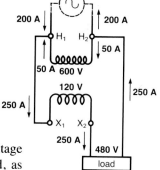

Figure 16-32
See Example 16-6.

4. Note that the 120 V winding is in series with the load. Because this winding can carry 250 A, the load can also draw 250 A. The power rating of the load can therefore be

   $$S = EI = 480 \times 250 = 120 \text{ kVA}.$$

   This load rating is four times the transformer rating, and yet the transformer is not overloaded.

5. The currents flowing in the circuit at full load are shown in Fig. 16-32.

   This example shows that when a conventional transformer is used as an autotransformer, it can sometimes carry a load that is far greater than the rated capacity of the transformer. In such cases, an autotransformer is the most economical way to make the voltage transformation.

16.18 Variable autotransformer

A variable autotransformer is often used when we wish to obtain a variable ac voltage from a fixed ac source. It is composed of a single-layer winding on a toroidal iron core (Fig. 16-33). A movable carbon brush in contact with the winding serves as a variable tap. The brush can be set in any position between zero and 330°.

Figure 16-33a
Variable autotransformer whose position can be set manually by means of a knob. The tap on the winding is seen on the left. (*Courtesy American Superior Electric*)

Figure 16-33b
Cutaway view of a 0-140 V, 15 A variable autotransformer showing (1) the laminated toroidal core; (2) the single-layer winding and (3) the movable brush. (*Courtesy American Superior Electric*)

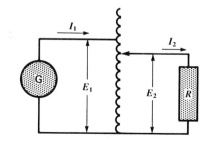

Figure 16-34
Schematic diagram of a variable autotransformer having a fixed 90 % tap.

As the brush slides over the bared portion of the windings, the secondary voltage E_2 increases in proportion to the number of turns swept out (Fig. 16-34). The input voltage E_1 is usually connected to a fixed 90 percent tap on the winding. This enables us to vary E_2 from zero to 110 percent of the input voltage.

Variable autotransformers are far more efficient than rheostats are, and they give much better voltage regulation under variable loads. The secondary line should always be protected by a fuse or circuit breaker so that the output current I_2 never exceeds the normal current rating of the autotransformer.

TRANSFORMERS IN THREE-PHASE SYSTEMS

To raise or lower the voltage of a three-phase transmission line we use transformers, as in the case of single-phase lines. The transformers may be three-phase, having three primary windings and three secondary windings mounted on a common core, as we saw in Fig. 16-12. Alternatively, they may be three single-phase transformers connected together to form a so-called three-phase transformer bank.

When single-phase transformers are used to transform a three-phase voltage, the windings can be connected in several ways. Thus, the primaries may be connected in delta and the secondaries in wye, or vice versa. As a result, the ratio of the input *line* voltage to the output *line* voltage depends not only upon the turns ratio of the transformers, but also upon how they are connected.

In making connections, it is important to observe transformer polarities. An error in polarity produces either a short-circuit or unbalanced line voltages.

16.19 Delta-delta connection

The three single-phase transformers P, Q, and R of Fig. 16-35 are connected in *delta-delta*. The figure shows the actual physical layout of the transformers; the corresponding schematic diagram is given in Fig. 16-36. Note the symmetrical way the connections between H_1 and H_2 and X_1 and X_2 are made.

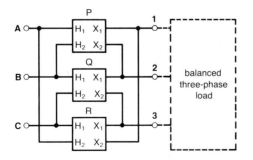

Figure 16-35
Three-phase transformer bank connected in delta-delta.

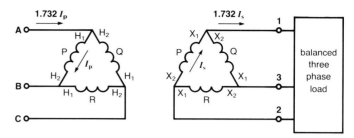

Figure 16-36

Schematic diagram of the delta-delta connection shown in Fig. 16-35. Primary and secondary windings that are drawn parallel to each other have voltages that are in phase.

The schematic diagram is drawn so that each secondary winding is parallel to the corresponding primary winding to which it is coupled. Because the primary and secondary voltages of a given transformer are in phase, E_{H1H2} of transformer P is in phase with E_{X1X2} of the same transformer. Similarly E_{H1H2} of transformer Q is in phase with E_{X1X2} of transformer Q.

Figure 16-36 shows that the incoming lines A, B, and C are directly connected to the primaries of the transformers. Furthermore, the voltages between outgoing lines 1, 2, and 3 are equal to the voltages induced in the respective secondary windings. If a balanced three-phase load is connected across lines 1, 2, and 3, the line currents on both the primary and secondary side of the transformer bank will be balanced.

As in any delta connection, the line currents are 1.732 times greater than the currents flowing in the respective windings.

16.20 Delta-wye connection

In a *delta-wye* connection, the primary windings of the three transformers are connected the same way as in Fig. 16-36. However, the secondary windings are connected with the X_2 terminals joined, to create a common neutral N (Fig. 16-37). In such a delta-wye connection, the primary voltage across each transformer is equal to the incoming line voltage. However, the outgoing line voltage is 1.732 times the

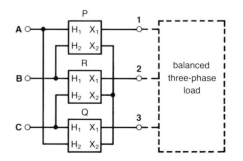

Figure 16-37

Three-phase transformer bank connected in delta-wye.

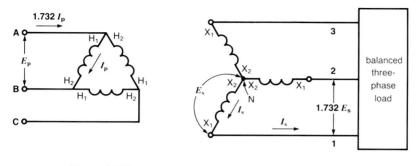

Figure 16-38
Schematic diagram of the transformer connections in Fig. 16-37.

secondary voltage across each transformer. The relative values of the corresponding currents in the transformer windings and transmission lines are given in Fig. 16-38. For example, the line currents in phases A, B, and C are 1.732 times the currents in the primary windings.

One of the important advantages of the wye connection is that it reduces the insulation requirements of the transformer. In effect, the HV windings have to be insulated for only 1/1.732 or 57.7 percent of the line voltage.

16.21 Wye-delta connection

The currents and voltages in a *wye-delta* connection are identical to those in the delta-wye connection of Section 16-19. The primary and secondary connections are simply interchanged. In other words, the H_2 terminals are connected to create a neutral, and the X_1, X_2 terminals are connected in delta.

16.22 Wye-wye connection

The *wye-wye* connection is used only when the neutral N_1 of the primary can be solidly connected to the neutral N_g of the source, usually by way of the ground (Fig. 16-39). When the neutrals are not joined, the line-to-neutral voltages become distorted (non sinusoidal).

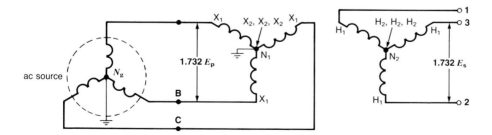

Figure 16-39
In a wye-wye connection, the neutral of the transformer bank is connected to the neutral of the source.

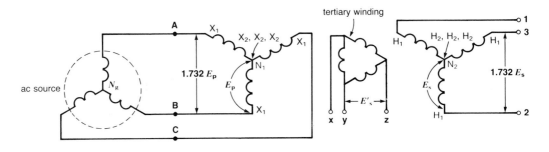

Figure 16-40
Wye-wye transformer connection with a tertiary winding.

A wye-wye connection can be used without joining the neutrals N_1 and N_g, provided that each transformer carries a third winding, called *tertiary winding*. The tertiary windings of the transformers must be connected in delta (Fig. 16-40). The tertiary windings eliminate voltage distortion, and maintain balanced line-to-neutral voltages even when the load is unbalanced. In addition, the tertiary windings often provide the substation service voltage where the transformers are installed.

16.23 Open-delta connection

We can transform the voltage of a three-phase line by using only two transformers connected in *open-delta*. The open-delta connection is identical to a delta-delta connection, but with one fewer transformer (Fig. 16-41). This is one of the advantages of a delta-delta connection because two transformers can continue to feed the load if one of them should become defective.

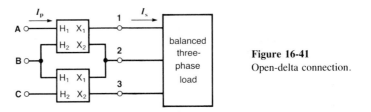

Figure 16-41
Open-delta connection.

In medium-and high-power installations, the open-delta connection is always temporary because the load capacity of the transformer bank is only 86.6 percent of the installed transformer capacity. For example, if the transformers in Fig. 16-41 are each rated at 100 kVA, the maximum possible load is not 200 kVA (as we might expect), but only 86.6% × 200 = 173 kVA.

16.24 Three-phase transformers

A transformer bank composed of three single-phase transformers may be replaced by a three-phase transformer. The magnetic core of such a transformer has three legs which carry the primary and secondary windings of each phase (Fig. 16-13). The windings are connected internally, either in wye or in delta (Fig. 16-37). Consequently, only six terminals have to be brought outside the tank. Figure 16-42 shows the three HV bushing of a large transformer.

For a given capacity, a three-phase transformer is always smaller and cheaper than three single-phase transformers. Nevertheless, single-phase transformers are sometimes preferred, particularly when an emergency standby unit is essential. For example, suppose a manufacturing plant absorbs 5000 kVA. To guarantee continued service, we can install a three-phase 5000 kVA transformer and keep a second one as a spare. Alternatively, we can install three single-phase transformers, each rated 1667 kVA, and one spare. The 3-phase transformer option is more expensive (total capacity 10 000 kVA) than the single-phase option (total capacity 6667 kVA).

Figure 16-42
Three-phase transformer rated 330 MVA, 735 kV/16 kV, 60 Hz. It serves to connect a 300 MVA synchronous capacitor to a 735 kV line. (*Courtesy Hydro-Quebec*)

16.25 Summary

In this chapter we learned that a transformer has the remarkable property of raising or lowering an ac voltage, depending upon the number of turns on its primary and secondary windings. The ratio of the voltages and that of the currents is related to the ratio of the number of turns.

The rating of a transformer is expressed in voltamperes rather then in watts because the heat generated in the iron core and the coils depends upon the apparent power of the load. Thus, although large transformers can have efficiencies as high as 99.5 percent, an adequate cooling system is needed.

We learned that the polarity of a transformer has a different meaning from the polarity of a battery, for example. It indicates which primary and secondary terminals have momentarily the same polarity. Polarity is very important when transformers are connected in parallel, or when they are connected to a three-phase system.

If a transformer had no resistance or reactance, its secondary voltage would remain fixed from no-load to full-load. However, a transformer is not perfect, and so it has a small internal impedance which causes the secondary voltage to change slightly when the transformer is loaded. The magnitude of this impedance is best expressed as a percentage of the rated load impedance of the transformer.

Voltage transformers and current transformers are used for metering and relay purposes, consequently they have to be very precise. In addition, for safety reasons, they must adequately isolate the secondary windings from the primary windings. Particular care must be taken to never open the secondary winding of a current transformer.

We also saw how an autotransformer can often carry a large load using a relatively small transformer. However, this advantage over conventional transformer is only possible when the primary and secondary voltages have about the same magnitude.

Finally, we learned that single-phase transformers can be used to create several types of three-phase transformer banks. This enabled us to put to use our knowledge of wye-and delta-connected systems.

TEST YOUR KNOWLEDGE

16-1   A single-phase transmission line delivers 120 A at a voltage of 14.4 kV. The apparent power it delivers is

a. 1728 kW   b. 120 W   c. 1728 kVA

16-2   The consumer in Fig. 16-1 absorbs 10.35 MVA of power. The current in the 69 kV single-phase line is

a. 15 A   b. 150 A   c. 21 562.5 A

16-3   The transformer ratio of T_1 in Fig. 16-1 is

a. 60.29   b. 16.586 V   c. 16.586

16-4   The primary winding of a transformer is *always* the winding that is connected to the source.

☐ true   ☐ false

16-5   In Fig. 16-2, E_p = 480 V and E_s = 7200. If the primary has 600 turns, the secondary has

a. 5760 turns   b. 40 turns   c. 9000 turns

16-6   In Fig. 16-5, suppose E_p = 480 V, E_s = 7200 V, and the load absorbs 180 kW. The value of I_p is

a. 375 A   b. 25 A   c. 180 kVA

16-7   In Fig. 16-9, if E_p = 12 470 V and E_s = 240 V, the peak value of E_s is

a. 480 V   b. 415.69   c. 339.4 V

16-8   A distribution transformer has a rating of 100 kVA, 14.4 kV to 480 V. If we apply a voltage of 7.2 kV to the primary side, the transformer will be able to deliver a maximum of

a. 50 kVA   b. 100 kVA   c. 100 kvar

16-9   If we know that a transformer has additive polarity, we don't need polarity marks to indicate where the respective H1 and X1 terminals are.

☐ true   ☐ false

16-10   In Fig. 16-12, if we apply 2400 V between terminals 4 and 1, the voltage between X_1 and X_2 will be

a. 120 V   b. 131.87 V   c. 109.2 V

16-11   A transformer has a rating of 15 kVA, 600 V/240 V, 60 Hz. What is the rated current on the 240 V side?

a. 16 A   b. 62.5 A   c. 625 A

What is the rated load impedance on the 240 V side?

d. 15 000 W   e. 15 Ω   f. 3.84 Ω

16-12   The transformer in Problem 16-11 has an impedance of 5 percent. What is the transformer impedance, in ohms, on the 240 V side?

a. 0.192 Ω   b. 76.8 Ω   c. 5.208 Ω

16-13   The secondary side of a VT or CT must *always* be grounded in order to

a.   get an accurate reading

b.   eliminate the capacitance between the primary and secondary windings

c.   reduce the danger of shock

16-14   A toroidal CT has a current rating of 1600/5 A. How many turns does it have on the primary?

a. 1 turn   b. 320 turns   c. none

16-15   In Fig. 16-36, if I_p = 600 A, what is the value of the line current in phase C assuming the system is balanced?

a. 1200 A   b. zero   c. 1039 A

16-16   In Fig. 16-37, the primary to secondary voltage rating of each transformer is 24 940 V/120 V. If the line voltage between phases A, B, and C is 24 000 V, the line voltage between phases 1, 2, and 3 is

a. 200 V   b. 115.48 V   c. 163.3 V

16-17   Each transformer in Fig. 16-41 has a rating of 4000 kVA. The maximum load they can carry is

a. 8000 kVA   b. 6928 kVA   c. 8000 kvar

QUESTIONS AND PROBLEMS

16-18   Name the principal components of a transformer.

16-19   What purpose does the no-load current of a transformer serve?

16-20   Explain how a voltage is induced in the secondary winding of a transformer.

16-21   The secondary winding of a transformer has twice as many turns as the primary. Is the secondary voltage higher or lower than the primary voltage?

16-22   State the voltage and current relationships between the primary and secondary windings of a transformer under load. The primary and secondary windings have N_1 and N_2 turns, respectively.

16-23   Name the losses produced in a transformer.

16-24   Which winding is connected to the load: the primary or secondary?

16-25   What conditions must be met in order to connect two transformers in parallel?

16-26   What is the purpose of taps on a transformer?

16-27   What is the difference between an autotransformer and a conventional transformer?

16-28   What is the purpose of a potential transformer? Of a current transformer?

16-29   Why must we never open the secondary of a current transformer?

16-30   The primary of a transformer is connected to a 600 V, 60 Hz source. If the primary has 1200 turns and the secondary has 240, calculate the secondary voltage.

16-31   The windings of a transformer respectively have 300 and 7500 turns. If the LV winding is excited by a 2400 V source, calculate the voltage across the HV winding.

16-32   A 6.9 kV transmission line is connected to a transformer having 1500 turns on the primary and 24 turns on the secondary. If the load across the secondary has an impedance of 5 Ω, calculate (a) the secondary voltage and (b) the primary and secondary currents.

16-33   The primary of a transformer has twice as many turns as the secondary. The primary voltage is 220 V and a 5 Ω resistor is connected across the secondary. Calculate (a) the power delivered by the transformer and (b) the primary and secondary currents.

16-34   A 3000 kVA transformer has a ratio of 60 kV to 2.4 kV. Calculate the rated current of each winding.

16-35   In Fig. 16-10, when 600 V is applied to terminals H_1 and H_2, 80 V is measured across terminals X_1 and X_2.

   a.   What is the voltage between terminals H_1 and X_2?

   b.   If terminals H_1X_1 are connected, calculate the voltage across terminals H_2X_2.

16-36   a.   Referring to Fig. 16-14, what would happen if we reversed terminals H_1 and H_2 of transformer B?

   b.   Would the operation of the transformer bank be affected if terminals H_1H_2 *and* X_1X_2 of transformer B were reversed? Explain.

16-37   What is meant by (a) transformer impedance and (b) percent impedance of a transformer?

16-38   A 2300 V line is connected to terminals 1 and 4 in Fig. 16-12. Calculate:

   a.   the voltage between terminals X_1 and X_2

   b.   the current in each winding if a 12 kVA load is connected across the secondary

16-39   A 66.7 MVA transformer has an efficiency of 99.3 % when it delivers full power to a load having a power factor of 100 percent.

   a.   Calculate the losses in the transformer under these conditions.

   b.   Calculate the losses and efficiency when the transformer delivers 66.7 MVA to a load having a power factor of 80 percent. (Hint: efficiency is always calculated on the basis of active power in and active power out)

Three-Phase Induction Motors

Introduction and Chapter Objectives

Three-phase induction motors are the motors most frequently encountered in industry. They are simple, rugged, low-priced, and easy to maintain. They run at essentially constant speed from no-load to full-load. However, the speed depends upon the frequency of the electric utility. Consequently, these motors are not easily adapted to speed control.

The importance of induction motors in industry can be appreciated from the fact that a modern manufacturing plant of 100 employees may contain as many as 500 motors.

In this chapter, we will cover the basic principles and construction of squirrel-cage and wound-rotor induction motors and linear motors.

17.1 Forces produced by a moving magnetic field

Before undertaking the detailed study of induction motors, we will cover the basic principles that enable us to build such machines. These principles depend upon (1) Faraday's law of electromagnetic induction, which we covered in Chapter 8, and (2) the force developed when a current-carrying conductor is immersed in a magnetic field, covered in Chapter 6.

Figure 17-1 shows a stationary copper plate and a movable permanent magnet. Copper is non-magnetic, and so the flux produced by the magnet passes through the plate as if it were not there. Thus, when the magnet and plate are both stationary, no forces are exerted and nothing happens.

But suppose we move the magnet to the right at a constant speed v_s. The flux will cut across the copper plate, and a voltage will be induced, which, in turn, will produce a current in the plate. The current is in the path of the flux created by the mag-

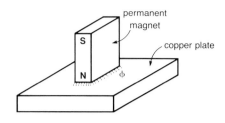

Figure 17-1
Stationary permanent magnet above a copper plate.

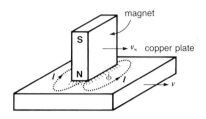

Figure 17-2
Moving the magnet to the right induces currents in the plate. The resulting force causes the plate to follow the magnet.

net (Fig. 17-2), and so a force acts on the plate. This force drives the plate in the same direction as the moving magnet. The plate is therefore dragged along with the magnet, at a certain speed v.

As the plate picks up speed, the flux lines cut across it less quickly, and so the induced voltage and the resulting induced current both begin to fall. The force dragging the plate along with the magnet therefore becomes smaller. If the plate moved at the same speed as the magnet, the flux lines would no longer cut across the plate. The induced current would become zero, and the force urging the plate to the right would disappear (Fig. 17-3). The magnitude of the induced current

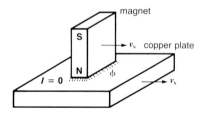

Figure 17-3
When the plate and magnet move at the same speed, the induced currents disappear and the force on the plate is zero.

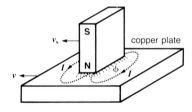

Figure 17-4
When the magnet moves to the left, it drags the plate with it.

depends therefore upon the *relative* speed $(v_s - v)$ between the magnet and the plate. The greater the difference between v_1 and v_2, the greater will be the current, and the resulting force.

If we move the magnet to the left, the direction of current flow will reverse, and the plate will be dragged toward the left (Fig. 17-4).

We conclude that the plate always tends to move in the same direction and at the same speed as the moving field.

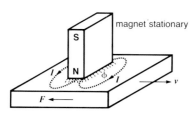

magnet stationary

Figure 17-5
A stationary magnet exerts a braking
force on the moving copper plate.

Suppose now that the magnet is stationary and that the *plate* is moving toward
the right at a speed v. The flux again cuts across the plate, and a current is again
induced (Fig. 17-5). The resulting force on the plate acts against the direction of
motion, and so a braking force F is exerted on the plate.

Thus, the magnet can cause the plate to move to the right or to the left,
depending upon the direction of v_s. Furthermore, if the magnet is stationary, it will
oppose the motion of the copper plate.

17.2 Torque produced by a rotating magnetic field

We can also produce rotary motion, based upon the principle of magnetic induction.
Figure 17-6 shows a permanent magnet N-S that can be rotated manually by means
of handle H. A rotor R, made of solid copper, is supported by bracket B, but is otherwise free to rotate. If we turn the handle clockwise at a speed n_s, the magnet will
induce voltages and currents in the rotor. As in the case of the copper plate, the rotor
will be dragged along by the revolving magnet. But as the rotor speed n increases,
the induced voltage will become smaller, as will the induced current. Thus, the
dragging force, or torque, will decrease as n approaches n_s. If n were to become
equal to n_s, the induced voltage (and consequent current) would fall to zero, and the
dragging force would disappear.

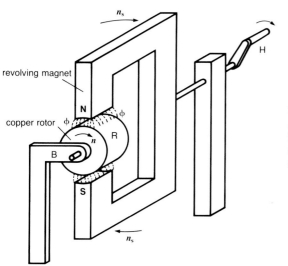

revolving magnet

copper rotor

Figure 17-6
A revolving permanent magnet
induces currents in the copper rotor,
causing the rotor to turn in the same
direction.

If we reverse the direction of rotation of the magnetic field, the rotor will turn in the opposite direction.

Finally, suppose that the magnetic field is stationary. If the rotor is turning, a braking force will act on it which will quickly bring it to a halt.

The behavior of a revolving rotor is therefore similar to that of a moving flat plate. This has given rise to two types of induction motors: standard (rotating) motors and linear motors.

17.3 Number of poles and construction of the rotor

The revolving magnetic field can have more than two poles. Figure 17-7 shows a field created by four poles that rotate as a group. Adjacent poles have opposite polarities N-S-N-S, as shown. The number of poles is always even because for every N pole there must be a corresponding S pole.

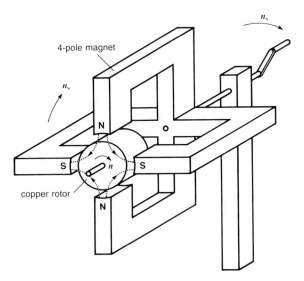

Figure 17-7
A revolving four-pole magnetic field causes the rotor to turn.

We mentioned that the rotor was made of solid copper. However, it could equally well be made of solid aluminum, or solid steel. It will still be carried along by the magnetic field. However, to develop the largest possible torque, we must first ensure that a strong field is produced by the N and S poles. For this reason, the rotor should be made of iron, which has a high permeability. In addition, the air gap between the rotor and the magnetic poles should be as small as possible.

Second, to obtain a really strong torque, the rotor currents must be as large as possible. This means that the rotor should be made of a metal having a low resistivity. The best choice is copper or aluminum. Thus, to obtain the highest possible torque, we use a rotor made of iron in combination with a good conductor, such as copper or aluminum. In practice, we use a laminated iron rotor that has slots around the circumference. Aluminum or copper bars are inserted in the slots, and the ends are short-circuited by so-called end rings. The resulting winding looks like a squirrel cage (Fig. 17-8), and the rotor is often called a squirrel-cage rotor.

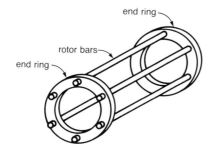

Figure 17-8
Elementary winding in a squirrel-cage rotor.

17.4 Slip and synchronous speed

Referring to Fig. 17-6 or 17-7, suppose the magnetic field is turning at a speed of n_s r/min, and the rotor is rotating in the same direction at a speed of n r/min. The relative speed is therefore $(n_s - n)$ r/min. The so-called *slip* of the rotor is given by the equation:

$$s = \frac{n_s - n}{n_s} \qquad (17\text{-}1)$$

The slip is zero when the rotor turns at the same speed and in the same direction as the magnetic field. On the other hand, the slip has a value of 1 when the rotor is stationary.

Example 17-1:

1. The field in Fig. 17-7 is turning cw at 1600 r/min, and the rotor is rotating cw at 1260 r/min. Calculate the slip.

2. Calculate the new value of slip if the rotor rotates cw at 1680 r/min while the field is still turning cw at 1600 r/min.

Solution:

1. $s = (n_s - n)/n_s = (1600 - 1260)/1600 = 240/1600 = 0.15$ or 15 %

2. $s = (n_s - n)/n_s = (1600 - 1680)/1600 = -80/1600 = -0.05$ or -5 %

The slip is negative when the rotor turns faster than the field.

17.5 Producing a revolving field

So far we have used mechanical means to produce a rotating field. The question arises, how can we produce a field that turns without having to use a mechanical drive?

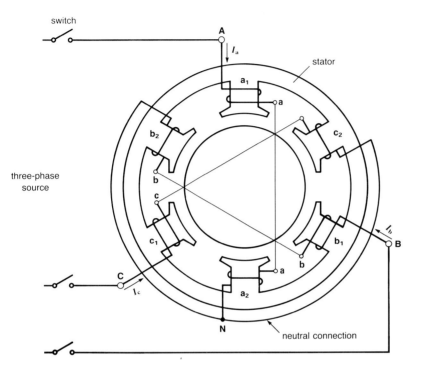

Figure 17-9

Simplified stator winding of a three-phase motor. The three phases are connected in wye forming the common neutral N. When the switch is closed, the three-phase currents are equal in magnitude and displaced by 120° in time. The arrows show the direction of current when the respective currents are positive.

Consider the special electromagnet of Fig. 17-9, which is called a *stator*. It has six salient poles carrying six identical coils. The coils are arranged in pairs — a_1, a_2; b_1, b_2; and c_1, c_2 — spaced at 120° to each other. Coils that are diametrically opposite (such as a_1 and a_2) are connected in series in such a way as to create one N and one S pole. Three of the six coil ends are joined together to form a common neutral N, and the remaining coil ends A, B, and C are connected to a three-phase source. The coil arrangement is perfectly symmetrical, and so the three-phase currents I_a, I_b, and I_c are identical but are displaced in time by 120° (Fig. 17-10). This means that the currents reach their maximum positive values at different times. Thus, when I_a is maximum positive, flux ϕ_a acts in line with coils a_1-a_2, as shown in Fig. 17-11. Then, 120 degrees later, when I_b reaches its maximum positive value, the flux ϕ_b acts in line with coils b_1-b_2 (Fig. 17-12). Another 120 degrees later, current I_c reaches its maximum positive value, and flux ϕ_c is oriented in line with coils c_1-c_2 (Fig. 17-13). After another 120 degrees (Fig. 17-14), I_a is again maximum positive, and the flux takes up the same position it had in Fig. 17-11.

The flux in Figs. 17-11 to 17-14 is obviously moving clockwise inside the six-pole stator. The sequence ϕ_a - ϕ_b - ϕ_c - ϕ_a - ϕ_b repeats continually, and so the flux rotates continually. We can reverse the direction of rotation by interchanging

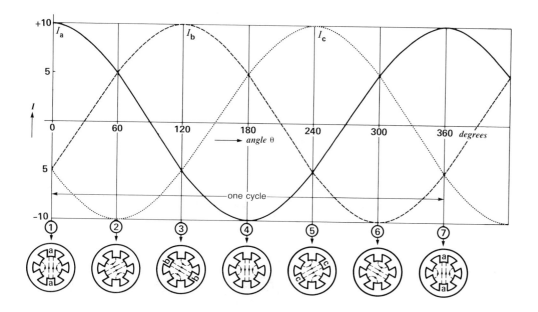

Figure 17-10
The currents in the windings vary sinusoidally. The resulting flux in the stator is shown at seven successive instants. The peak current is assumed to have a value of 10 A.

Figure 17-11
Flux produced by coils $a_1 - a_2$ when I_a is $+10$ A.

Figure 17-12
Flux produced by coils $b_1 - b_2$ when I_b is $+10$ A.

Figure 17-13
Flux produced by coils $c_1 - c_2$ when I_c is $+10$ A.

Figure 17-14
Flux produced by coils $a_1 - a_2$ when I_a is again $+10$ A.

any two stator leads. As we learned in Section 15-10, this reverses the phase sequence at the motor terminals.

In comparing Figs. 17-11 and 17-14, it is evident that the flux makes a complete turn in the time it takes I_a to go from one maximum positive value to the next. In other words, the flux makes a complete turn during each cycle. If the frequency of the source is 60 Hz, the flux makes one turn in 1/60 s. This corresponds to 60 revolutions per second, or 3600 r/min. The speed of rotation is therefore synchronized with the frequency of the source.

Although the stator has six poles, it only has two poles per phase. Thus, the three phases together produce a revolving magnetic field that has two poles. The shape of the field at successive intervals is also shown in Fig. 17-10.

17.6 Effect of the number of poles

We can increase the number of poles per phase from 2 to 4 to 6 (or to any higher even number) by changing the design of the stator. Increasing the number of poles reduces the speed of rotation of the magnetic field. For example, in Fig. 17-15, the 4-pole stator (four poles per phase) produces a 4-pole revolving field that turns at exactly half the speed of a 2-pole stator. The four poles rotate together like the spokes of a wheel. Consequently, the rotating field is similar to that shown in Fig. 17-7.

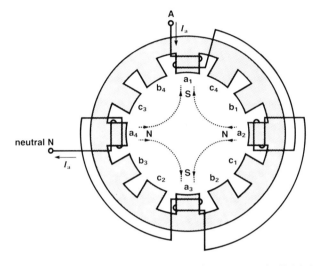

Figure 17-15
Flux produced by the coils of phase A in a 4-pole stator when I_a is maximum positive. Note that the coil connections produce alternate N and S poles as we move around the stator. The coils of phases B and C are wound the same way but are shifted in space.

The speed of rotation of the magnetic field depends therefore upon the frequency of the source and the number of poles on the stator. Its value is given by the equation:

$$n_s = \frac{120\,f}{p} \qquad (17\text{-}2)$$

where

n_s = synchronous speed, in revolutions per minute (r/min)

f = frequency of the source, in hertz (Hz)

p = number of poles per phase

The speed of rotation is called *synchronous speed* because it is exactly in step with the frequency.

Example 17-2:

A three-phase induction motor has 16 poles on the stator. If it is connected to a special source that generates a frequency of 5 Hz, calculate the speed of rotation of the magnetic field.

Figure 17-16
Stator core of a medium-power (50 hp) three-phase motor. The large number of slots permits a distributed winding that produces a better motor performance than a salient-pole winding does.

Solution:

$$n_s = 120 \, f/p = (120 \times 5) \, / \, 16 = 37.5 \text{ r/min} \qquad \text{Eq. 17-2}$$

The synchronous speed of the motor is therefore 37.5 r/min.

17.7 Construction of a squirrel-cage induction motor

Having covered the general principles, we will now examine the practical construction of an induction motor. A three-phase induction motor has two main parts: a stationary stator and a revolving rotor.

The stator produces the revolving field in a manner similar to the stator of Fig. 17-9 or 17-15. However, instead of having salient poles, it consists of a hollow cylindrical core made up of stacked laminations. A number of evenly spaced slots are punched out of the internal circumference. The slots provide space for the stator winding (Figs. 17-16, 17-17, and 17-18). The laminated core is supported inside a steel frame that carries mounting feet and lifting lugs and also serves to accurately position the end-bells. End-bells are the two disk-shaped ends of the motor that support the rotor shaft (Fig. 17-20).

The rotor is also made of punched laminations. The rotor slots are carefully lined up during the stacking process to provide space for the rotor winding. The winding is composed of copper bars that are inserted into the slots and then welded together at each end by means of two copper end rings. For motors below 100 hp, the rods and end rings are usually made of die-cast aluminum. In the manufacturing process, molten aluminum under high pressure is forced into a mold that surrounds the iron

Figure 17-17
Placing pre-formed coils into the slots of a large-power motor. Note the symmetrical distribution of the winding. (*Courtesy General Electric*)

rotor. The squirrel-cage winding is thus formed in a fraction of a second. Figures 17-19 to 17-21 show typical rotor contruction and motor assembly.

The completed rotor is mounted on a steel shaft and then turned down in a lathe to create a small gap between the rotor and the stator. The length of the air gap depends upon the size of the motor, but it usually ranges from 0.4 mm to about 5 mm for very large machines. The air gap is made as small as mechanical tolerances will permit.

17.8 Locked-rotor conditions

At the instant a three-phase motor is connected to the line, the rotor is still at rest. However, the stator immediately creates a revolving field. This field cuts across the

Figure 17-18
The stator is dipped in varnish and baked in order to protect the windings against the environment. The varnish penetrates the windings thereby preventing vibration in the slots at the same time as it improves the transfer of heat from inside the machine. The coil ends are firmly secured to resist the powerful electromagnetic forces when the motor is starting. (*Courtesy General Electric*)

rotor bars, inducing a small voltage in them. However, large currents begin to flow in the rotor bars because they are short-circuited together. The bars are immersed in the magnetic field created by the stator, and so a strong force ($F = BLI$) acts on each bar. The sum of the forces on all the bars produces a strong locked-rotor torque which acts in the same direction as the revolving field.

Under locked-rotor* conditions, the motor acts like a three-phase transformer whose secondary winding is short-circuited (remember that all the rotor bars are short-circuited together). Consequently, the rotor and stator windings begin to heat

---

*   The term locked-rotor is sometimes abbreviated to LR.

Figure 17-19
Squirrel-cage rotor of a large-power induction motor. The slots are just beneath the smooth outer surface and are therefore not visible. The thirteen radial ducts permit cooling along the length of the rotor. Additional cooling is provided by the two fans at each end. The very robust shaft is designed to resist mechanical shock and to prevent the rotor from coming in contact with the stator. (*Courtesy General Electric*)

up very quickly because of the large I^2R losses. The current in the three lines of the stator is between 5 and 6.5 times the rated full-load current. Consequently, the I^2R losses are between 25 and 42 times their normal value. Furthermore, the power factor is usually below 40 percent.

Because of the large power that is drawn from the source and because of the high heating rate, we must never leave a motor in the locked-rotor condition for more than a few seconds.

Figure 17-20
Exploded view of a squirrel-cage induction motor having copper rotor bars. The illustration shows the ball bearings mounted on each end of the shaft, the two end-bells next to the stator, the ventilating fan, and the protective shield on the extreme left. The windings are fed from the three terminals in the terminal box. (*Courtesy Brook Crompton Parkinson*)

17.9 Acceleration of a squirrel-cage motor

If the rotor is free to turn, the locked-rotor torque will accelerate the rotor in the same direction as the rotating field. But as the motor picks up speed, the rotating field cuts across the rotor bars at a lower speed, and so the induced voltage begins to fall. This, in turn, causes the rotor current to decrease. Although the rotor torque may decrease, it continues to accelerate the rotor.

How fast will the motor turn? The rotor can never reach synchronous speed because if it did, it would be moving at the same speed as the revolving field. The induced voltage would then be zero, the currents in the rotor bars would disappear, and the torque would become zero. Therefore, a three-phase induction motor always runs at slightly less than synchronous speed. That is why it is sometimes

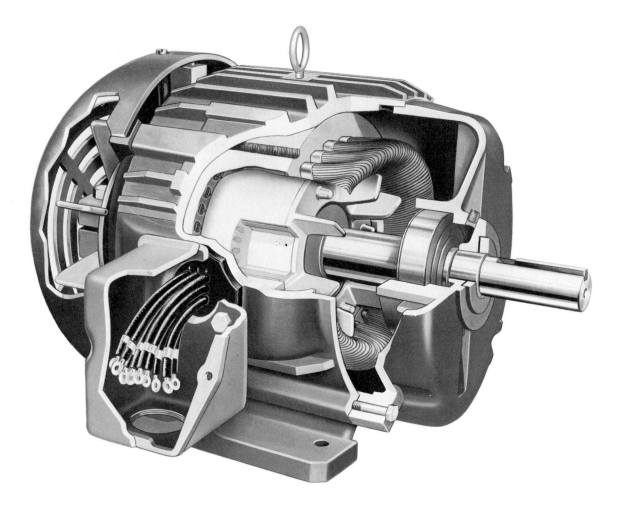

Figure 17-21
Cutaway view of a 5 hp squirrel-cage induction motor having a die-cast aluminum rotor. This duel-voltage 230 V/460 V motor has nine leads coming out of the terminal box. Note that the cooling fins are die-cast integrally with the rotor bars and end rings. (*Courtesy Siemens-Allis*)

called an *asynchronous motor*. The difference between the rotor speed and synchronous speed amounts to a slip of less then one percent.

While the rotor is accelerating, a frequency effect also takes place. When the rotor is locked, the frequency of the voltage and current in the rotor bars is the same as the line frequency. But as the rotor picks up speed, the frequency decreases progressively. For example, when the rotor is turning at one-half synchronous speed, the stator flux cuts across the rotor bars at half the speed it did when the rotor was locked. As a result, the frequency of the voltage and current in the rotor bars falls to one-half the line frequency.

By the same reasoning, if the rotor turns at 90 percent of synchronous speed (slip = 10 percent), the frequency of the voltage and current in the rotor bars will be only 10 percent of line frequency. The rotor frequency is therefore related to the slip by the equation:

$$\overline{f_r = sf} \qquad (17\text{-}3)$$

where

f_r =   rotor frequency, in hertz (Hz)

s =   slip

f =   frequency of the source, in hertz (Hz)

The progressive drop in rotor frequency causes the inductive reactance of the rotor to decrease with increasing speed. This tends to increase the torque during the acceleration period. On the other hand, the decrease in rotor current with increasing speed tends to decrease the torque. These two opposing effects produce the typical torque-speed curves shown in Fig. 17-22.

17.10 Squirrel-cage motor at no-load

When a squirrel-cage motor is operating at no-load, it runs very close to synchronous speed. For example, a 100 hp motor having a synchronous speed of 1800 r/min

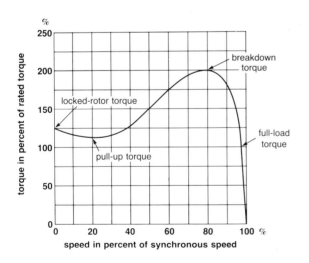

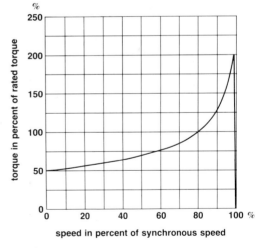

Figure 17-22a
Typical torque-speed characteristic of a 2 hp, 3-phase squirrel-cage induction motor. The curve illustrates the interpretation of locked-rotor, pull-up, breakdown and full-load torques.

Figure 17-22b
Typical torque-speed characteristic of a 2000 hp, 3-phase squirrel-cage induction motor. Note the low starting torque, and almost constant speed before the breakdown torque is reached.

has a typical no-load speed of 1799 r/min. This represents a slip of (1800 − 1799)/1800 = 0.00055, or about 0.05 percent. Under these conditions, the current in the rotor bars is very small. Thus, as far as the stator is concerned, it is as if the rotor bars were open-circuited. Consequently, the no-load current drawn by the stator is mainly that needed to produce the revolving magnetic field. Thus, at no-load the motor draws mainly reactive power from the line, and so the power factor is low. The no-load current is typically 30 percent to 50 percent of the full-load current, and the power factor is about 20 percent.

17.11 Squirrel-cage motor under load

Suppose an induction motor is initially running at no-load. If we uddenly apply a mechanical load, the motor will begin to slow down. This causes the magnetic flux to cut across the rotor bars more quickly. The voltage induced in the rotor bars increases progressively, as does the rotor current. As a result, the motor develops more and more torque. Eventually, when the motor has slowed down enough, the torque it develops becomes equal to the load torque. The speed then remains stable (see Section 1-15).

The full-load speed is only slightly less than synchronous speed. Thus, a 1/4 hp motor having a synchronous speed of 1800 r/min has a typical full-load speed of 1725 r/min. Similarly, a 5000 hp motor has a typical full-load speed of 1780 r/min. Thus, induction motors are often called *constant-speed* machines because the change in speed from no-load to full-load is very small.

Between no-load and full-load, the magnetic field in the motor is practically constant, and so the reactive power (kvar) needed to produce it is practically constant. However, the active power (kW) absorbed by the motor increases in almost direct proportion to the mechanical load. Consequently, the power factor of an induction motor improves as the load increases. Figure 17-23 shows the typical power factor and efficiency curve of a 2 hp motor as the load increases from zero to rated output.

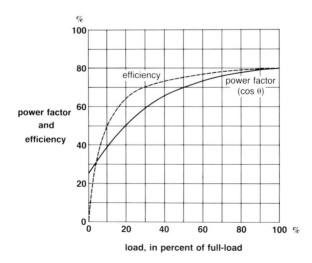

power factor and efficiency

load, in percent of full-load

Figure 17-23
Typical efficiency and power factor of a 2 hp, 3-phase induction motor.

This curve shows that we should never select a 100 hp motor to do a 20 hp job. The reason is that when a motor operates below its rated output, the power factor and efficiency are always below their best values.

17.12 Squirrel-cage motor under overload

If we increase the mechanical load beyond the rated capacity of the motor, both the rotor and stator begin to overheat, which can reduce the useful life of the motor. However, as long as the excessive temperature does not last too long, no harm is done.

There is a limit, however, to the overload we can apply. If the load torque exceeds the so-called *breakdown torque* of the motor, the machine will stop and the circuit breakers will trip because of the resulting high current. The breakdown torque (Fig. 17-22) ranges between 1.5 and 2.5 times rated torque, depending upon the design of the motor.

17.13 Effect of rotor resistance

The resistance of the rotor has an important effect on the torque-speed characteristic of a three-phase induction motor. Fig. 17-24 shows the torque-speed curve of a standard 5 hp, 1760 r/min, 208 V motor having copper conductors in the rotor. If brass is used instead of copper, the torque-speed curve changes considerably, as shown. The locked-rotor torque increases from 30 N.m to 60 N.m and, at the same time, the locked-rotor current *decreases* from 90 A to 75 A. This is a big advantage because we get more torque while drawing less current. Unfortunately, at rated torque, the speed falls from 1760 r/min to 1600 r/min when brass is used. In most cases, this decrease in speed is a disadvantage. Furthermore, the efficiency drops from 85 percent to 80 percent.

Thus, it would be desirable to have a motor whose rotor resistance is high under locked-rotor conditions and low when the motor runs at rated speed. This objective can be met by using a wound-rotor induction motor.

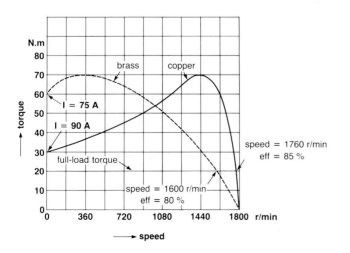

Figure 17-24
Effect on the torque-speed characteristic when brass instead of copper is used in the rotor winding.

17.14 Wound-rotor induction motor

A wound-rotor induction motor (Figs. 17-25 and 17-26) is identical to a squirrel-cage motor except that the rotor contains a three-phase winding. The winding is similar to the stator winding and is usually connected in wye. The rotor winding is placed in the rotor slots but is insulated from the iron core. The ends of the winding are connected to three slip rings mounted on the shaft. Brushes ride on the slip rings so that stationary external resistors can be connected in series with the rotor winding (Fig. 17-27).

The motor is started with the three-phase rheostat in its position of maximum resistance. As the motor accelerates, the resistance is gradually reduced, and finally made zero when the motor reaches rated speed.

If required, the speed of the motor can also be varied under load by changing the external resistance. However, this is done only in relatively small machines (below 100 hp) because the high I^2R losses produce a low efficiency.

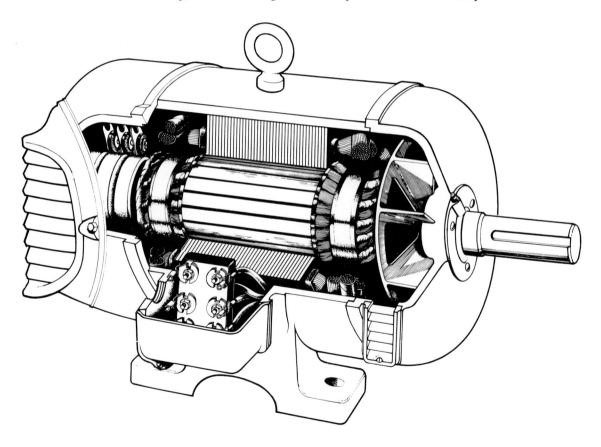

Figure 17-25
Cutaway view of a wound-rotor induction motor. The stator is identical to that in a squirrel-cage motor. (*Courtesy Brook Crompton Parkinson*).

One of the important uses of wound rotor motors is to drive high-inertia loads. Such loads take a long time to come up to speed, and an ordinary squirrel-cage motor would either burn out its windings or melt the rotor bars during start-up. With a wound-rotor motor, we can limit the starting current to a low value and still obtain a good starting torque. The reduced current prevents the motor from overheating. Once the high-inertia load is up to speed, the external resistors are cut out of the circuit.

17.15 Sector motor

Consider a standard 3-phase, 4-pole induction motor having a synchronous speed of 1800 r/min. Let us cut the stator in half so that half the stator winding is removed and only two complete poles (per phase) are left. Let us connect the three phases in wye without making any other change to the existing coil connections. This will

Figure 17-26
Exploded view of a wound-rotor induction motor. The three slip rings can be seen on the left-hand side of the rotor. The brushes are mounted inside the left-hand end bell. (*Courtesy Brook Crompton Parkinson*).

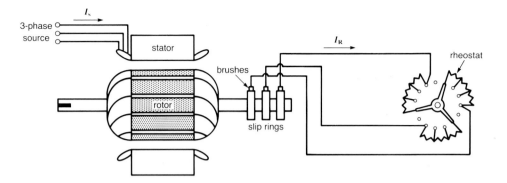

Figure 17-27
Starting and controlling the speed of a wound-rotor induction motor using an external rheostat.

give us a so-called *sector stator*, shown in Fig. 17-28. Let us mount the original rotor above this sector stator, leaving a small air gap.

If we connect the stator terminals to a 3-phase, 60 Hz source, the rotor will again turn at slightly below 1800 r/min. The reason is that the sector motor produces a "revolving" flux that moves at the same linear speed as the flux in the original three-phase motor. However, instead of making a complete turn, the field simply travels continuously from one end of the stator to the other.

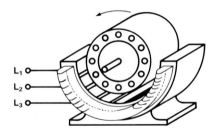

Figure 17-28
Sector stator causes the squirrel-cage rotor to rotate.

17.16 Linear induction motor

The sector stator could be laid out flat without affecting the linear speed of the magnetic field. Such a flat stator produces a field that moves at constant speed in a straight line.

If a flat plate is brought near the flat stator, the traveling field drags the plate along with it, as was explained in Section. 17-1. In practice, we generally use a simple aluminum or copper plate as a "rotor" (Fig. 17-29). The combination of flat stator and flat rotor is called a *linear induction motor*. The direction of the motor can be reversed by interchanging any two stator leads.

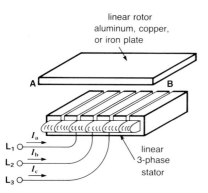

linear rotor
aluminum, copper,
or iron plate

A
B

L_1
I_a
L_2
I_b
L_3
I_c

linear
3-phase
stator

Figure 17-29
Linear induction motor produces a moving
field that drives the motor in a straight line.

Example 17-3:

The stator of a 600 hp, 3-phase, 60 Hz induction motor has six poles per phase and an internal diameter of 460 mm. Calculate the speed at which the flux travels along the internal surface of the stator.

Solution:

a. The synchronous speed of the flux is:
    $$n_s = 120 \, f/p = (120 \times 60)/6 = 1200 \text{ r/min}$$ Eq. 17-2

b. the length L of the internal circumference is
    $$L = \pi d = 3.1416 \times 460 = 1445 \text{ mm} = 1.445 \text{ m}$$

c. the flux makes 1200 turns per minute, and so it covers a distance of $1200 \times 1.445 = 1734$ meters per minute, or $(1734 \times 60) = 104\,050$ m/h

d. knowing that 1609 m = 1 mile, the linear speed v_s is
    $$v_s = 104\,050 / 1609 = 64.7 \text{ miles per hour.}$$

This means that if this particular stator were laid out flat, it would become a linear motor having a synchronous speed of 64.7 miles per hour. By changing the design of the windings, it is possible to obtain linear synchronous speeds of several hundred miles per hour.

17.17 Summary

In this chapter, we saw another practical application of Faraday's law ($E = BLv$) and of the force on a conductor ($F = BLI$). In effect, whenever a flux moves across a metal surface, the induced voltage E produces a current I. Because the current is itself in the path of the flux, a force is produced that tends to drag the metal along with the flux.

The moving flux is conveniently obtained by making use of a three-phase source. Thus, the induction motor embodies some of the most useful applications of the three-phase theory we have covered so far. It is called an induction motor because power flows from the stator to the rotor by electromagnetic induction across the air gap. Unlike the dc motor, the rotor is not connected to the power supply. As a result, we obtain a motor that has no brushes, no commutator, and a much simpler winding.

We saw that there are two basic types of induction motors: squirrel-cage and wound-rotor. The latter is more expensive, but it permits us to vary the resistance in the rotor circuit so as to obtain a high starting torque and high efficiency at full-load.

Finally, we learned that a linear induction motor is a particular case of an induction motor that has a flat stator. In the next chapter, we will investigate some of the practical applications and torque-speed features of induction motors.

TEST YOUR KNOWLEDGE

17-1   If a magnet moves swiftly across a sheet of non-magnetic material it will drag the sheet along with it.

☐   true   ☐   false

17-2   The field in a motor rotates at 3600 r/min, and the rotor turns in the same direction at 2700 r/min. The slip is

a. 0.33   b. 0.25   c. 2.5 %

17-3   The field in a motor rotates at 3600 r/min, and the rotor turns in the *opposite* direction at 2700 r/min. The slip is

a. − 1.75   b. − 0.25   c. 1.75

17-4   In Fig. 17-9, if the frequency of the source is 120 Hz, the speed of the rotating field is

a. 1800 r/min   b. 7200 r/min   c. 2400 r/min

17-5   A three-phase motor having eight poles per phase is connected to a source whose frequency is 50 Hz. The revolving field turns at

a. 250 r/min   b. 750 r/min   c. 750 r/s

17-6   The locked-rotor current relates to the condition in which the rotor is not turning.

☐   true   ☐   false

17-7   A squirrel-cage motor rated at 440 V, 3-phase, 60 Hz runs at a no-load speed of 1798 r/min. If the applied voltage is reduced to 220 V, the new speed will

a. drop to about 900 r/min   b. decrease only slightly
c. not change at all

17-8   A 3-phase, 60 Hz induction motor has a full-load speed of 252 r/min. The number of poles per phase is

a. 28   b. 29   c. 30

17-9   The copper bars of a squirrel-cage rotor are replaced by aluminum. The locked-rotor torque will

a. decrease   b. increase   c. remain the same

QUESTIONS AND PROBLEMS

17-10   Name the principal components of an induction motor

17-11   Explain how a revolving field is set up in a three-phase induction motor.

17-12   Why does the rotor of an induction motor turn slower than the revolving field?

17-13   Describe the principle of operation of a linear induction motor.

17-14   How can we change the direction of a three-phase motor?

17-15   One of the bars in a squirrel-cage rotor carries a current of 600 A. The bar is 23 cm long, and it is momentarily in a field whose flux density is 0.5 T. Calculate the force on the bar, in newtons and in pounds force.

17-16   a. Calculate the synchronous speed of a 3-phase, 14-pole induction motor excited by a 60 Hz source.

b. What is the full-load speed if the slip is 6 percent?

c. What is the frequency of the current in the rotor bars at full-load?

CHAPTER 18

Selection and Application of Three-Phase Induction Motors

Introduction and Chapter Objectives

When we select a three-phase induction motor for a particular application, we often discover that several types can fill the need. Consequently, we have to make a choice. The selection is generally simplified because the manufacturer of the lathe, fan, pump, and so forth indicates the type of motor best suited to drive the load. Nevertheless, it is useful to know something about the special construction and characteristics of the various types of three-phase induction motors on the market.

In this chapter, we will discuss the more important kinds of induction motors and how they operate under both normal and abnormal conditions.

18.1 Standardization and classification of induction motors

Industrial motors under 500 hp all have standardized dimensions. Thus, a 60 hp, 1725 r/min, 60 Hz motor of one manufacturer can be replaced by that of any other manufacturer without having to change the mounting holes, shaft height, or type of coupling. The standardization covers not only frame sizes, but also establishes minimum requirements on starting torque, locked-rotor current, overload capacity, and temperature rise. These standards are set by organizations such as ANSI (American National Standards Institute), IEEE (Institute of Electrical and Electronics Engineers), and NEMA (National Electrical Manufacturers Association)*.

---

*   In Canada, the CSA (Canadian Standards Association) is largely responsible for setting standards.

18.2 Classification according to environment and cooling methods

Motors are grouped into several distinct categories, depending upon the environment in which they have to operate. We limit our discussion to five important classes.

1. *Drip-proof motors*. The frame protects the windings against liquid drops and solid particles that fall at any angle from 0 to 15 degrees from the vertical. The motors are cooled by a fan directly coupled to the rotor. Cool air is drawn into the motor through vents in the frame, blown over the windings, and then expelled. Drip-proof motors can be used in most locations (Figs. 18-1 and 18-2).

Figure 18-1
Drip-proof induction motor rated 450 hp, 1770 r/min, 3-phase, 60 Hz. (*Courtesy Gould*)

Figure 18-2
Drip-proof induction motor in a particularly clean industrial environment. (*Courtesy General Electric*)

2. *Splash-proof motor*. The frame protects the windings against liquid drops and solid particles that fall at any angle from 0 to 100 degrees from the vertical. Cooling is similar to that in drip-proof motors. These motors are mainly used in wet locations.

3. *Totally enclosed, non-ventilated motors*. These machines have closed frames which prevent the free exchange of air between the windings and the outside atmosphere. They are designed for very wet or dusty locations. Such motors are usually rated below 10 hp because it is difficult to get rid of the heat. The motor losses are dissipated by natural convection and radiation from the frame.

4. *Totally enclosed, fan-cooled motors*. Medium- and high-power motors that are totally enclosed are usually cooled by an external blast of air. An external fan, directly coupled to the shaft, blows air over the ribbed motor frame. A concentric outer shield prevents physical contact with the fan and serves to channel the air stream over the frame (Figs. 18-3 and 18-4).

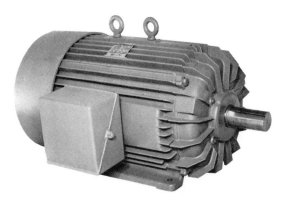

Figure 18-3
Totally enclosed fan-cooled induction motor rated 350 hp, 1765 r/min, 60 Hz. Air blows over the ribbed frame from the rear to the shaft-end of the motor. The ribs improve the heat transfer from the frame. (*Courtesy Gould*)

Figure 18-4
Totally enclosed fan-cooled induction motor rated 25 hp, 3515 r/min, 230 V/460 V, 3-phase, 60 Hz for a vertical pump. (*Courtesy General Electric*)

5. *Explosion-proof motors*. These motors are used in highly inflammable or explosive surroundings such as in coal mines, oil refineries, and grain elevators. The motors are totally enclosed (but not airtight) and the frames are designed to withstand the enormous pressure that may build up inside the motor if an internal short-circuit should occur and to prevent the arc from igniting the surrounding atmosphere (Figs. 18-5 and 18-6).

Figure 18-5
Explosion-proof induction motor, totally enclosed, fan cooled. (*Courtesy Siemens-Allis*)

Figure 18-6
This 50 hp totally-enclosed, fan cooled, explosion-proof motor has given decades of service in a dusty and hazardous environment. (*Courtesy Canadian General Electric*).

18.3 Classification according to electrical and mechanical properties

In addition to special enclosures, three-phase squirrel-cage motors can be designed to produce various electrical and mechanical characteristics, as listed below.

1. *Motors with standard locked-rotor torque (NEMA Design B).* Most induction motors belong to this group. The locked-rotor torque depends upon the size of the motor and ranges from 130% to 70% of full-load torque as the power increases from 20 hp to 200 hp (15 kW to 150 kW). The corresponding locked-rotor current should not exceed 6.4 times the rated full-load current. These general-purpose motors are used to drive fans, centrifugal pumps, machine-tools, and so forth.

Example 18-1:

A 75 hp, 460 V, 3-phase, 60 Hz, 880 r/min squirrel-cage induction motor has a NEMA Design B rating. The full-load efficiency is 92 % and power factor is 81.5 %. Calculate:

1. the slip at full-load

2. the full-load torque, in ft.lbf

3. the active power consumed by the motor

4. the apparent power absorbed by the motor

5. the full-load current

6. the reactive power absorbed by the motor

Solution:

1. a) The synchronous speed of the motor must be close to its rated speed of 880 r/min

    b) The nearest synchronous speed corresponds to that of an 8-pole motor:
        $n_s = 120 \, f/p = 120 \times 60/8 = 900$ r/min.   Eq. 17-2

    c) The full-load slip is $s = (n_s - n)/n_s$ Eq. 17-1
        $$= (900 - 880)/900 = 0.0222, \text{ or } 2.22 \text{ \%.}$$

2. The full-load torque is:
    $$T = 5252 \, P/n = (5252 \times 75)/880 = 447.6 \text{ ft.lbf} \qquad \text{Eq. 1-10}$$

3. The active power P_i consumed by the motor is
    $$P_i = 100 \, P_o/\text{eff} = (100 \times 75 \times 746)/92 = 60 \, 815 \text{ W} = 60.8 \text{ kW}$$

    Eq. 1-11

4. The apparent power S absorbed by the motor is
    $$S = P_i/\cos \theta = 60.8/0.815 = 74.6 \text{ kVA} \qquad \text{Eq. 15-2}$$

5. The full-load current drawn by the motor is
$$I = S/(1.732E) = 74\ 600/(1.732 \times 460) = 93.6\ \text{A} \qquad \text{Eq. 15-3}$$

6. The reactive power Q absorbed by the motor is
$$Q = \sqrt{S^2 - P_i^2} = \sqrt{74.6^2 - 60.8^2} = 43.2\ \text{kvar} \qquad \text{Eq. 14-20}$$

This reactive power creates the magnetic field in the motor.

2. *High starting-torque motors (NEMA Design C).* These motors are used when starting conditions are difficult. Pumps and piston-type compressors that have to start under load are two typical applications. In general, these motors have a special double-cage rotor. In the range from 20 hp to 200 hp, the locked-rotor current should not exceed 6.4 times the rated full-load current.
    The excellent torque-speed characteristic of a double-cage rotor (Fig. 18-7) is based upon these facts:

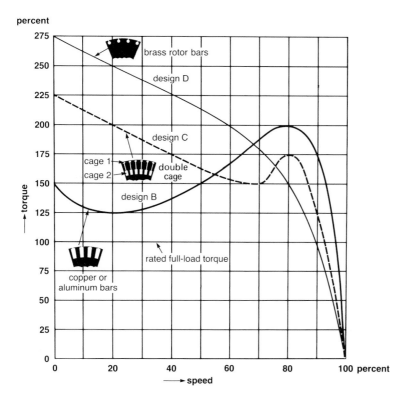

Figure 18-7
Torque-speed curves of three types of squirrel-cage induction motors.

a) the frequency of the rotor current diminishes as the motor speeds up and

b) a conductor that lies close to the rotor surface (cage 1) has a lower inductive reactance than one buried deep inside the iron core (cage 2).

When the motor is starting up, most of the rotor current flows in cage 1 because of its relatively low reactance. Thus, cage 2 is effectively out of service during this period, and because the small rotor bars of cage 1 produce a relatively high rotor resistance, the starting torque is high. But as the motor approaches synchronous speed, the frequency of the current in the rotor falls. As a result, the reactance becomes negligible, and current begins to flow in cage 2, which has a much lower resistance then cage 1. Thus, the effective rotor resistance at rated speed is much lower than at standstill. For this reason, the double-cage rotor develops both a high starting torque, and a low slip at full-load (see Section 17.13).

Example 18-2:

A 3 hp, 575 V, 3-phase, 60 Hz, 1080 r/min induction motor has the NEMA Design C torque-speed characteristic shown in Fig. 18-7. Calculate the actual torque (in N.m) and speed at the 40 percent speed point.

Solution:

a) The synchronous speed that is closest to 1080 r/min corresponds to a 6-pole motor, for which n_s = 1200 r/min

b) The 40 percent speed corresponds to (40/100) × 1200 = 480 r/min

c) The full-load torque is given by Eq. 1-9:
$T = 9.55\ P/n = (9.55 \times 3 \times 746)/1080 = 19.8$ N.m.

d) From the curve, the torque is 200 percent at 40 percent speed; the torque at 480 r/min is therefore
$T = (200/100) \times 19.8 = 39.6$ N.m

3. *High-slip motors (NEMA Design D).* The rated speed of high-slip motors lies between 85% and 95% of synchronous speed. These motors develop high starting torques and can be used to drive high-inertia loads (such as centrifugal dryers) which take a relatively long time to reach full speed. The high-resistance squirrel cage is made of brass, and the motors are usually designed for intermittent operation to prevent overheating. The large drop in speed with increasing load is also ideal to drive impact machine tools, such as shears and punch presses equipped with a flywheel. The flywheel gradually stores up mechanical energy during the idling period and releases it in a surge of power when the machine-tool impacts. The power delivered by the flywheel during the short impact period is usually many times the nameplate horsepower rating of the motor.

The graphs of Fig. 18-7 enable us to compare the torque-speed characteristics of these motors. The rotor construction is also shown, and it can be seen that the distinguishing properties are obtained by changing the rotor design.

Example 18-3:

A 1.5 hp, 200 V, 3-phase, 60 Hz, 1620 r/min high-slip motor has the following characteristics:

full-load current = 5.1 A
locked-rotor current = 46 A
locked-rotor torque = 250 percent

We want to rewind this motor so that it can function on a 460 V, 3-phase, 60 Hz line. Calculate:

1. how the properties of the rewound motor compare with the original machine.

2. the winding changes that must be made.

Solution:

1. a) The horsepower rating, the efficiency, power factor, and torque-speed characteristic are not affected when a motor is rewound for a different voltage.

    b) It follows that the apparent power at full-load is constant. Thus, when the voltage rating increases, the motor current must decrease in direct proportion. The full-load current at 460 V is

    $$I_{FL} = 5.1 \times (200/460) = 2.22 \text{ A}$$

    c) The new locked-rotor current is

    $$I_{LR} = 4 \times (200/460) = 20 \text{ A}$$

2. a) The full-load current is only 2.22 A, compared with the original value of 5.1 A. Consequently, the conductor size of the motor windings can be reduced by a factor of 5.1/2.22 = 2.3. For example, if the original motor was wound with No. 17 ½ wire (0.924 mm^2) the new winding would be made with No. 21 wire (0. 411 mm^2).

    b) The second important change is to increase the number of turns per coil in the ratio of the voltages, that is, 460/200 = 2.3. Thus, if the coils in the original motor were made of 13 turns of No. 17 ½ wire, the new coils would have (13 × 2.3) = 30 turns of No. 21 wire. In effect, the volts per turn in the rewound motor must be the same as before.

18.4 Choice of motor speed

The synchronous speed of induction motors changes by jumps, depending upon the frequency and number of poles. Consequently, the choice of motor speed is rather limited. For example, it is impossible to build an efficient 2400 r/min, 60 Hz induction motor. For good efficiency, the motor must run close to synchronous speed; thus, at either 1800 or 3600 r/min.

The speed of a motor is obviously determined by the required speed of the load. At low speeds, it is often preferable to use a high-speed motor and a gear box instead of directly coupling a low-speed motor to the load. There are several advantages to the gear-box approach:

1. For a given power output, a high-speed motor is smaller and costs less than a low-speed motor. Furthermore, its efficiency and power factor are higher.

2. High-speed motors always have a greater locked-rotor torque (as a percentage of full-load torque) than that of low-speed motors of similar type and power.

When equipment has to operate at very low speeds (100 r/min or less), a gear box is unavoidable. The gears are often an integral part of the motor, making for a very compact unit (Fig. 18-8).

On a 60 Hz system, a gear box is also required when equipment has to run above 3600 r/min. For example, in one industrial application, a large gear unit is used to couple a 5000 r/min centrifugal compressor to a 3000 hp, 3560 r/min induction motor.

Figure 18-8
Three-phase gear motor rated 3 hp, 125 r/min, 460 V, 3-phase, 60 Hz.
The motor speed is 1750 r/min. (*Courtesy Reliance*).

Example 18-4:

A 1 hp, 1740 r/min induction motor is coupled to a gear box having a gear ratio of 200. Calculate the full-load output torque and speed.

Solution:

a) the shaft output speed is 1740/200 = 8.7 r/min

b) the output power is slightly less than 1 hp because of the losses in the gear box. If we assume a loss of 10 percent, the output power at the low-speed shaft is 0.9 hp.

c) The output torque is
$$T = 5252 \, P/n = (5252 \times 0.9)/8.7 = 543 \text{ ft.lbf} \qquad \text{Eq. 1-10}$$

Note that this torque exceeds the torque developed by the 75 hp motor in Example 18-1.

In some applications (Fig. 18-9), it is convenient to use a motor that can operate at two speeds. One obvious way to accomplish this result is to place two separate three-phase windings in the stator. For example, if a 60 Hz motor has a 2-pole and a 6-pole winding it can operate at either 3600 or 1200 r/min. It is clear that for a

Figure 18-9
Multi-speed induction motors are often used to drive hoists and cranes in port facilities, and on in ships equipped with ac power. (*Courtesy Brown Boveri*)

given horsepower rating, such a two-winding motor will be bigger than a conventional single-speed motor.

However, when the speed ratio is 2:1, it is possible to use a single winding, and to obtain the required pole change by simply reconnecting a few external leads. Because there is never an idle winding, such two-speed machines are smaller and therefore cheaper. The single winding is sometimes called a Dahlander winding.

In two-speed blower applications, a speed ratio of 2:1 is too large a spread. In such cases, a so-called *pole-amplitude modulated* (PAM) winding can be reconnected to produce either 8 or 10 poles. Such PAM motors are built in sizes up to several hundred horsepower.

18.5 Plugging an induction motor

If we want to bring an induction motor to a rapid stop, we can simply interchange two stator leads. This causes the revolving field to suddenly turn in the opposite direction and therefore opposite to the rotation of the rotor (Fig. 18-10). During this so-called *plugging* period, the motor acts as a *brake*.

It absorbs kinetic energy from the still-revolving load, causing its speed to fall. The energy is entirely dissipated as heat in the rotor. Unfortunately, the rotor also

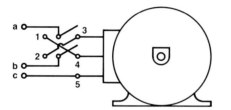

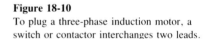

Figure 18-10
To plug a three-phase induction motor, a switch or contactor interchanges two leads.

continues to receive power from the stator, and this power is also dissipated as heat. Consequently, plugging produces very high I^2R losses in the rotor. They are even greater than the I^2R losses when the rotor is locked. Motors should not be plugged too frequently because high rotor temperatures may melt the rotor bars or overheat the stator winding.

Example 18-5:

A 50 hp, 575 V, 3-phase, 60 Hz, 1765 r/min motor is repeatedly plugged, as shown in Fig. 18-11. The motor is insulated with class B (120°C) insulation, for which the maximum permissible temperature rise (by resistance) is 80°C. The winding is made of copper.

To determine the temperature of the winding after 10 plugging cycles, the resistance of the winding was measured between terminals 4 and 5 immediately after the last reversal. The result gave 0.94 Ω, and the ambient temperature was 26°C.

The motor was then disconnected from the line and allowed to cool to room temperature during a 24-hour period. The resistance was again measured (0.62 Ω), as was the ambient temperature ($t_c = 23°C$). Calculate the temperature of the winding after the plugging period, and state whether the temperature rise is excessive.

Solution:

a) The resistance of the copper winding at 0°C would be:

$$R_o = R_t/(1 + \alpha t_c) \qquad\qquad \text{Eq. 4-3}$$
$$= 0.62/(1 + 0.00427 \times 23) = 0.564 \ \Omega$$

b) the temperature t_H of the winding after the plugging test is given by:

$$R_t = R_o (1 + \alpha t_H)$$
$$0.94 = 0.564 (1 + 0.00427 \ t_H)$$
$$t_H = 156°C$$

c) the ambient temperature at the time of the plugging test was 26°C. Thus, the temperature *rise* is $(156 - 26) = 130°C$. This temperature rise far exceeds the 80°C limit for Class B insulation. The motor is definitely running too hot.

18.6 Effect of inertia

High-inertia loads put a heavy strain on induction motors because they prolong the starting period. The starting current in both the stator and rotor is high during this interval, and so overheating is a major problem. To limit the temperature rise, induction motors are often started on reduced voltage. This reduces the power and current which the motor draws from the line. Unfortunately, it also increases the time it takes to bring the load up to speed, but this is usually not too important.

When very high inertia loads have to be accelerated (or brought to a rapid stop), wound-rotor motors are recommended. The reason is that most of the heat produced in the rotor circuit is dissipated by the external resistors. Furthermore, we can maintain a consistently high torque by gradually varying the external rotor resistance during the acceleration or deceleration periods.

18.7 Braking with direct current

An induction motor and its high-inertia load can also be brought to a quick stop by disconnecting the three-phase line and circulating dc current in the stator winding (Fig. 18-11). Any two stator terminals can be connected to the dc source.

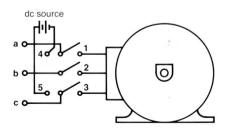

Figure 18-11
A three-phase induction motor can be braked by connecting the stator to a dc source.

The direct current produces stationary N and S poles in the stator. The number of poles created is equal to the number of poles the motor develops normally. Thus, a 3-phase, 4-pole induction motor produces four dc poles no matter how the motor terminals are connected to the dc source.

As the rotor sweeps past the stationary field, a voltage is induced in the rotor bars. The voltage produces a current, and the resulting rotor I^2R losses are dissipated at the expense of the kinetic energy stored in the revolving parts. The motor finally comes to rest when all the kinetic energy has been dissipated as heat in the rotor. (this is a practical application of the principle discussed in Section 17-1.)

The advantage of dc braking is that it produces far less heat than does plugging. In effect, the energy dissipated in the rotor is only equal to the original kinetic energy stored in the revolving masses. The energy dissipated in the rotor is independent of the magnitude of the dc current. However, a smaller dc current increases the braking time.

18.8 Abnormal conditions

Abnormal motor operation may be due to internal causes (short-circuit in the stator, overheating of the bearings, etc.) or to external disturbances. External problems may be caused by:
1. mechanical overload,
2. supply voltage changes,
3. single phasing, and
4. frequency changes.

We will now examine the effect of these external disturbances.

18.9 Mechanical overload

Although standard induction motors can develop at least 1.5 times their rated power output for short periods, they should not be allowed to run continuously beyond their rated capacity. Overloads produce stator and rotor currents that are above the rated values. The consequent overheating deteriorates the insulation and reduces the useful life of the motor. In practice, the higher motor current causes thermal overload relays to trip, bringing the motor to a stop before its temperature rises too much.

Some drip-proof motors are designed to carry a continuous overload of 15 percent. This overload capacity is indicated on the nameplate by the so-called *service factor* 1.15.

18.10 Line voltage changes

According to national standards, a motor shall operate successfully on any voltage within $\pm 10\%$ of the nominal voltage and for any frequency within $\pm 5\%$ of the nominal frequency.

A change in line voltage has an important effect on the torque-speed curve of the motor because the torque at any speed is proportional to the *square* of the applied voltage. Thus, if the line voltage drops 10%, the torque will fall 20%. Such voltage drops are often produced during start-up because of the large starting current.

On the other hand, if the line voltage is too high, the magnetic flux in the motor rises above its normal level. This increases both the iron losses and the magnetizing current, with the result that the temperature increases slightly, and the power factor is somewhat reduced.

Finally, a slight imbalance of the three-phase line voltage can produce a serious unbalance of the three line currents. This condition increases the stator and rotor losses, yielding a higher temperature. A voltage imbalance of as little as 3.5% can cause the temperature to increase by 15°C. The utility company should be notified whenever the phase-to-phase line voltages differ more than 1 percent.

18.11 Single-phasing

If one line of a three-phase line is accidentally opened, or if a fuse blows while the motor is running, the machine will continue to run as a *single-phase motor*. The current drawn from the remaining two lines will almost double, and the motor will begin to overheat. The thermal relays protecting the motor will eventually trip the circuit breaker, thereby disconnecting the motor from the line.

The torque-speed curve is seriously affected when a three-phase motor operates on single phase. The breakdown torque decreases to about 40% of its rated value, and the motor develops no starting torque at all. Consequently, a fully loaded three-phase motor may simply stop if one of its feeder lines is accidentally opened.

18.12 Frequency variation

Important frequency changes never take place on a large electric utility system, except during a major disturbance. However, the frequency may vary significantly on isolated systems in which electrical energy is generated by diesel engines or gas turbines. The emergency supply in a hospital, the electrical system on a ship, the generators in a lumber camp, etc., are examples of this type of supply.

The most important consequence of a frequency change is the resulting change in speed: if the frequency drops 20%, the speed of the induction motor drops 20%.

Machine tools and other motor-driven devices imported from countries where the frequency is 50 Hz may cause problems when they are connected to a 60 Hz system. Everything runs 20% faster than normal, and this may not be acceptable in some applications. In such cases, we have to gear down the motor speed or supply an expensive auxiliary 50 Hz source.

A 50 Hz motor operates well on a 60 Hz line, but its terminal voltage should be raised to 60/50 (or 120%) of the nameplate rating. The new breakdown torque is then equal to the original breakdown torque, and the starting torque is only slightly reduced. Power factor, efficiency, and temperature rise remain satisfactory.

A 60 Hz motor can also operate on a 50 Hz line, but its terminal voltage should be reduced to 50/60 (or 83%) of its nameplate rating. The breakdown torque and starting torque are then about the same as before, and the power factor, efficiency, and temperature rise remain satisfactory.

18.13 Typical applications of induction motors

The following list of typical applications show how induction motors are used in industry:

1. In Montana, a 900 hp, 2300 V, 3-phase, 60 Hz, 3600 r/min squirrel cage induction motor is used to drive a pipeline pump. The motor is weather-protected and is designed to operate in a 40°C ambient with a maximum temperature rise of 80°C by resistance.

2. In Sydney, Australia, a 12 000 hp, 11 kV, 3-phase, 50 Hz, 1490 r/min wound-rotor induction motor drives the feedwater pump in a thermal generating station. The motor is totally enclosed, and the closed-circuit cooling air gives up its heat to an air/water heat exchanger.

3. In Stuttgart, West Germany, a 3300 kW, 5000 V, 3-phase, 425 to 595 r/min wound-rotor induction motor pumps water from Lake Constance to meet the needs of the city. The motor is totally enclosed, and a closed-circuit air/water heat exchanger ensures adequate cooling.

4. In Cleveland, Ohio, an overhead crane in a plant is propelled along its supporting I-beams by two linear three-phase motors. The stators are mounted on the crane a few millimeters away from the flat I-beam, which acts as the stationary rotor. The motors develop a thrust of 300 lbf.

5. The crane on a ship is driven by a 3-speed pole-changing induction motor rated at 19/7/3.3 hp, with synchronous speeds of 1800/900/450 r/min. A single winding is used for the 1800/900 r/min speeds and a second winding for the 450 r/min. The motor is totally-enclosed, non-ventilated, and integrated with a gear box.

6. In Ontario, Canada, twelve 1500 hp, 3600 r/min squirrel-cage induction motors drive the heat transport pumps of the Pickering nuclear generating station. The 4.16 kV, 3-phase, 60 Hz motors are totally enclosed with an air/water heat exchanger. The motors are equipped with a heavy flywheel to ensure a brief period of continued operation in the event of a momentary power failure.

18.14 Summary

This chapter has given us a better understanding of the practical application of three-phase induction motors. It has also shown us how the theories of previous chapters fit into the total picture. Thus, we saw that the mechanical aspects reviewed in Chapter 1 become an integral part of the motor. Furthermore, the concepts of active, reactive, and apparent power begin to take on a more interesting form.

We also learned that there are many industrial applications of induction motors, both rotary and linear.

TEST YOUR KNOWLEDGE

18-1   An explosion-proof motor is designed to withstand *external* explosions.

☐ true   ☐ false

18-2   A 500 hp, 3-phase, 2.4 kV squirrel-cage motor has a rated current of 125 A. The locked-rotor current is probably between

a. 700 A and 900 A   b. 600 A and 800 A
c. 40 A and 70 A

18-3   When a flywheel is part of a punch press, the driving motor must have:

a. a high locked-rotor torque   b. a high power factor
c. a high slip at full-load

18-4   Two drip-proof NEMA design B, three-phase induction motors have the following ratings: motor X: 200 hp, 1760 r/min, 2.4 kV; motor Y: 60 hp, 450 r/min, 460 V. motor X is bigger than motor Y.

☐ true   ☐ false

18-5   If a three-phase induction motor has to be plugged frequently, it is best to use

a. a NEMA design D motor   b. a wound-rotor induction motor
c. a wound-rotor induction motor with
external resistors

18-6   A 40 hp, 460 V, 3-phase, 60 Hz induction motor has a full-load rated current of 45 A. The rated full-load current of an identical motor, but rated at 575 V is:

a. 56.25 A   b. 36 A   c. 28.8 A

18-7   A 40 hp, 460 V, 3-phase, 60 Hz induction motor develops a locked-rotor torque of 74 ft.lbf. The torque expressed in newton meters is

a. 100.3 N.m   b. 54.57 N.m   c. 16.64 N.m

18-8   A 5 hp, 460 V, 3-phase, 60 Hz induction motor has a rated speed of 837 r/min, and a locked-rotor torque of 114 N.m. When the motor is connected across-the-line, the terminal voltage falls to 410 V. The locked-rotor torque under these conditions is:

a. 101.6 N.m   b. unchanged   c. 90.56 N.m

18-9   A 50 Hz squirrel-cage induction motor has a rated speed of 419 r/min. It has

a. 16 poles   b. 14 poles   c. 15 poles

18-10   A 200 hp, 460 V, 3-phase, 60 Hz induction motor has to be installed on a 50 Hz system. The recommended 50 Hz voltage is

a. 383 V   b. 552 V   c. 460 V

QUESTIONS AND PROBLEMS

18-11   Explain why a NEMA design B motor is not the best choice to drive a pump.

18-12   Is it possible for a three-phase motor to run on a single-phase line?

18-13   What type of ac induction motor should you recommend for a variable-speed pump?

18-14   A three-phase induction motor can be brought to a quick stop by plugging it. Why *must* the line be disconnected when the motor has come to a halt?

18-15   If a dc current flows in the winding of a three-phase stator, the rotor is very hard to turn. Why?

18-16   Referring to Fig. 18-7, a 60 hp NEMA design B motor has a rated speed of 1152 r/min. Calculate

a.   the rated full-load torque, in N.m

b.   the locked-rotor torque

c.   the breakdown torque and corresponding breakdown speed

d.   the pull-up torque and corresponding speed

e.   the mechanical power developed when the motor is running at 600 r/min

18-17   Draw the typical torque-speed curve of a NEMA design C motor rated at 25 hp, 60 Hz, 1750 r/min. Show the torque in newton meters versus the speed in revolutions per minute.

18-18   The resistance between two terminals of a stator is 3 Ω when the winding temperature is 22°C. Knowing that the winding is made of copper, calculate the resistance for a winding temperature of 95°C.

18-19   A 1.5 hp, 460 V, 3-phase, 60 Hz, 870 r/min induction motor has a full-load efficiency of 77 %, and a power factor of 62 %. Calculate:

    a.   the active power drawn by the motor

    b.   the reactive power absorbed by the motor

    c.   the rated line current

18-20   In Problem 18-19, calculate the impedance per phase at full-load knowing that the stator is connected in wye.

CHAPTER 19

Synchronous Motors

Introduction and Chapter Objectives

Electric motors and generators whose speeds are exactly synchronized with the line frequency are called *synchronous* machines. Synchronous machines have special properties which become particularly useful when the machines are large. Thus, most synchronous motors are rated above 200 hp, while synchronous generators may range up to 1500 MVA.

A synchronous machine can operate as either a motor or a generator, and the transition from one to the other takes place very smoothly without requiring any change in electrical connections. In this chapter, we will study the synchronous machine when it operates as a motor.

19.1 Construction of synchronous motors

A synchronous motor is composed of a rotor and a stator. The stator is identical to that of a three-phase induction motor. Consequently, it is wound to produce 2, 4, 6 or more poles and may be connected in either delta or wye. When the stator is connected to a three-phase source, a revolving field is set up, as in the case of an induction motor. The speed of the revolving field is again given by the equation

$$n_s = \frac{120f}{p}$$ Eq. 17-2

where

n_s = synchronous speed, in revolutions per minute (r/min)

f = frequency, in hertz (Hz)

p = number of poles, per phase

Figure 19-1
Stator of synchronous motor rated 600 hp, 1200 r/min, 2300 V, 3-phase, 60 Hz. (*Courtesy H. Roberge*)

Figure 19-1 shows the typical construction of the stator of a synchronous motor.

The rotor is composed of a set of salient poles that carry identical coils connected in series (Fig. 19-2). A dc current I_x flows through the coils, and the connections are made so that adjacent poles have opposite magnetic polarities. The rotor behaves therefore like a revolving dc field. The exciting current I_x is fed into the rotor by means of a set of brushes and two slip rings.

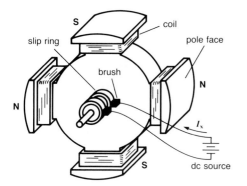

Figure 19-2
Schematic diagram of the rotor of a four-pole synchronous motor. The four coils are connected in series to two slip rings.

Figure 19-3 shows the construction of a 6-pole rotor. The number of poles on the rotor is equal to the number of poles per phase on the stator. Thus, a 6-pole rotor is always associated with a 6-pole stator.

The rotor also carries a squirrel-cage winding, similar to that in an induction motor. This winding is composed of rotor bars that pass through slots in the pole faces. The bars are short-circuited together by means of end rings (Fig. 19-4). In some motors, the squirrel-cage winding does not form a continuous circle around the rotor. The rotor bars are still short-circuited together, but only pole by pole (Fig. 19-6).

When the rotor is mounted inside the stator, a small air gap separates the two. The pole face on the rotor is shaped so that the length of the air gap increases gradually as we move away from the center (Fig. 19-5). The variable length ensures that the ac voltages and currents in the stator remain as sinusoidal as possible.

19.2 Acceleration of a synchronous motor

We will first consider what happens when a 4-pole synchronous motor is connected, at standstill, to a 3-phase, 60 Hz source. The dc winding is not excited but is connected to a low external resistance R (Fig. 19-6). The stator has four poles per phase, and so a 4-pole revolving field will be created that turns at 1800 r/min. Let us assume it turns clockwise.

As the field cuts across the squirrel-cage rotor bars, the resulting induced current will produce a torque, and the rotor will quickly accelerate to a speed close to 1800 r/min. A current is also induced in the dc winding, and so it contributes to some of the accelerating torque. During this starting period, the machine behaves like an ordinary three-phase induction motor.

When the rotor reaches a speed of, say, 1795 r/min, the stator field is moving across the rotor poles at a relative speed of only (1800 − 1795) = 5 r/min. At this low speed, the currents induced in the rotor bars and the dc winding are small, and so the squirrel-cage torque is much weaker than before.

However, as the stator flux slowly moves across the iron pole face, it develops a

Figure 19-3
Partially-assembled rotor of a synchronous motor rated 600 hp, 1200 r/min, 2300 V, 3-phase, 60 Hz. The six salient poles are dovetailed into the central core so as to resist the centrifugal forces. (*Courtesy H. Roberge*)

Figure 19-4
Closeup of the salient pole in Fig. 19-3, showing the rotor bars and end rings. The end rings of adjacent poles are later bolted together to form a closed circle. (*Courtesy H. Roberge*)

Figure 19-5
Relative position of the rotor and stator laminations of a large synchronous motor. The air gap in the center of the pole is smaller than near the pole tips. The deep stator slots carry pre-formed armature coils.

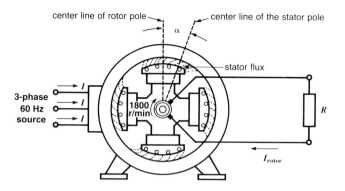

Figure 19-6
Synchronous motor connections during start-up and immediately after it has attained synchronous speed. The induced ac rotor current has fallen to zero and the rotor is carried along by the revolving stator flux because of the reluctance torque. Note that the two brushes ride on two slip rings.

strong force of magnetic attraction. The magnetic force produces a reluctance torque that pulls the rotor into synchronism with the field.* Once pull-in is achieved, the rotor runs at the same speed as the rotating field, namely at 1800 r/min. Thus, the induction motor has become a synchronous motor.

Once the motor attains synchronous speed, the stator flux no longer cuts across the squirrel-cage winding or the dc winding. As a result, the induced voltage (and current) in both windings is zero. However, the rotor poles are carried along by the field because of the reluctance torque. The lines of force are slanted at an angle, as shown in Fig. 19-6, because the windage and friction losses impose a drag on the rotor. This causes the center line of the rotor poles to lag slightly behind the center line of the stator poles. The mechanical angle α between the two center lines increases with increasing mechanical load. However, at no-load, α is only a few degrees.

19.3 Effect of dc excitation

When the motor runs at no-load with no dc excitation, the ac current I drawn from the three-phase line is large, and lags almost 90° behind the voltage. The reason is that a lot of reactive power is needed to create the magnetic field. Consequently, the no-load current I in Fig. 19-6 may be greater than the rated full-load current of the machine.

If we circulate a dc current in the rotor winding in such a way as to create a flux in the same direction as the stator flux, a remarkable thing begins to happen. The stator current I begins to fall. In effect, the mmf created by the rotor reduces the mmf the stator has to develop, in order to produce the required magnetic field. Thus, the more we increase the mmf of the rotor, the less the stator mmf has to be.

---

*   The reluctance torque is produced in exactly the same way as was discussed in Section 5-19.

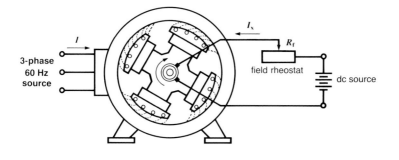

Figure 19-7
The stator current first decreases and drops to a minimum as the dc exciting current is increased. The current then increases again as I_x is further increased. The field rheostat permits control of I_x. Note that the center line of the rotor poles always lags slightly behind the center line of the stator poles.

Consequently, the stator current I decreases as the dc exciting current I_x is raised (Fig. 19-7).

As we continue to raise I_x, less and less reactive power has to be supplied by the stator to produce the required flux. Finally, when I_x is high enough, the stator needs to produce no reactive power at all. Under these conditions, the stator current I is in phase with the stator voltage because only active power is being supplied. The active power keeps the motor running and also supplies its iron and copper losses.

What happens if we continue to raise the dc excitation? The magnetic field tends to become greater than it should be and, to keep it at its original value, the stator mmf must now *oppose* the excess mmf created by the dc field. As a result, as I_x increases, current I again begins to increase. However, I acts in the opposite direction to what it did before, when I_x was small. Consequently, the reactive power becomes *negative*. This means that the synchronous motor is actually feeding reactive power back into the line! Consequently, the motor is now acting like a capacitor as far as the three-phase line is concerned.

Figure 19-8 shows the change in stator current I as the dc exciting current I_x is varied. When $I_x = 0$, $I = 100$ A, and the current lags almost 90° behind the voltage. But as we raise I_x, the stator current begins to fall and eventually reaches a minimum of 2 A when I_x is 5 A. The power factor of the motor is then 100 percent. As we continue to raise I_x, the stator current begins to increase, eventually reaching a value of 50 A when $I_x = 10$ A. The current now leads the voltage by almost 90°.

When the dc current is less than 5 A, the motor draws reactive power from the line. The synchronous machine is said to be *under-excited*, and it behaves like a three-phase inductor. Conversely, when the dc current is greater than 5 A, the synchronous machine feeds reactive power into the line. The machine is said to be *over-excited*, and it behaves like a capacitor.

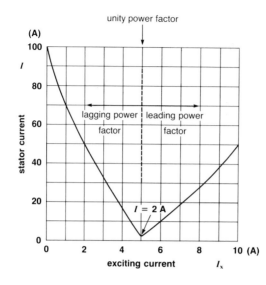

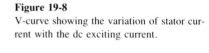

Figure 19-8
V-curve showing the variation of stator current with the dc exciting current.

A synchronous motor that runs at no-load on a three-phase line can therefore act as a continuously variable three-phase inductor or capacitor. The amount of reactive power absorbed (or delivered) by the machine depends upon the level of dc excitation. When the excitation is "normal" (5 A in Fig. 19-8), the machine neither delivers nor absorbs reactive power.

Example 19-1:

A synchronous motor operating on a 3-phase, 4 kV line has the V curve characteristic shown in Fig. 19-8. Calculate:

1. the magnitude of the reactive power for dc excitation currents of 1 A, 5 A, and 10 A.

2. the equivalent capacitance of the motor when $I_x = 10$ A

Solution:

1. a) Referring to Figs. 19-7 and 19-8, for $I_x = 1$ A, $I = 70$ A

    and from

    $$S = 1.732 \, EI$$ Eq. 15-3

    we have

    $$S = 1.732 \times 4000 \times 70 = 485 \text{ kVA}$$

At this level of excitation, the machine is under-excited, and so it draws nearly 485 kvar of reactive power from the line. (The motor is operating at no-load, therefore the active power is negligible compared with the apparent power S of 485 kVA.

Thus, the reactive power is nearly equal to the apparent power S.)

b) For $I_x = 5$ A, we find $I = 2$ A, and so

$$S = 1.732 \ EI = 1.732 \times 4000 \times 2 = 13.85 \text{ kVA}$$

Because I is at its minimum value, the machine absorbs only active power from the line. This power (13.85 kW) is mainly used to overcome the friction, windage, and iron losses (The apparent power S is numerically equal to the active power at unity power factor.)

c) When $I_x = 10$ A, we find $I = 50$ A; therefore

$$S = 1.732 \ EI = 1.732 \times 4000 \times 50 = 346 \text{ kVA}$$

The machine is over-excited, and so it *delivers* nearly 346 kvar to the three-phase line. (The reactive power is again nearly equal to S because the motor absorbs very little active power.)

2. We can calculate the equivalent capacitance of the motor as follows (Fig. 19-9).

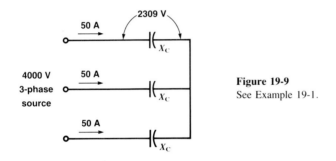

Figure 19-9
See Example 19-1.

a) current in each "capacitor" = 50 A

b) voltage across each "capacitor" = 4000/1.732 = 2309 V

c) capacitive reactance per phase is

$$X_c = E/I = 2309/50 = 46.18 \ \Omega.$$

d) from $X_c = \dfrac{1}{2\pi fC}$ Eq. 13-2

$$C = \frac{1}{2\pi fX_C} = \frac{1}{2\pi \times 60 \times 46.18} = 57.4 \times 10^{-6} = 57.4 \ \mu F$$

Thus, as far as the three-phase line is concerned, the synchronous motor has exactly the same effect as a three-phase, wye-connected capacitor bank composed of three 57.4 μF capacitors. That is why such a machine is sometimes called a synchronous capacitor (Fig. 19-10).

Figure 19-10a
Synchronous capacitor whose reactive power can be varied from − 200 Mvar (supplying
reactive power) to + 300 Mvar (absorbing reactive power). The 8-pole machine operates
at no-load on a 16 kV, 3-phase, 60 Hz line. (*Courtesy Hydro-Québec*)

Figure 19-10b
The synchronous capacitor is used to regulate the voltage of a 735 kV transmission line. It
is enclosed in a steel shell that contains hydrogen to provide an efficient cooling system.
(*Courtesy Hydro-Québec*)

19.4 Synchronous motor under load

Consider the synchronous motor of Fig. 19-11 running at no-load on a three-phase line. The rotor current is adjusted so that the ac line currents are minimum. This means that the machine is drawing only active power from the line, just enough to keep the machine running. We assume that the field created by the stator is turning clockwise.

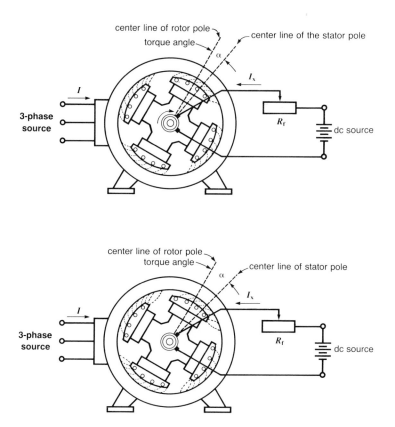

Figure 19-11
Synchronous motor operating at no-load and turning clockwise. The torque angle and stator current are small.

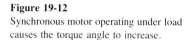

Figure 19-12
Synchronous motor operating under load causes the torque angle to increase.

The poles of the rotor are slightly behind the poles of the stator, and the angular displacement between them is α degrees. The displacement at no-load amounts to only a fraction of a degree. Thus, we can assume that the center lines of the rotor poles and stator poles are lined up.

What happens when a mechanical load is suddenly applied to the rotor? The motor begins to slow down, causing the rotor poles to fall further behind the stator poles. Angle α increases, causing the flux lines between the rotor and stator poles to become more "stretched", and this produces a larger torque. When the torque developed by the motor is equal to the load torque, the so-called *torque angle* α again remains constant. However, its value (Fig. 19-12) is now greater than at no-load. In this new condition, current I drawn by the stator is also greater than before. However, the machine continues to run at synchronous speed because the rotor and stator poles remain locked together. Consequently, the machine still operates as a synchronous motor.

However, there is a limit to the mechanical load that can be applied. As we increase the load torque, the rotor poles lag farther and farther behind the stator poles, and current I becomes greater and greater. However, at a certain critical torque, called *pull-out torque*, the rotor poles break away from the stator poles and the motor starts running below synchronous speed. The motor is said to have "lost synchronism". When this happens, the average torque developed by the motor is only that due to the squirrel-cage winding. In effect, when the motor runs below synchronous speed, it behaves like an induction motor. The stator current increases greatly, and the circuit breakers protecting the machine will immediately trip.

In a 4-pole machine, the pull-out torque occurs at an angle α of about 40°. This condition is shown in Fig. 19-13. Note how the flux lines between the rotor and stator poles are "stretched". They behave like rubber bands between the N and S pole faces (rotor and stator), tending to keep the poles together. However, in the critical position of Fig. 19-13, the "rubber bands" are about to snap.

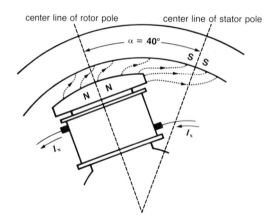

Figure 19-13
Position of rotor poles when a 4-pole synchronous motor has reached its pull-out torque.

19.5 Relationship between torque and torque angle

We have seen that the torque developed by the synchronous motor depends upon the torque angle α. The relationship between the two is given by the graph of Fig. 19-14. Curve 1 applies for a level of dc excitation that produces unity power factor at rated load. In the case of our 4-pole motor, the pull-out torque T_m is 1.6 times the rated torque, and it occurs at an angle of about 40°. The motor continues to develop a positive torque when α exceeds 40°. However, the torque is smaller than T_m, and it falls to zero when $\alpha = 90°$.

The mechanical angle α_m at which pull-out occurs depends upon the number of

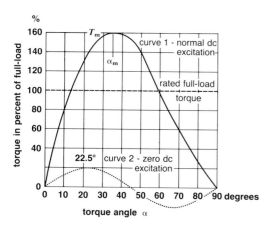

Figure 19-14

Typical torque versus torque angle curves for a 4-pole synchronous motor. Curve 2 is the reluctance torque developed by the motor when the poles are not excited.

poles on the synchronous machine. Thus, in a 2-pole machine α_m is about 90°, while in a 20-pole α_m machine it is about 9°.

Curve 2 shows the torque developed in the 4-pole motor when there is no dc excitation. This reluctance torque reaches a maximum of 20 percent of rated torque, and it occurs at an angle of 22.5°.

19.6 Excitation of synchronous machines

There are several ways to provide the dc excitation for a synchronous machine. One of the simplest is to use a dc generator G called an *exciter* (Fig. 19-15). The exciter output current I_x is varied by changing the exciter field excitation I_o, using a field rheostat R. The main exciting current I_x flows through a set of brushes and slip rings that are mounted on the rotor of the synchronous machine.

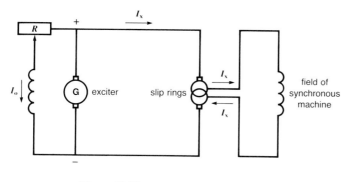

Figure 19-15

The dc excitation is provided by an exciter.

Another means of excitation is to use two dc generators G_1 and G_2 (Fig. 19-16). Generator G_1 is called the *pilot* exciter because it provides the field excitation for the main exciter. Such excitation systems are used when generator G_2 has a rating of several hundred kilowatts. By using a pilot exciter, the small control power E_oI_o associated with I_o can be used to control the much larger power E_xI_x associated with I_x.

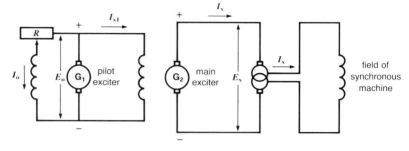

Figure 19-16
Large synchronous machines are sometimes excited by two exciters. The small pilot exciter requires little control power.

A third method of excitation is to use a three-phase generator (also called exciter), and rectify its ac output (Fig. 19-17). The magnitude of I_x can be controlled by varying the dc excitation I_o of the exciter. The advantage of this system is that the brushes formerly associated with the dc generator are eliminated. But we still have the brushes riding on the slip rings of the synchronous machine.

Brushes are a source of trouble because they have to be replaced. Furthermore, in large machines, the wear of dozens of brushes causes carbon dust to be deposited

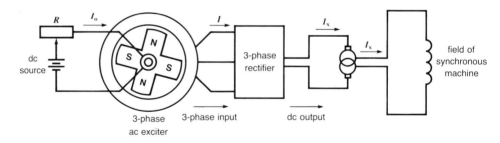

Figure 19-17
Exciting a synchronous machine with an ac exciter and rectifier.

at various places. Such conducting dust creates a serious hazard because it may cause flashovers and short-circuits. As a result, many synchronous machines today are equipped with so-called *brushless* excitation systems.

Figure 19-18 illustrates the basic components of such a system. A three-phase exciter is mounted on the shaft of the synchronous machine, along with a three-phase rectifier. A dc control source (1) provides the dc excitation current I_{x1} for the exciter. The poles (2) of the exciter are stationary, and the armature (3) rotates. The three-phase output of the exciter is delivered to the rectifier (5) by means of cable (4). The dc output I_x of the rectifier is fed directly into the rotor of the synchronous machine without having to flow through brushes and slip rings.

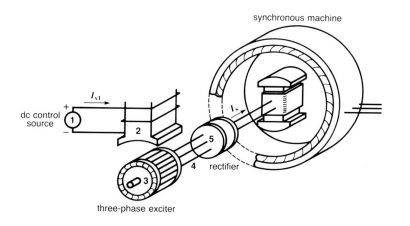

synchronous machine

I_{x1}

dc control
source
1

2

5

4   rectifier

3

three-phase exciter

Figure 19-18
Schematic diagram of a brushless excitation system for a synchronous motor or
generator.

Because of the high reliability of modern rectifiers, most synchronous motors
and virtually all synchronous generators are now equipped with brushless excitation
systems (Fig. 19-19). To appreciate the excitation requirements of synchronous
machines, it is useful to refer to Table 19A.

TABLE 19A   EXCITATION REQUIREMENTS OF SYNCHRONOUS MACHINES

Function of the synchronous machine	Power rating	DC excitation (rated values)		
		power	voltage	I_x
motor	2 hp	120 W	120 V	1 A
motor	4000 hp	50 kW	250 V	200 A
generator	500 MVA	2400 kW	400 V	6000 A
generator	1500 MVA	6720 kW	600 V	11 000 A

19.7 Hunting of synchronous machines

Synchronous motors tend to oscillate for several seconds whenever they are subjected to sudden load changes. The reason is that the rotor and stator poles are held
together by the "elastic" magnetic field. Thus, when a sudden load is applied to a
synchronous motor, the rotor poles start moving toward a new position. However,

Figure 19-19a
Synchronous motor rated 4000 hp, 200 r/min, 3-phase, 6.9 kV, 60 Hz, 80% power factor
designed to drive an ore crusher. The brushless exciter is mounted on the overhung shaft and
is rated 50 kW, 250 V. (*Courtesy General Electric*)

in doing so, they briefly overshoot the mark and therefore the rotor oscillates around
the new position until it finally comes to "rest". This temporary action of hide and
seek is called *hunting*. The frequency of the oscillations is typically one or two per
second. The frequency depends upon the mechanical inertia of the revolving parts
and the electrical properties of the synchronous machine.

Hunting is particularly important in synchronous motors that drive pulsating
loads, such as reciprocating compressors. If the load pulsates at close to the natural
hunting frequency of the synchronous machine, the oscillations may build up,
causing the machine to eventually lose synchronism. The oscillations also may be
transmitted to the floor and through the building, causing damage to foundations
and structures. However, such problems are usually foreseen (and corrected) in the
design stage.

Figure 19-19b
Closeup of the 50 kW exciter showing the armature winding and five of the six diodes used to rectify the ac current.

19.8 Rating of synchronous motors

Synchronous motors are rated according to horsepower, speed, line voltage, and power factor. The power factor rating is necessary because synchronous motors are often designed to operate at leading power factor. When they do, they deliver reactive power to the factory in which they are installed. The amount of reactive power is controlled by varying the dc excitation. The reactive power is absorbed by induction motors in the factory, and so the power factor of the entire plant can be maintained at the desired level.

Synchronous motors may be designed to operate at a leading power factor of as much as 80 % (Fig. 19-19). However, if there is no need for reactive power, a unity power factor motor can do the job just as well. It is slightly smaller and cheaper than the 0.8 power factor motor because less exciting current is needed. Consequently, the field windings are smaller. Furthermore, the rated stator current is smaller, which reduces the size of the stator windings.

Example 19-2:

A 3000 r/min, 400 V, 60 Hz, 3-phase synchronous motor has an efficiency of 96 % and a power factor rating of 0.9. Calculate the maximum reactive power it can deliver to the electrical system.

Solution:

a)   Power output of the motor is

$$P_o = 3000 \text{ hp} \times 746 = 2\,238\,000 \text{ W} = 2238 \text{ kW}$$

b)   Active power absorbed by the motor is

$$P_i = \frac{100\, P_o}{\text{eff}} = \frac{100 \times 2238}{96} = 2331 \text{ kW} \qquad \text{Eq. 1-11}$$

c)   Rated apparent power of the motor is

$$S = P \,/\, \text{power factor} = 2331 \,/\, 0.9 = 2590 \text{ kVA}$$

d)   Reactive power the motor can deliver is

$$Q = \sqrt{S^2 - P^2} \qquad \text{Eq. 14-20}$$
$$= \sqrt{2590^2 - 2331^2} = 1129 \text{ kvar}$$

19.9 Starting synchronous motors

Special starters are used for synchronous motors. Many are designed for across-the-line starting, but large motors may require reduced-voltage starting. These starters are considered in Chapter 22, so for now we will discuss only the general principles of starting synchronous motors.

During the acceleration phase, the dc field winding is connected to an external "discharge" resistor (Fig. 19-6). The presence of the discharge resistor improves the starting torque of the motor. It also limits the high voltage that would otherwise appear across the field terminals while the motor speed is low. When the motor is running at close to synchronous speed, the dc excitation is suddenly applied. The timing is done automatically, so as to develop the highest possible pull-in torque. We recall that the field produced by the stator sweeps slowly past the salient poles when the rotor is running at close to synchronous speed. Consequently, when dc excitation is applied, we must be sure that the resulting N and S poles on the rotor are facing the opposite S and N poles on the stator, so as to maximize the force of attraction.

If the load has high inertia, the pull-in torque may not be big enough to pull the machine into synchronism. In such cases, the synchronous motor must be synchronized alone, and the load gradually brought up to speed by a using a magnetic clutch.

19.10 Application of synchronous motors

Synchronous motors are used to drive pumps, compressors, fans, pulp grinders, and other heavy industrial equipment. They are preferred to induction motors whenever the speed of rotation is low. At low speeds, induction motors tend to have a low power factor whereas synchronous motors can operate at unity power factor.

At the opposite end of the power spectrum, we find tiny synchronous motors that are used in electric clocks and timing devices. They are single-phase devices having typical outputs of 0.75 W (1/1000 hp) at efficiencies of 15 percent. Motors that are geared down to 1 r/min have typical efficiencies of 1 percent.

19.11 Summary

In this chapter, we learned that one of the main advantages of the synchronous motor is that its magnetic field can be produced by a dc current flowing in the rotor windings. The motor therefore does not have to draw reactive power from the ac line, and it can operate at unity power factor. Furthermore, we saw that if the motor is over-excited, it will supply reactive power to the line to which it is connected. Thus, it behaves, in part, like a capacitor.

A synchronous motor also runs at constant speed, in step with the frequency. But this is usually only a marginal advantage, except in the case of clock motors.

Compared with large squirrel-cage induction motors, the only disavantage of a synchronous motor is its limited pull-in torque. This may prevent the motor from pulling into synchronism, particularly if the load has a high inertia.

TEST YOUR KNOWLEDGE

19-1   A synchronous motor can pull into step even in the absence of dc excitation.

☐ true   ☐ false

19-2   The dc excitation of a synchronous motor must not be applied until the motor is running at close to synchronous speed.

☐ true   ☐ false

19-3   A 500 hp, 2200 V, 3-phase, 60 Hz, 30 pole synchronous motor is connected to a line whose frequency is 60.3 Hz. Its speed is

a. 240 r/min   b. 720 r/min   c. 241.2 r/min

19-4   A 28-pole synchronous motor draws an exciting current of 80 A from a 125 V dc source. The resistance of a rotor field coil is

a. 43.75 Ω   b. 55.8 mΩ   c. 1.56 Ω

19-5   When a synchronous motor is under-excited,

a. its field coils overheat

b. its speed drops slightly

c. it absorbs reactive power from the ac line

19-6   In Fig. 19-8, the field current needed to obtain a leading power factor and a stator current of 40 A is

a. 2.5 A   b. 9 A   c. 5.196 A

19-7   A 4500 hp, 6900 V, 3-phase, 60 Hz, 1200 r/min, 0.8 power factor synchronous motor drives a centrifugal compressor in a coal mine. If it has an efficiency of 97.0 %, calculate the ac line current when the motor operates at its rated output.

a. 362 A   b. 627 A   c. 209 A

QUESTIONS AND PROBLEMS

19-8   Compare the construction of a synchronous motor with that of a wound-rotor induction motor.

19-9   Why does the speed of a synchronous motor remain constant even under variable load?

19-10   What is meant by a synchronous capacitor?

19-11   Explain what causes a synchronous motor to hunt.

19-12   If we over-excite a synchronous motor, will it deliver more power to the load?

19-13   A synchronous motor driving a pump operates at unity power factor. If the dc excitation is increased, what happens to (a) the stator current and (b) the temperature of the motor?

19-14   A 10 000 hp, 6.6 kV, 3-phase, 60 Hz, 327 r/min, 0.9 power factor synchronous motor drives a pulpwood grinder in a paper mill. If it has a full-load efficiency of 97.8 %, calculate the losses in the machine.

19-15   In Table 19A, calculate the ratio of excitation power to the power rating of (a) the 2 hp motor, (b) the 4000 hp motor, and (c) the 1500 MVA generator.

CHAPTER 20

Synchronous Generators

Introduction and Chapter Objectives

Three-phase synchronous generators are the primary source of all the electrical energy we consume. They are also the largest energy converters in the world, converting mechanical energy into electrical energy in powers ranging up to 1500 MW. In this chapter, we will study the construction and characteristics of these large, modern ac generators.

Among other things we will describe the behavior of a generator when it is connected in parallel with dozens of other generators. This will be compared with the performance when the generator is connected alone to an isolated load. We will also show the steps that must be taken to synchronize the machine with another generator. Finally, we will develop an equivalent circuit of the synchronous generator which can be used to predict its performance.

The performance of synchronous generators is closely related to that of synchronous motors. Consequently, we will occasionally return to the theory of synchronous motors covered in the previous chapter.

20.1 Stationary-field ac generator

Commercial ac generators are built with either a stationary or a rotating dc field. A stationary-field ac generator (Fig. 20-1) has the same outward appearance as a dc generator. The salient poles create the dc field, which is cut by a revolving armature. However, the armature has a three-phase winding whose terminals are connected to three slip-rings mounted on the shaft. A set of brushes, riding on the slip-rings, permits the armature to be connected to an external three-phase load. The armature is driven by a gasoline engine, water turbine, or some other source of motive power. As we learned in Section 15-2, a three-phase voltage is induced whose value depends upon the speed of rotation and the magnitude of the dc exciting current in the stationary poles.

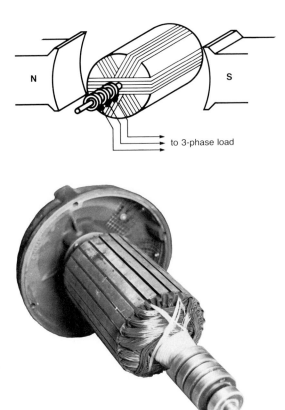

Figure 20-1
Three-phase stationary-field ac generator.

Figure 20-2
Rotor of a stationary-field ac generator rated 20 kVA, 3600 r/min, 115/230 V, single-phase, 60 Hz unity power factor. The armature is 200 mm long and has a diameter of 143 mm. The two single-phase windings are brought out to four slip rings. *(Courtesy Electro-Mecanik)*

Stationary-field generators are satisfactory when the output power is less than 50 kVA (Fig. 20-2). However, for greater outputs, it is cheaper, safer, and more practical to employ a stationary armature and a revolving dc field.

20.2 Revolving-field ac generator

From an electrical standpoint, the stator of a revolving-field ac generator is identical to that of a three-phase synchronous motor. However, the windings are always connected in wye and the neutral is grounded. We prefer a wye connection to a delta connection because the voltage per phase is only 1/1.732 or 58 percent of the voltage between the lines. This means that the highest voltage between a stator conductor and the grounded stator core is only 58 percent of the line voltage. Thus, by using a wye connection, we can reduce the amount of insulation in the slots. Figures 20-3 and 20-4 show the typical construction of large ac generators.

The rated voltage of a synchronous generator depends upon its power rating. In general, the greater the power, the higher the voltage. However, the rated voltage seldom exceeds 25 kV.

Revolving-field generators are built with two types of rotors: *salient-pole* rotors and smooth, *cylindrical* rotors. Salient-pole rotors are used in low-speed generators, such as those driven by hydraulic turbines. Cylindrical rotors are used in high-speed generators driven by steam turbines. The turbines or driving engines are called *prime movers*.

Figure 20-3
Stator of a steam turbine generator rated 722 MVA, 3600 r/min, 19 kV, 60 Hz, 3-phase. The windings are water-cooled and the entire stator will later be enclosed in a metal housing (see background). The housing contains hydrogen under pressure to further improve the cooling. (*Courtesy Brown Boveri*)

20.3 Number of poles

The number of poles on a synchronous generator depends upon the speed of rotation and the frequency we wish to produce. The frequency is given by the basic equation:

$$f = \frac{pn_s}{120}$$
Eq. 17-2

Figure 20-4a
The stator of a large, slow-speed synchronous generator is composed of steel laminations that are stacked at the generator site. This picture shows the stator of a 500 MVA, 200 r/min, 15 kV, 60 Hz, 3-phase generator designed to operate at a lagging power factor of 95 %. The internal diameter is 9250 mm, and the axial length is 2500 mm. (*Courtesy Marine Industrie*)

where

f =   frequency of the induced voltage (Hz)

p =   number of poles on the rotor

n_s =   speed of the rotor (r/min)

Figure 20-4b
The 376 pre-formed coils are placed in the 376 slots and interconnected to produce a 3-phase
wye-connected winding. The current per phase is 19 250 A. (*Courtesy Marine Industrie*)

Example 20-1:

A hydraulic turbine turning at 150 r/min drives a synchronous generator rated at 400 MVA, 14 kV, 3-phase, 60 Hz.

1. how many poles does the rotor have?

2.   What is the rated armature current per phase?

Solution:

1. $p = 120\, f/n_s = 120 \times 60/150 = 48$ poles   Eq. 17-2

2.   From Eq. 15-3 we have $S = 1.732\, EI$

therefore $I = S/1.732 = \dfrac{400 \times 10^6}{14\,000 \times 1.732} = 16\,496$ A

20.4 Salient-pole rotors

Most hydraulic turbines must turn at low speeds (between 50 and 300 r/min) in order to extract the maximum power from a waterfall. Because the rotor is directly coupled to the waterwheel, and because a frequency of 60 Hz is required, a large number of poles must be placed on the rotor. The salient poles are mounted on a circular steel frame that is fixed to a revolving vertical shaft (Fig. 20-5).

In addition to the dc winding, we often add a squirrel-cage winding embedded in the pole faces (Fig. 20-6). Under normal conditions, this winding does not carry any current because the rotor turns at synchronous speed. However, when the load on the generator changes suddenly, the rotor begins to hunt, producing momentary speed variations above and below synchronous speed. This induces a voltage in the squirrel-cage winding, causing a large current to flow. The current reacts with the magnetic field of the stator, producing forces that dampen the oscillation of the rotor. For this reason, the squirrel-cage winding is sometimes called a *damper winding*.

20.5 Cylindrical rotors

Steam turbines are smaller and more efficient when they run at high speed. The same is true of synchronous generators. Thus, there is an economic advantage to using high-speed equipment. However, to produce the desired frequency, we cannot use less than two poles, and this fixes the highest possible speed. On a 60 Hz system, it is 3600 r/min. The next lower speed is 1800 r/min, corresponding to a 4-pole machine. Consequently, these so-called *steam-turbine generators* have either two or four poles. The rotor is a long solid-iron cylinder which has slots milled lengthwise out of the cylindrical mass (Fig. 20-7a). Concentric field coils, firmly wedged into the slots, serve to create the dc poles (Figs. 20-7b and 20-8).

The high speed of rotation produces strong centrifugal forces which impose an upper limit on the diameter of the rotor. For this reason, cylindrical rotors have to be very long when the power output of the generator is in the 1000 MW range.

20.6 AC generator under load

The performance of an ac generator depends upon the type of load it has to supply. Although there are many types of loads, they can all be grouped into two basic categories:

1.   isolated loads, powered by a single generator, and

2.   a fixed-voltage, fixed frequency infinite bus.

From the standpoint of power generation, the second category is by far the most important, so we will discuss it first.

20.7 AC generator on an infinite bus

A large electric utility system, fed by dozens of big generators, is so powerful that the frequency and line voltage are, to all practical purposes, fixed. Thus, when a

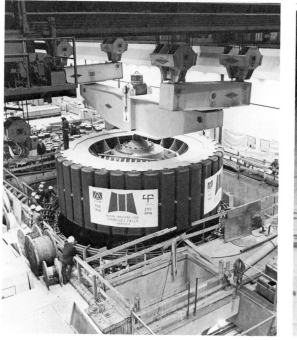

Figure 20-5
Salient-pole rotor of a 500 MVA, 200 r/min synchronous generator being lowered inside its stator. The diameter is 30 feet.
(*Courtesy Marine Industrie*)

Figure 20-6
Salient-pole of a low-speed, 250 MVA synchronous generator showing the twelve slots that will carry the squirrel-cage damper winding.

Figure 20-7a
Cylindrical rotor of a steam-turbine generator showing the slots being milled out along its length. They will carry the dc winding. The generator is rated 1530 MVA, 1500 r/min, 27 kV, 50 Hz; it has an effective pole length of 7490 mm and a diameter of 1800 mm. (*Courtesy Allis-Chalmers Power Systems Inc., West Allis, Wisconsin*).

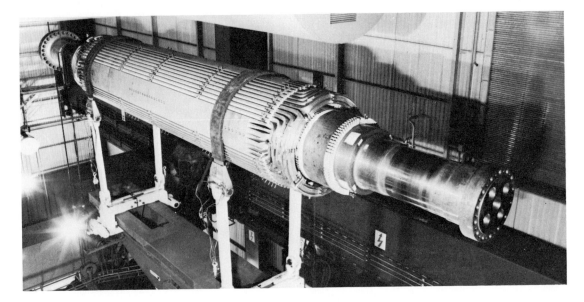

Figure 20-7b
Cylindrical rotor with its 4-pole dc winding. The total mass is 225 tons and the exciting current of 11 200 A is supplied by a 600 V dc brushless exciter. The air gap between the revolving rotor and stationary stator is 120 mm.

Figure 20-8
End view of a two-pole cylindrical rotor. Each pole is composed of eight concentric coils that carry the dc current. Flat, bare conductors are placed at the top of each slot and extend outwards in artistic display. They will be flattened down over the dc winding, and short-circuited to create the damper winding. (*Courtesy Brown Boveri*)

single generator is added or disconnected from the system, it cannot affect either the voltage or the frequency of the system. Such a system is called an *infinite bus*.

Once connected to an infinite bus, an ac generator becomes part of a network comprising dozens (sometimes hundreds) of other generators that deliver power to millions of loads. It is impossible, therefore, to specify the nature of the load (large or small, resistive, inductive, or capacitive) connected to the terminals of this particular generator. What, then, determines the power the machine delivers?

Part of the answer lies in what we learned in the chapter on synchronous motors. We recall that a synchronous motor is effectively connected to an infinite bus because the voltage and frequency of the line are constant. When a *load* torque is applied to the motor, it *absorbs* active power from the line. Similarly, if a prime mover exerts a *driving* torque on a generator, it will *deliver* active power to the line.

In effect, whenever a generator is connected to an infinite bus, we can raise the active power (kilowatts) it delivers to the system by increasing the torque of the turbine. On the other hand, if we want to vary the amount of reactive power (kilovars) we simply vary the dc excitation. These two basic methods — torque control and excitation control — are the way in which the generator output is adjusted. Because the kW and kvar output can be varied independently, the power factor and the magnitude of the generator load are actually determined by the generator and its prime mover.

We will now examine in more detail what actually happens when an ac generator is connected to an infinite bus. Such an ac generator is often called a *synchronous generator* because its average speed has to be directly proportional to the frequency of the system to which it is connected.

20.8 Synchronous machine operating as a generator

In order to observe the smooth transition from motor action to generator action, let us begin our study with a synchronous motor operating at no-load (Fig. 20-9). Because the voltage and frequency of the ac source are fixed, the motor is actually connected to an infinite bus. The rotor is turning clockwise, and it lags behind the stator poles by a small angle α. The excitation is adjusted so that the motor operates at unity power factor; the line current *I* is therefore small.

Instead of applying a mechanical load to the shaft (thus tending to slow the machine down), let us apply a driving torque, tending to speed it up. This can be done by connecting a gasoline engine or other prime mover to the motor shaft (Fig. 20-10). As the driving torque is increased, angle α will gradually decrease and eventually become zero when the rotor and stator poles are perfectly lined up. When this happens, the windage and friction losses are entirely supplied by the driving engine, and no more power is drawn from the three-phase line. Consequently, line current *I* becomes zero (Fig. 20-11).

Let us continue to increase the driving torque by opening the throttle of the engine. The rotor poles will pull ahead of the stator poles, and so the flux between them will have the shape shown in Fig. 20-12. The magnetic "rubber bands" exert a pull on the rotor that acts *against* the direction of rotation. Consequently, the engine has to exert a torque to maintain this position of the poles. The engine is therefore delivering mechanical power to the synchronous machine. What happens to this power?

It is entirely converted into electric power, which is returned to the three-phase line. As a result, the synchronous machine now operates as a generator. It is called a synchronous generator because its frequency is determined by the three-phase line to which the generator is connected.

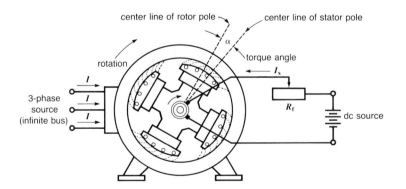

Figure 20-9
Synchronous motor connected to an infinite bus and running at no-load. The arrows show the positive direction of the current in each phase. The dc current is adjusted so that the motor operates at unity power factor.

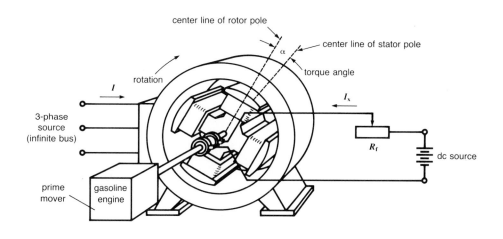

Figure 20-10
When a prime mover, such as gasoline engine, is coupled to the shaft, the torque angle can gradually be reduced to zero.

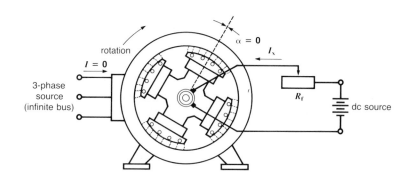

Figure 20-11
When the torque angle is zero, the prime mover supplies all the power needed to drive the rotor. The ac line current becomes zero. Note that the flux is now symmetrically distributed over the pole face.

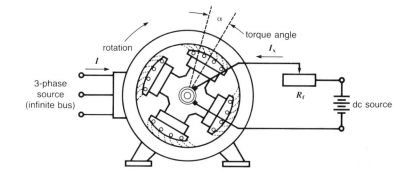

Figure 20-12
When the torque of the prime mover is increased, the rotor poles pull ahead of the stator poles. The poles continue to attract each other, therefore a continuous braking torque is exerted on the rotor as it rotates.

As we continue to increase the driving torque of the engine, angle α increases, as does the power delivered to the three-phase system. The stator current I also increases. However, there is a limit to the driving torque that can be applied to the rotor. If we exceed the *pull-out torque* of the generator, the rotor poles will suddenly break away from the stator poles, and the generator is said to lose synchronism. When this happens, the net power delivered to the three-phase line falls abruptly to zero. The line current I increases dramatically, and the circuit breakers protecting the machine will trip*.

The pull-out torque T_m of the synchronous generator has the same magnitude as when the machine was operating as a synchronous motor (see Fig. 19-14). Furthermore, the critical angle α_m where pull-out occurs is also the same. The only difference is that the torque acts in the opposite direction, and α_m is positive instead of negative.

In summary, we can increase the active power (kilowatts) delivered to an infinite bus by increasing the mechanical power input to the synchronous generator. The increased mechanical power is obtained by raising the torque of the turbine or driving engine.

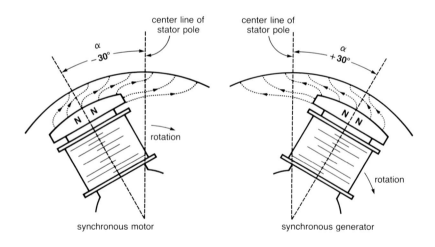

Figure 20-13
Comparison of the rotor position and flux pattern when a synchronous machine is connected to an infinite bus and running as a motor or a generator.

---

*   The sudden disappearance of any opposing torque causes the driving engine to start racing unless automatic means are used to limit its speed. Such overspeeds can be dangerous because the centrifugal forces acting on the poles may tear the rotor apart, literally causing a mechanical explosion inside the machine.

As a final point of comparison, Fig. 20-13 shows the flux pattern between the rotor poles and stator poles when the machine is acting as a motor ($\alpha = -30°$), and as a generator ($\alpha = +30°$). Note that the rotor poles are always ahead of the stator poles when the machine runs as a generator.

20.9 Active and reactive power flow

We have seen that when a synchronous machine is connected to a three-phase line, it can operate as either a motor or a generator, depending upon the torque that is exerted on the shaft. When the machine operates as a motor, it *draws* active power P from the line. When it operates as a generator, it *delivers* active power to the line. Thus, if a 3-phase wattmeter is connected between the synchronous machine and the line, it will give a positive reading when power flows into the machine (Fig. 20-14). Conversely, it will give a negative reading when power flows out of the machine, in which case it is acting as a synchronous generator (Fig. 20-15). Note that the dc exciting current I_x is the same in both cases. We conclude that the magnitude of the active power received or delivered by the synchronous machine depends only upon the magnitude of the torque exerted on the shaft.

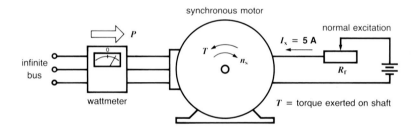

Figure 20-14
Wattmeter reading and active power flow when the synchronous machine is operating as a motor. The torque exerted on the shaft is opposite to the rotation.

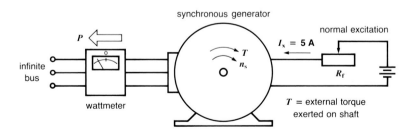

Figure 20-15
Wattmeter reading and active power flow when the synchronous machine is operating as a generator. The torque exerted on the shaft is in the direction of rotation.

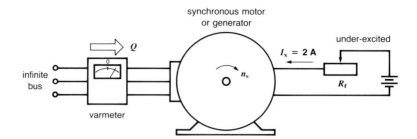

Figure 20-16
Varmeter reading and reactive power flow when a synchronous motor (or generator) is under-excited.

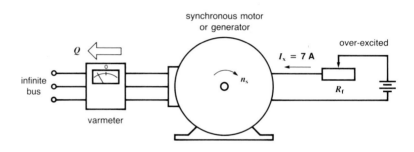

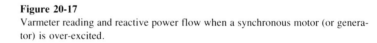

Figure 20-17
Varmeter reading and reactive power flow when a synchronous motor (or generator) is over-excited.

What happens if we vary the dc excitation? This will vary the amount of reactive power that flows into (or out of) the machine.

If we decrease the excitation below its "normal" value, the synchronous machine will absorb reactive power from the line. **This is true whether the machine operates as a motor or as a generator**. If a 3-phase varmeter is connected into the line, it will give a positive reading (Fig. 20-16).

Conversely, if we raise the excitation above its normal value, the machine will deliver reactive power to the line, whether it is acting as a motor or as a generator. The varmeter will give a negative reading (Fig. 20-17).

20.10 Synchronization of an ac generator

On a large electric utility system, dozens of synchronous generators are effectively connected in parallel to supply the enormous power requirements of a geographical region. As the demand builds up during the day, generators are successively added to the system to provide the extra power. Later, as the power demand falls, selected machines are successively disconnected from the system until the utility load again builds up the following day. Synchronous generators are therefore continually being connected and disconnected from a power grid in response to consumer demand. However, before connecting a generator to a system, it must first be *synchronized*. We cannot just throw it on the line by closing a switch. A generator is said to be synchronized when it meets the following conditions:

1. the generator frequency is equal to the system frequency,

2. the generator voltage is equal to the system voltage,

3. the generator voltage is in phase with the system voltage, and

4. the generator phase sequence is the same as the system phase sequence.

To synchronize a generator, we proceed as follows:

1. Adjust the speed of the turbine (or other prime mover) so that the generator frequency is very close to the system frequency.

2. Adjust the dc excitation so that the generator terminal voltage E_o is equal to the system voltage E (Fig. 20-18).

3. Observe the phase angle between E_o and E by means of a *synchroscope*. This instrument has a pointer that indicates the phase angle between the two voltages, covering the entire range from zero to 360°. Although the degrees are not

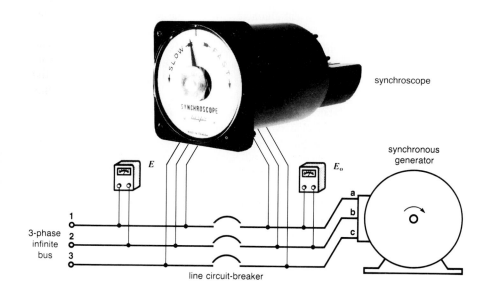

Figure 20-18
Method of synchronizing a generator. (*Courtesy Lab-Volt*)

shown, the dial has a zero marker to indicate when the voltages are in phase. In practice, when the frequency of the generator is close to system frequency, the pointer rotates slowly as it tracks the changing phase angle between the generator and system voltages. If the generator frequency is higher than the system frequency, the pointer rotates clockwise. Otherwise, it rotates counterclockwise.

4.   Carefully adjust the speed regulator of the turbine so that the pointer of the synchroscope just barely creeps across the dial.

5.   Make a final check to see that the generator voltage is equal to the system voltage.

6.   Then, the moment the pointer crosses zero, close the line circuit breaker, thus connecting the generator to the system.*

The same synchronizing procedure is followed in isolated systems when a generator has to be connected in parallel with another generator. In modern generating stations, synchronization is done automatically.

20.11 Equivalent circuit of an ac generator

In the study of dc machines, we found it useful to employ an equivalent circuit showing the essential components of the armature and field. It enabled us to predict the voltages and currents when the machine was under load. In the same way, it is useful to have an equivalent circuit of a synchronous machine. It should represent, as simply as possible, the electrical characteristics of the armature and field.

The three armature (stator) windings of a synchronous machine have identical electrical properties. The only difference between them is that the voltages and currents are displaced (in time) by 120°. Consequently, we can use one winding on the stator to represent all three. Referring to Fig. 20-19, a voltage E_o is induced in this winding by the revolving flux ϕ created by the rotor. The winding has a resistance R_s and a reactance X_s. These elements are in series with the induced voltage E_o. The external terminals are labeled A and N, where N is the neutral of the three-phase machine. The corresponding terminal voltage is E.

The revolving field is connected in series with a field rheostat R_f to a dc source E_x. The resulting current I_x produces the flux ϕ. The external terminals of the field are marked 1 and 2, but the slip rings are not shown.

The equivalent circuit shown in Fig. 20-18 represents with good accuracy the electrical properties of a synchronous machine. In practice, R_s is much smaller than X_s, and so we usually neglect it. The reactance X_s is called the *synchronous reactance* of the machine.

---

*   If the phase sequence 1-2-3 of the infinite bus is not the same as the phase sequence a-b-c of the "incoming" generator, it will be impossible to observe the slow rotation of the synchroscope pointer. If the circuit breaker is closed, a short-circuit will result.

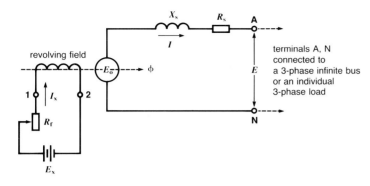

Figure 20-19
Equivalent circuit of a 3-phase generator, showing only phase A, and the dc exciting circuit.

20.12 Phasor diagram of a synchronous generator

To illustrate the use of the equivalent circuit, let us draw the phasor diagram when the synchronous machine operates as a generator on a three-phase line (infinite bus). We assume that I_x is adjusted so that the generator operates at unity power factor. This means that I is in phase with the terminal voltage E. (Fig. 20-20).

Next, because the machine is operating as a synchronous generator, we know that the rotor poles *lead* the stator poles by a certain mechanical angle α (see Fig. 20-13). As a result, voltage E_o that is induced in the stator by the rotor must *lead* E by a corresponding *electrical* angle δ. Finally, the IX_s drop must be at 90° to I, because of the inductive nature of X_s. The complete phasor diagram for the synchronous generator is shown in Fig. 20-20.

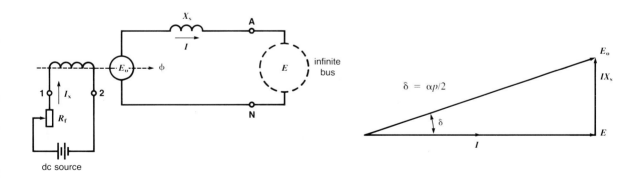

Figure 20-20
Circuit and phasor diagram when a 3-phase generator is connected to an infinite bus and operating at unity power factor.

It can be proved that the electrical torque-angle δ is related to the mechanical torque-angle α by the equation:

$$\delta = \alpha p/2 \tag{20-1}$$

where p is the number of poles on the synchronous machine.

Suppose we raise I_x so that the generator is over-excited. How will this affect the phasor diagram, knowing that the mechanical power remains the same? The theory tells us that over-excitation causes the generator to deliver reactive power to the line. Thus, as far as the generator is concerned, the load across its terminals is inductive. This means that I lags behind the terminal voltage E. However, the component I_p that is in phase with E must be the same as before. The reason is that the active power delivered by the generator is unchanged because the mechanical power

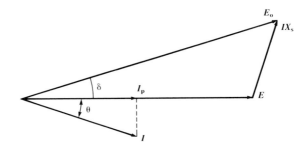

Figure 20-21
Phasor diagram when the generator excitation in Fig. 20-20 is increased without changing the power input of the prime mover.

input is the same. The resulting phasor diagram is shown in Fig. 20-21. Note that IX_s is again 90° ahead of I. The power factor of the synchronous generator is equal to cos θ, and the torque angle δ is slightly smaller than in Fig. 20-20.

Example 20-2:

A 150 MVA, 15 kV, 3-phase, 200 r/min, wye-connected synchronous generator has a synchronous reactance of 1.4 Ω per phase. We want it to deliver 120 MW of active power and 60 Mvar of reactive power to the 15 kV infinite bus to which the generator is connected. Calculate:

1. the current per phase and the line-to-neutral voltage

2. the phase angle between the line current and line-to-neutral voltage

3. the internal voltage drop due to the synchronous reactance

4. the excitation voltage E_o that is required

5. the electrical torque angle α

6. the mechanical torque angle δ

7. the mechanical torque exerted by the prime mover

Solution:

This may appear to be a complicated problem, but if we attack it step by step, it is fairly easy to solve.

1. a) the apparent power delivered by the generator is

$$S = \sqrt{P^2 + Q^2} = \sqrt{120^2 + 60^2} = 134 \text{ MVA}$$

b) the current per phase is

$$\cdot I = S/1.732E = 134 \times 10^6/(1.732 \times 15\ 000) = 5158 \text{ A}$$

c) the line-to-neutral voltage is $E/1.732 = 15\ 000/1.732 = 8660$ V

2. a) the power factor of the generator is

$$\cos \theta = P/S = 120/134 = 0.895$$

b) the phase angle between I and the line-to-neutral voltage is

$$\theta = \arccos 0.895 = 26.4°$$

3. the internal IX_s drop is $IX_s = 5158 \times 1.4 = 7221$ V

Note that the internal voltage drop is almost as great as the terminal voltage of the generator: this is quite common in synchronous machines. However, unlike in a dc generator, this voltage drop produces no heat because it is associated with a reactance.

4. a) to determine the excitation voltage E_o, we draw a phasor diagram of the values we have calculated so far (see Fig. 20-22). The IX_s drop is at 90° to current I, and so the phase angle between 7221 V and 8660 V is $(90 + 26.4) = 116.4°$.

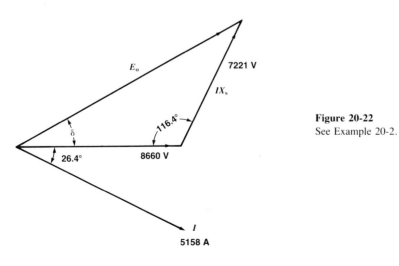

Figure 20-22
See Example 20-2.

b)   the magnitude of E_o can be found using the cosine law (Section 1-20)

$$E_o^2 = 8660^2 + 7221^2 - 2 \times 8600 \times 7221 \cos 116.4 \qquad \text{Eq. 1-15}$$

from which the required excitation voltage is found to be $E_o = 13\ 518$ V

5.   the electrical torque angle δ can also be found using the cosine law. Thus

$$7221^2 = 13518^2 + 8660^2 - 2 \times 13518 \times 8660 \cos \delta$$

from which $\cos \delta = 0.878$, and so $\delta = \text{arcos } 0.878 = 28.6°$

6.   a)   the number of poles on the generator is
$p = 120f/n_s = 120 \times 60/200 = 36$ Eq. 17-2

b)   the mechanical torque angle is
$\alpha = 2\ \delta/p = 2 \times 28.6/36 = 1.59$ Eq. 20-1

The center line of the rotor poles is therefore only 1.59° ahead of the stator poles. Thus, it takes a very small mechanical displacement to produce the 120 MW output.

7.   the torque of the prime mover is:
$T = 9.55P/n = 9.55 \times 120 \times 10^6/200 = 5.73$ MN.m   Eq. 1-9

20.13 AC generator feeding an isolated load

When an ac generator feeds an isolated load, the terminal voltage does not remain constant, as in the case of an infinite bus. The voltage regulation from no-load to full-load can be quite large, depending upon the power factor of the load. In order to maintain a constant line voltage, the dc excitation has to be varied whenever the load varies. Furthermore, the speed of the prime mover also has to be regulated to maintain a constant frequency.

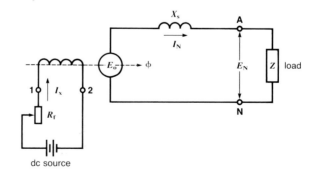

Figure 20-23
Equivalent circuit of a three-phase generator feeding an isolated load. Only phase A is shown, and the dc exciting circuit.

Figure 20-23 shows a generator feeding in isolated load Z. The generator is operating at full load, delivering rated current I_N at rated voltage E_N. We wish to determine the magnitude of the excitation voltage E_o for various load power factors with the generator delivering its rated output.

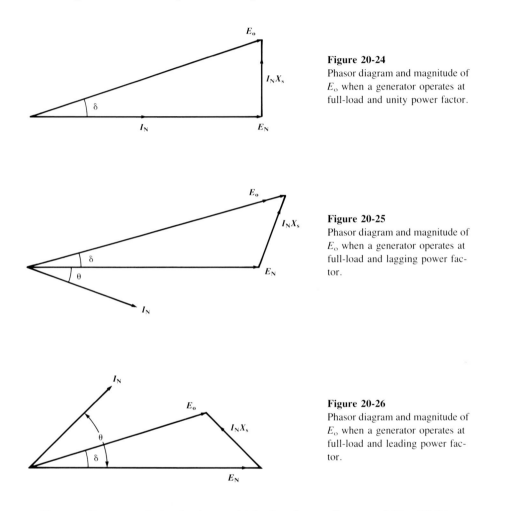

Figure 20-24
Phasor diagram and magnitude of E_o when a generator operates at full-load and unity power factor.

Figure 20-25
Phasor diagram and magnitude of E_o when a generator operates at full-load and lagging power factor.

Figure 20-26
Phasor diagram and magnitude of E_o when a generator operates at full-load and leading power factor.

For a unity power factor load, we obtain the phasor diagram of Fig. 20-24.

For a lagging power factor load, current I_N lags the terminal voltage E_N, and the resulting phasor diagram is shown in Fig. 20-25. Note that E_o is now greater than in Fig. 20-24. This means that the dc exciting current I_x has to be raised at lagging power factors in order to obtain the same terminal voltage.

For a leading power factor, current I_N leads E_N, and the resulting phasor diagram is shown in Fig. 20-26. The induced voltage E_o is now *less* than the terminal voltage E_N. This surprising result is due to the partial resonance effect between the capacitance of the load and the synchronous reactance X_s.

20.14 Voltage regulation of an isolated generator

The previous phasor diagrams enable us to predict the full-load to no-load voltage regulation of an ac generator when it supplies an isolated load. However, the power factor of the load must be specified.

The voltage regulation in percent can be calculated from the equation

$$\text{percent regulation} = \frac{E_o - E_N}{E_N} \times 100 \qquad (20\text{-}2)$$

where

E_N = rated voltage of the alternator

E_o = open-circuit voltage (at the same speed and excitation level)

Example 20-3:

A 3-phase generator having a rating of 72 MVA, 21 kV has a synchronous reactance of 5 Ω per phase. Calculate

1. the rated current I_N of the generator

2. the induced voltage E_o that is required at rated load, 80 percent power factor lagging

3. the no-load to full-load voltage regulation, in percent, at this load

Solution:

1. $I_N = \dfrac{S}{1.732E} = \dfrac{72 \times 10^6}{1.732 \times 21\ 000} = 1980 \text{ A}$ Eq. 15-1

2. a) Referring to Fig. 20-27, the line-to-neutral voltage is

   $E_N = \dfrac{21\ 000}{1.732} = 12\ 124 \text{ V}$

   b) The current lags behind E_N by an angle arcos $0.8 = 36.9°$

   c) The internal voltage drop due to the synchronous reactance of the machine is

   $IX_s = 1980 \times 5 = 9900 \text{ V}$

   This voltage drop is at 90° to phasor I_N, as shown. Consequently, angle $\theta' = 90 + \theta = 90 + 36.87 = 126.9°$.

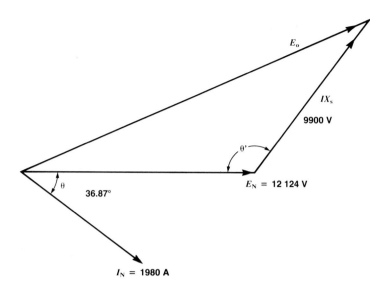

Figure 20-27
See Example 20-3.

d) From the cosine law we have

$$E_o{}^2 = 9900^2 + 12\ 124^2 - 9900 \times 12\ 124 \cos 126.87 \qquad \text{Eq. 1-15}$$

$$= 98.01 \times 10^6 + 147 \times 10^6 + 72 \times 10^6 = 317 \times 10^6$$

hence $E_o = \sqrt{317\ 000\ 000} = 17\ 804$ V

3. If the load is disconnected, the induced voltage E_o will appear across the generator terminals. The voltage regulation is

$$\text{percent regulation} = \frac{E_o - E_N}{E_N} \times 100 \qquad \text{Eq. 20-3}$$

$$= \frac{17\ 804 - 12\ 124}{12\ 124} \times 100 = 46.85\ \%$$

Figure 20-28 shows the typical voltage regulation curves for a generator having a rated voltage of 1500 V and a rated current of 100 A. The curves show how the terminal voltage varies with the load current at unity power factor, at 80 percent power factor lagging, and at 80 percent power factor leading. In each case, the dc excitation current was adjusted to give rated voltage E_N at the rated current I_N of the machine. Once adjusted, the current was then kept fixed.

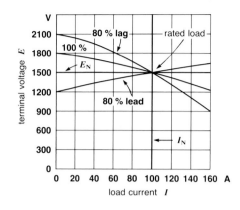

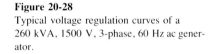

Figure 20-28
Typical voltage regulation curves of a 260 kVA, 1500 V, 3-phase, 60 Hz ac generator.

20.15 Summary

Synchronous generators are similar to synchronous motors and they can be represented by the same equivalent circuit diagram. We learned that when a generator is connected to an infinite bus, we can control the active power output by varying the torque of the prime mover. Furthermore, we can control the reactive power output by varying the dc excitation. When the excitation is "normal" the generator neither delivers nor receives reactive power, and thus it operates at unity power factor. However, if we raise the excitation above the unity power factor level, the generator will deliver reactive power to the infinite bus. Similarly, if we under-excite the generator, it will absorb reactive power. This behavior was also observed in the synchronous capacitor, covered in the previous chapter.

We also learned how to bring a generator "on line" by following a logical synchronizing procedure. Then, we used the equivalent circuit diagram to solve some practical problems. We saw that a generator has a high internal impedance, but because it is mainly reactive, it does not produce a corresponding amount of real losses. As a result, synchronous generators, like other electrical machines, have high efficiencies.

Finally, we used many of the basic principles covered in previous chapters to understand how synchronous generators operate. Thus, we used electromagnetic induction, Ohm's law, magnetic forces, phasor diagrams and mechanical relationships to obtain a coherent understanding of these machines.

TEST YOUR KNOWLEDGE

20-1   A wye-connected synchronous generator has a rated line voltage of 19 kV. The *peak* voltage between a conductor in a slot and the stator core is
a. 15.5 kV   b. 10.97 kV   c. 19 kV

20-2   A 400 Hz aircraft generator is driven at 12 000 r/min. The number of poles is

a. 30   b. 4   c. 2

20-3   A steam-turbine generator is rated at 722 MVA, 19 kV, 3-phase, 60 Hz. The rated current per phase is

a. 38 A   b. 38 kA   c. 21940 A

20-4   Figure 20-13 represents the mechanical angle α of a 4-pole synchronous machine. The corresponding electrical angle delta is

a. 60°   b. 15°   c. 30°

20-5   When a synchronous generator is under load, the rotor poles are always ahead of the center-line of the stator poles.

☐ true   ☐ false

20-6   If a synchronous generator is connected to an infinite bus, it will absorb reactive power if it is over-excited

☐ true   ☐ false

20-7   If a synchronous motor is connected to an infinite bus, it will absorb reactive power if it is over-excited.

☐ true   ☐ false

20-8   A synchronous machine neither delivers nor receives active power when the center lines of the rotor and stator poles are lined up

☐ true   ☐ false

20-9   A synchronous generator can deliver active power at the same time that it *absorbs* reactive power

☐ true   ☐ false

20-10   When a synchronous generator delivers only reactive power to an infinite bus, it has the same effect on the bus as a capacitor.

☐ true   ☐ false

20-11   When a synchronous generator delivers only reactive power to a load, the prime mover must exert a large torque.

☐ true   ☐ false

20-12   In Fig. 20-28, the voltage regulation for an 80 percent power factor lagging load is

a. 23.09 %   b. 40 %   c. 28.6 %

20-13   In Fig. 20-28, the dc exciting current is greatest when the load has a leading power factor.

☐ true   ☐ false

20-14   In a revolving-field ac generator, the armature is the stator.

☐ true   ☐ false

QUESTIONS AND PROBLEMS

20-15   What are the advantages of using a stationary armature in large synchronous generators?

20-16   State the main differences in construction between salient-pole and steam-turbine generators. For a given power output, which of these machines requires the bigger torque? Which machine is larger?

20-17   In analyzing a hydropower site, it is found that the turbines should turn at close to 350 r/min in order to develop the maximum mechanical power. If a directly coupled generator must produce a frequency of 60 Hz, calculate (a) the number of poles on the rotor, and (b) the exact turbine speed.

20-18   An ac generator delivers power to an isolated load having a lagging power factor. If the no-load line voltage is 600 V, must the excitation be increased or decreased to maintain the same voltage under load? Draw a phasor diagram to explain your answer.

20-19   State the procedure to synchronize a generator with an infinite bus.

20-20   An ac generator turning at 1200 r/min generates a no-load line voltage of 6.9 kV, 60 Hz. How will the terminal voltage be affected if the following loads are connected to its terminals?

(a) resistive load
(b) inductive load
(c) capacitive load

20-21   In Problem 20-20, calculate the no-load voltage and frequency if the speed is (a) 100 r/min and (b) 5 r/min.
The dc exciting current is held constant.

CHAPTER 21

Single-Phase Motors

Introduction and chapter objectives

Single-phase motors outnumber by far every other type of electric motor. The reason is that single-phase motors are used in all domestic appliances and tools, such as washers, fans, saws, clocks, and so forth.

The principle of operation of single-phase motors is not as simple as that of three-phase motors. Although there are many types of single-phase motors we will limit our study to the more common types, particularly the single-phase induction motor, which is the most frequently used. We shall then go on to explain hysteresis motors, synchros, and stepper motors. We also will study the special instrument motor found in watthourmeters that measure the energy in millions of homes and factories.

21.1 Single-phase induction motor

If one of the three lines feeding a three-phase induction motor is disconnected, the motor will continue to rotate and will develop a substantial mechanical torque. This is proof that an induction motor can run on single-phase power. However, if the motor is initially at rest, it will not start when single-phase power is applied. Thus, a single-phase induction motor is not self-starting.

But if we spin the shaft in one direction or the other, the motor will immediately pick up speed until it runs at close to synchronous speed. The synchronous speed is given by the same equation as that for three-phase motors, namely

$$n_s = 120f/p \qquad (21\text{-}1)$$

where

n_s = synchronous speed, in revolutions per minute (r/min)

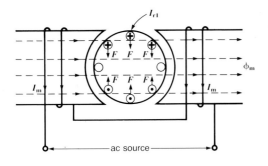

Figure 21-1
The forces in the rotor of a stationary single-phase motor cancel each other and so the starting torque is zero.

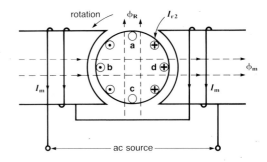

Figure 21-2
When the rotor turns, the induced currents I_{r2} produce a rotor flux that is at right angles to the stator flux. The two fluxes together produce a revolving field.

f = frequency of the single-phase source, in hertz (Hz)

p = number of poles on the stator.

Figure 21-1 is the schematic diagram of a 2-pole single-phase induction motor. It is composed of a stator and a squirrel-cage rotor. When an ac voltage is applied to the stator winding, the resulting current I_m produces a flux ϕ_m. This is an ac flux which increases, decreases, and reverses periodically, in step with the line frequency. However, it is not a rotating flux, and so the rotor does not turn. The ac flux induces large ac currents I_{r1} in the rotor bars of the stationary rotor. Consequently, the rotor bars are subjected to a force F because the currents are immersed in the flux ϕ_m. Unfortunately, the sum of these forces is zero because some of them tend to make the rotor turn cw, while others tend to make it turn ccw (Fig. 21-1). For this reason, the motor cannot start by itself.

21.2 Theory of the revolving field

If the motor is started, it unexpectedly develops a torque in the same direction of rotation, and it rapidly picks up speed. What produces this driving torque when the motor starts turning?

One of the theories explaining the torque of a single-phase motor may be stated as follows. As soon as the motor begins to rotate, a weak revolving field is set up in the machine. This field is caused by the combined action of the mmf of the stator and the mmf produced by the currents in the rotor. As the motor picks up speed, the rotating field becomes stronger and stronger. It reaches its maximum strength when the rotor runs at synchronous speed. The rotating field is then similar to that produced by a three-phase motor. Indeed, if we placed a conventional 3-phase winding on the stator (in addition to the single-phase winding), we would discover that the voltages induced in the three windings are displaced at 120° to each other. This is perhaps the best proof that a revolving field is produced in a single-phase motor when it runs at close to synchronous speed. But how is this field produced?

Referring to Fig. 21-2, when the motor is turning, a voltage is induced in every conductor in sections abc and adc of the rotor because these conductors are cutting across the flux lines ϕ_m. The induced voltages produce rotor currents I_{r2} which together develop a mmf that produces a flux ϕ_R. This flux did not exist when the rotor was stationary. It so happens that ϕ_R reaches its maximum value 90° after ϕ_m is maximum. The combined action of this 90° time delay and the 90° mechanical displacement between ϕ_m and ϕ_R produces the revolving field.

21.3 Means of starting single-phase motors

To make a single-phase motor self-starting, we add an extra winding to the stator. This *auxiliary* winding, or *starting* winding, has the same number of poles as the main winding. However, the poles are shifted in space from those of the main winding. In most cases, they are placed exactly midway between the poles of the main winding (Fig. 21-3).

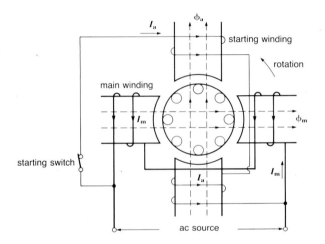

Figure 21-3
To make a single-phase motor self-starting, a starting winding is added.

When the main and auxiliary windings are connected to a single-phase source, the main winding produces a flux ϕ_m, and the auxiliary winding a flux ϕ_a. If the two fields are out of phase, a revolving field is produced. It rotates at a fixed, synchronous speed given by Eq. 21-1. We obtain a perfect rotating field when ϕ_m and ϕ_a are equal in magnitude and out of phase by 90°. However, a strong rotating field and good starting torque can still be produced for phase shifts as small as 20° between ϕ_m and ϕ_a.

In order for ϕ_m and ϕ_a to be out of phase, currents I_m and I_a in the main and auxiliary windings must be out of phase. The desired phase shift can be obtained by using different R/X ratios for the two windings, where R and X are the resistance and reactance of the windings. In some cases, we connect a capacitor in series with the auxiliary winding so that I_m leads I_a. These phase-shift methods go under the general term "split-phase".

Once the motor has reached about 80 percent of synchronous speed, the auxiliary winding is disconnected from the source by an automatic switch. The motor then continues to run as a true single-phase machine, with only the main winding in service.

The automatic switch is usually actuated by the centrifugal force that acts on spring-loaded weights mounted on the rotor shaft. When the motor approaches its rated speed, the centrifugal force causes the weights to suddenly fly out, which opens the circuit of the auxiliary winding. In some motors, the automatic switch consists of a current-sensitive relay.

Figures 21-4 to 21-10 illustrate the components and construction of a single-phase motor.

Figure 21-4
The construction of a single-phase motor is very similar to that of a three-phase motor. The laminated stator on the left has treated-paper slot liners to insulate the windings from the core. The squirrel-cage rotor on the right is identical to that of a 3-phase induction motor. The rotor bars, end rings and fan blades are made of die-cast aluminum. *(Courtesy Lab-Volt)*

Figure 21-5
The main coils of this 4-pole motor each consist of four concentric coils connected in series. The connections are made so that adjacent poles have opposite magnetic polarities. The two wires are the terminals of the main winding. Note that the laminated core is supported by a rolled iron frame that will also support the end bells. (*Courtesy Lab-Volt*)

Figure 21-6
The next step consists of threading the auxiliary winding into the slots. The four poles of the auxiliary winding are arranged to straddle the poles of the main winding. The adjacent poles of the auxiliary winding are connected in series and produce alternate N and S poles. (*Courtesy Lab-Volt*).

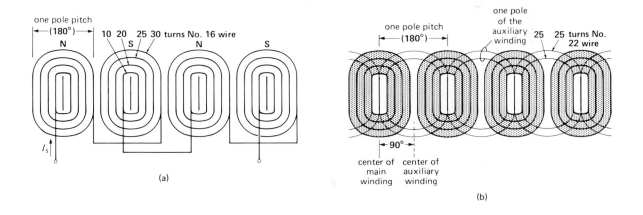

Figure 21-7
Schematic diagram showing how (a) the main and (b) the auxiliary winding are connected in a 4-pole motor. The turns per coil and wire size are typical for a 1/4 hp, 120 V, 60 Hz motor.

Figure 21-8
After assembly, the centrifugal switch is mounted on the end bell, and the spring-loaded weights are mounted on the shaft. When the motor is at rest, the spring pushes a plastic collar against the switch, thus closing the circuit of the starting windings.

Figure 21-9
When the motor approaches synchronous speed, the weights suddenly fly outward, as shown above. The plastic collar moves to the left, thus opening the starting circuit. In most commercial motors the action of the centrifugal switch cannot be observed because it is mounted inside the frame. (*Courtesy Lab-Volt*)

Figure 21-10
Cutaway view of a 5 hp, 1725 r/min, 230 V, 60 Hz single-phase motor. The centrifugal switch mechanism can be seen between the fan and the die-cast end ring. (*Courtesy Gould*)

21.4 Resistance split-phase motor

In a *resistance split-phase* motor (Fig. 21-11), the main winding has many turns of relatively large wire. Its inductive reactance is therefore large, and the resistance is low. As a result, current I_m lags considerably behind the applied voltage E.

On the other hand, the starting winding has relatively few turns of small wire. Consequently, it has a high resistance compared with its inductive reactance. The current I_a in the starting winding lags therefore only slightly behind voltage E.

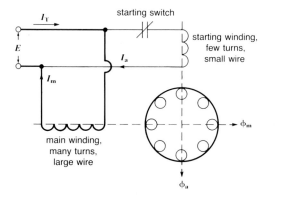

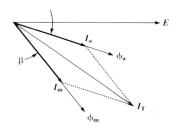

Figure 21-11

Schematic diagram of a resistance split-phase motor.

Figure 21-12

Phasor diagram of a resistance split-phase motor under locked-rotor conditions.

Figure 21-12 shows the resulting phase angle β between I_m and I_a. This angle is obviously the same as that between ϕ_m and ϕ_a. The two fluxes produce the rotating field that starts the motor. The locked-rotor current I_T is the phasor sum of I_m and I_a. It is usually 6 to 7 times the rated full-load current of the motor.

Because the wire used in the starting winding is small, it heats up very quickly. To prevent excessive temperatures, the motor should reach its running speed in 2 to 3 seconds. If the starting period exceeds 5 s, the auxiliary winding will start to smoke, and unless the protective devices trip, the winding will burn out.

A resistance split-phase motor is therefore not adapted to frequent starts and stops. Furthermore, it cannot be used to bring high-inertia loads up to speed. When heavy-duty starting conditions are encountered, it is preferable to use a capacitor-start motor.

21.5 Capacitor-start motor

The *capacitor-start* motor (Fig. 21-13) is identical to the resistance split-phase motor except that the auxiliary winding has about the same number of turns as the main winding. A capacitor is connected in series with the auxiliary winding and a starting switch again disconnects it when the speed reaches about 80 percent of synchronous speed.

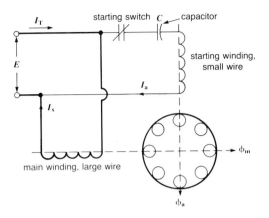

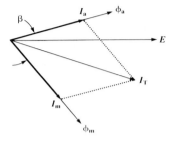

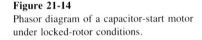

Figure 21-13
Schematic diagram of capacitor-start motor.

Figure 21-14
Phasor diagram of a capacitor-start motor under locked-rotor conditions.

The capacitive reactance of the capacitor is high enough so that current I_a in the auxiliary winding is *ahead* of the applied voltage E (Fig. 21-14). However, current I_m in the main winding lags behind the voltage by the same angle as before. It follows that the phase angle between currents I_m and I_a (and between fluxes ϕ_m and ϕ_a) is greater than the angle in the resistance split-phase motor. Consequently, the locked-rotor torque is greater. Furthermore, if currents I_m and I_a have the same magnitude as those of the resistance split-phase motor, the total current I_T drawn from the line will be smaller.

Thus, a capacitor-start motor has the double advantage of producing a larger locked-rotor torque with a smaller locked-rotor current. The current is only 4 to 5 times the nominal full-load current. This means that the starting time can be longer than in the case of a resistance split-phase motor without causing the auxiliary winding to overheat. However, it is only the starting characteristics that are better; under normal running conditions capacitor-start, and resistance split-phase motors have exactly the same properties. The reason is that only the main windings are then in service, and they are the same in both machines.

The increasing use of capacitor-start motors is due to the excellent characteristics and low cost of modern *electrolytic* capacitors. These capacitors have relatively large microfarad ratings compared to their size. Although they cannot remain in continuous service, they are well adapted for intermittent use. Before the development of these capacitors, repulsion-induction motors were used when starting conditions were severe. Repulsion-induction motors with their commutators and brush gear are still in use, but they are becoming obsolete.

21.6 Properties of split-phase induction motors

The efficiency and power factor of split-phase motors are usually low. Thus, at full load, a 1/4 hp motor has a power factor and efficiency of about 60 %, compared with values of 70 % for three-phase motors of equal power.

Figure 21-15
Capacitor-run motor supported in a resilient mount
to prevent the vibrations from being transmitted to the
base. (*Courtesy Brook Crompton Parkinson*)

Considerable reactive power is needed to create the magnetic field. As a result, the no-load current may be as high as 80 percent of full-load current. Thus, even at no-load, these motors have a temperature rise approaching its full-load value.

Another disadvantage of single-phase motors is the mechanical vibration they produce. The mechanical frequency is exactly twice the electrical frequency. For example, any single-phase motor that operates on a 60 Hz source will vibrate at a mechanical frequency of 2 x 60 = 120 Hz. If the motor is mounted on a rigid base, the vibration may be amplified, producing a loud noise. To eliminate the noise, we can mount the motor on a special cradle support. The end bells are supported on rubber rings that dampen the 120 Hz vibrations so they are not transmitted to the base (Fig. 21-15).

Three-phase motors do not produce this vibration and so they are quieter.

The speed of split-phase motors is essentially constant. Thus, a 1/4 hp motor having a synchronous speed of 1800 rpm will have a full-load speed of about 1730 rpm. The direction can be reversed by interchanging the terminals of either the main winding or the auxiliary winding. However, the direction cannot be reversed while the motor is running. Unlike a three-phase motor, a single-phase motor cannot be plugged.

As in the case of three-phase motors, the torque of a single-phase motor varies as the square of the voltage across its terminals. Thus, at a given speed, if the line voltage is 10 percent below the rated voltage of the motor, the corresponding torque will be 20 percent lower. This rule also applies when the motor is starting up, with both windings in service.

Example 21-1:

A 1/4 hp, 120 V, 60 Hz, 1725 r/min resistance split-phase motor develops a locked-rotor torque of 1.5 N.m.
1. What is the magnitude of the rated locked-rotor torque, in percent?
2. If the line voltage drops to 112 V, what is the value of the locked-rotor torque?

Solution:

1. a) the rated torque of the motor is

$$T = 9.55\ P/n \hspace{3cm} \text{Eq. 1-9}$$

$$= 9.55 \times 0.25 \times 746/1725 = 1.032\ \text{N.m}$$

   b) the L.R. torque in percent $= \dfrac{\text{locked-rotor torque}}{\text{rated torque}} \times 100$

$$= \dfrac{1.5}{1.032} \times 100$$

$$= 145\ \%$$

2. When the line voltage falls from 120 V to 112 V, the locked-rotor torque drops from 1.5 N.m to

$$T = 1.5 \times (112/120)^2 = 1.31\ \text{N.m}$$

Example 21-2:

The motor in Example 21-1 has a full-load efficiency of 61 % and a power factor of 58 %. Calculate:

1. the rated full-load current

2. the reactive power absorbed by the motor

3. the slip

Solution:

1. a) the active power input to the motor is:

$$P_i = 100\ P_o/\text{eff} \hspace{3cm} \text{Eq. 1-11}$$

$$P_i = 100 \times (0.25 \times 746)/61 = 305.7\ \text{W}$$

   b) the apparent power of the motor is:

$$S = P/\text{power factor} = 305.7/0.58 = 527\ \text{VA} \hspace{2cm} \text{Eq. 14-22}$$

   c) the full-load current is $I = S/E = 527/120 = 4.39\ \text{A}$

2. the reactive power absorbed by the motor is:

$$Q = \sqrt{S^2 - P^2} = \sqrt{527^2 - 305.7^2} = 429.3\ \text{var} \hspace{1.5cm} \text{Eq. 14-20}$$

3. the slip is found the same way as for a of 3-phase motor:

$$s = (n_s - n)/n_s \hspace{3cm} \text{Eq. 17-1}$$

$$s = (1800 - 1725)/1800 = 0.0417, \text{ or } 4.17\ \%.$$

21.7 Capacitor-run motor

The capacitor-run motor is basically a two-phase motor. It is composed of a main winding and an auxiliary winding that is *permanently* connected to the line in series with a capacitor. These motors are used in hospitals and studios because they are particularly quiet. Their mechanical construction is simpler than that of split-phase motors because they do not contain a starting switch. The starting torque is inherently low, but because the motors are often used to drive fixed, low starting-torque loads such as fans and pumps, this presents no problem. The power factor at full-load is about 90 percent, because the capacitor is always in service.

21.8 Shaded-pole motor

Shaded-pole induction motors are often used when the power output is less than 1/20 hp. These small squirrel-cage motors have salient poles on the stator, and the main winding is a simple coil. The auxiliary winding is composed of a single turn of heavy copper that is short-circuited on itself. The turn, or ring, is wrapped around the "shaded" portion of each salient pole as shown in Fig. 21-16.

Part of the flux ϕ_m created by the coil links with the ring, inducing a large current I_a. This current produces a flux ϕ_a. Because of the inductance of the ring, I_a, and therefore ϕ_a, lag behind ϕ_m. At the pole face, the mechanical displacement between ϕ_m and ϕ_a, together with the phase lag between them, causes a revolving field to be produced. This weak revolving field produces a weak torque which starts the motor going.

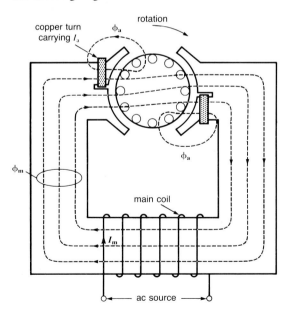

Figure 21-16
Shaded-pole motor and the fluxes and currents it produces.

Figure 21-17
Shaded-pole motor rated 5 millihorsepower, 2900 r/min, 115 V, 60 Hz. (*Courtesy Gould*)

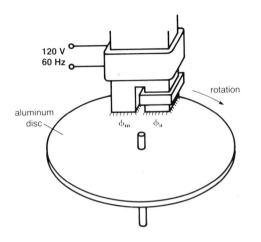

Figure 21-18
Shaded-pole motor in which the phase shift between ϕ_m and ϕ_a causes an aluminum disc to turn.

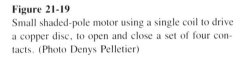

Figure 21-19
Small shaded-pole motor using a single coil to drive a copper disc, to open and close a set of four contacts. (Photo Denys Pelletier)

Although the starting torque, efficiency, and power factor are very low, the simplicity of the windings and the absence of a starting switch give this motor a distinct advantage. The direction of rotation cannot be changed; it is fixed by the location of the short-circuiting rings. In effect, the time lag between ϕ_a and ϕ_m causes the flux to continually shift from the unshaded toward the shaded portion of the pole face. This continual shift determines the direction of rotation of the motor. For example, the direction of rotation of the motor in Fig. 21-16 is clockwise.

Figure 21-18 shows a slow-speed shaded-pole motor in which the rotor is a simple aluminum disc. The phase shift between ϕ_m and ϕ_a produces a revolving field that drives the disc in the direction shown. Figure 21-19 shows such a single-pole shaded-pole motor driving a set of contacts that open and close in sequence to create an incandescent-lamp display.

21.9 Series motor

The single-phase series motor (Fig. 21-20) is similar to a dc series motor, and so it is equipped with a commutator. However, both the field and armature are laminated because they carry an ac flux. An ac series motor can therefore operate on both dc and ac, and that is why it is often called a *universal* motor.

The main advantage of fractional-horsepower series motors is their high speed, which may range from 5 000 to 25 000 r/min. Unlike induction motors, their speed does not depend upon the frequency or the number of poles. They are used to drive vacuum cleaners and small, powerful hand tools, such as hand drills and circular saws.

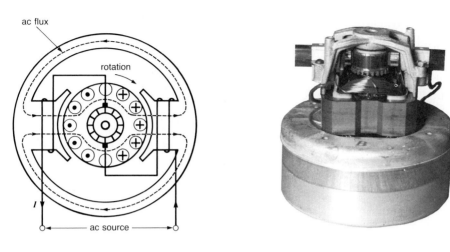

Figure 21-20
Series single-phase motors can run at very high speeds and will operate on both dc and ac power. The photograph shows a 1/3 hp, 120 V, 10 000 r/min series motor used in a domestic vacuum cleaner.

When an ac voltage is applied to the terminals of a series motor, both the armature current and the flux produced by the poles change direction simultaneously. As a result, the torque developed by the rotor always acts in the same direction. There is no revolving field in such a motor; the operating principle is the same as that of a dc series motor.

The starting torque of a series motor is high, and its speed decreases rapidly from no-load to full-load. This is an advantage in such tools as circular saws and electric drills for which we want the speed to drop significantly as the torque increases.

21.10 Choice of single-phase motors

Because of its relatively low cost, the split-phase resistance motor is the one most commonly used. However, it is applied to drives in which the starting periods are infrequent. It is mainly built in ratings between 1/20 hp and 1/4 hp. Motors of this type are used to drive fans, centrifugal pumps, washing machines, oil burners, and small machine tools, such as lathes, drill presses, and grinding wheels.

The capacitor-start motor is used for applications that require a high or prolonged starting torque. It is built in ratings between 1/8 hp and 10 hp. Such motors are used on compressors, large fans, and centrifugal pumps.

Single-phase series motors are manufactured in many types and sizes, ranging from small toy motors to large traction motors of several hundred horsepower. They operate on frequencies ranging from 60 hertz to 16 2/3 hertz.

Finally, there is a great variety of tiny synchronous motors, rated below 1/50 hp. They drive electric clocks and timing devices, and will be covered in Sections 21-12 and 21-13.

21.11 Synchros

In remote control systems, we sometimes want to vary the position of a small rheostat that is one or two meters away. This problem is easily solved by using a flexible shaft. But if the rheostat is 100 m away, the flexible-shaft solution becomes impractical. We then use an "electrical shaft" to move the rheostat. How does such a shaft work?

Consider two conventional three-phase synchronous motors in which the three stator windings are connected together, as shown in Fig. 21-21. The rotor windings are connected in parallel but instead of carrying a dc current, they are energized from a single-phase source. When the machines are connected this way, they are called *synchros*. The remarkable feature about this arrangement is that the rotor on one machine automatically tracks with the rotor on the other. Thus, if we turn one rotor cw through 29 degrees, the other rotor will move cw through 29 degrees. Such a system enables us to control the position of a device at any remote location because the wires connecting the two synchros can be threaded through holes, around corners, and over any obstacle.

Two small synchros called *transmitter* and *receiver*, are required. In Fig. 21-21, the transmitter is driven by a control knob, and the receiver is coupled to a rheostat. The 5-conductor cable linking the transmitter and receiver is the electrical shaft.

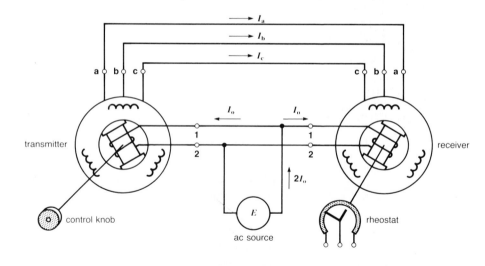

Figure 21-21
Synchro system composed of two 3-phase synchronous motors.

The behavior of this synchro control system is explained as follows. Assume that the transmitter and receiver are identical and that the rotors are stationary. When the rotors are excited by the ac source, they behave like the primaries of two transformers, inducing voltages in the stator windings. The single-phase voltages induced in the three stator windings of a synchro are always unequal because the windings are displaced by 120°. The magnitude of the voltages depends upon the position of the rotor.

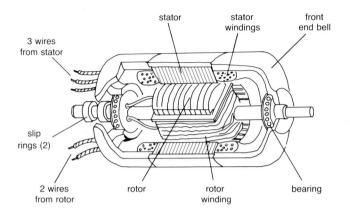

Figure 21-22
Cutaway view of a synchro.

Nevertheless, no matter what the respective stator voltages may be, they are identical in both machines when the rotors occupy the same position. As a result, the stator voltages balance each other and, consequently, no current flows in the lines connecting the stators. Thus, referring to Fig. 21-21, $I_a = I_b = I_c = 0$. The rotors, however, carry a small exciting current I_o.

If we rotate the rotor of the transmitter slightly, its three stator voltages will change. They will no longer balance the stator voltages on the receiver; consequently, currents I_a, I_b and I_c will flow in the lines connecting the two devices. These currents produce a torque on both rotors, tending to line them up. Because the rotor of the receiver is free to move, it will line up with the transmitter. As soon as the rotors are aligned, the stator voltages are again in balance, and the torque-producing currents disappear.

Synchros are often used as indicators to show the position of an antenna, valve, or gun turret. The torque required to drive a pointer is much smaller than that needed to drive a rheostat. Such transmitters and receivers are built with watch-like precision to ensure they will track with as little error as possible. Figure 21-22 shows the construction of a two-pole synchro. It is built exactly like a miniature two-pole three-phase synchronous motor. Nevertheless it is a single-phase device.

21.12 Hysteresis motors

The operation of a hysteresis motor is based upon a revolving field and the remanent magnetism in a rotor. The revolving field is created by a single-phase stator in one of the various split-phase methods we have just learned. The rotor is composed of a permanent-magnet (PM) material. In order to eliminate eddy-current effects, suppose it is made of a ceramic PM material having very high resistivity (Section 5-6).

The revolving field created by the stator can be represented by two magnets N, S

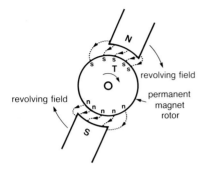

Figure 21-23

The revolving field produced by a stator creates poles of opposite polarity in the permanent magnet rotor. As a result, the rotor is always subjected to a driving torque even when it runs at the same speed as the revolving field.

Figure 21-24

Rotor and stator of a 32-pole, 115 V, 60 Hz single-phase hysteresis motor. The rim of the rotor is made of a ferrite ceramic material.

that rotate around the PM rotor (Fig. 21-23). As previously explained in Sections 5-21 and 5-22, poles of opposite polarity are continually being induced in the rotor under the N and S poles of the stator. A torque is therefore produced which drags the rotor along with the field. Thus, a hysteresis motor produces its torque in a fundamentally different way from that in an induction motor.

Because the torque is due to the magnetic attraction between the rotor and stator poles, it does not depend upon the relative speed of the rotor with respect to the stator. In other words, the torque is the same whether the N, S poles barely creep around the rotor, or whether they move quickly.

Thus, the main advantage of a hysteresis motor is that it produces a *constant* torque up to, and including, synchronous speed. This means that a hysteresis motor can bring high-inertia loads, such as tape decks and heavy turntables, up to synchronous speed. Then, because the frequency of the electric utility system is constant, the speed remains fixed like a clock.

Hysteresis motors are synchronous motors, and they are used in millions of electric clocks and timing devices. On large electric utility systems, the frequency is kept very constant, although not with quartz-watch precision. However, the frequency is monitored and adjusted so that during a 24-hour day on a 60 Hz system, the number of cycles generated is exactly 24 x 3600 x 60 = 5 184 000. Thus, although electric clocks may run slow or fast during certain times of the day, they are absolutely precise over any 24-hour period. (This of course is true only if the system has not suffered an outage).

Figure 21-24 shows a 32-pole shaded-pole hysteresis motor used to drive a kitchen clock.

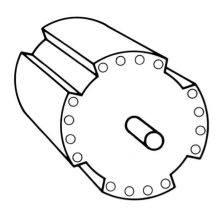

Figure 21-25
A conventional squirrel-cage rotor with four salient poles will operate as a synchronous motor when placed inside a 4-pole single-phase stator.

21.13 Reluctance-torque motor

Single-phase *reluctance-torque* motors have a salient-pole rotor that also carries a squirrel-cage winding (Fig. 21-25). The stator may be a split-phase or shaded-pole type. When it is energized, the squirrel-cage winding brings the motor nearly up to synchronous speed. The reluctance torque then comes into play, and the rotor is pulled into synchronism by virtue of the salient poles. This behavior was described in Section 19-2 for a three-phase synchronous motor. The same explanation applies to single-phase motors. Reluctance-torque motors can bring only low-inertia loads up to synchronous speed because their pull-in torque is limited. However, they are considerably cheaper than hysteresis motors.

21.14 Stepper motors

Stepper motors have salient poles on both the rotor and stator. The stator poles carry windings that are excited in sequence, either slowly or in rapid succession. The reluctance torque causes the rotor to advance in response to these excitation pulses. Stepper motors are not really single-phase motors because they are often powered by a dc source. The source is connected in sequence to individual windings that are equally spaced around the stator. This causes the rotor to advance in steps.

Stepper motors are made for digital-control systems. They enable an object to be positioned with high precision. The object (tracing pen, drill, or punch) is coupled to the motor by gears, and so each step of the motor corresponds to a precisely known distance, or angular movement.

One of the simplest stepper motors consists of a 3-pole stator and 2-pole iron rotor (Fig. 21-26). The rotor is shown at rest in the vertical position. Current I_1

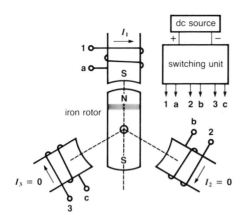

Figure 21-26

Stepper motor with rotor in the vertical position because of the flux produced by I_1. The band across the rotor is merely to identify the pole.

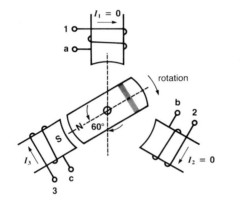

Figure 21-27

The rotor turns cw through 60° when I_1 is switched off and I_3 is switched on.

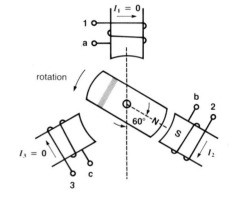

Figure 21-28

The direction of rotation can be reversed by turning off I_1 and turning on I_2.

flows so as to keep the rotor locked in this position. To make the rotor turn cw, we interrupt I_1 and cause I_3 to flow, by means of the switching unit. This will cause the rotor to step cw by an angle of precisely 60°. Current I_3 is then maintained for as long as we want the rotor to stay in this new position (Fig. 21-27).

If we want the rotor to turn ccw, we energize winding 2 instead of winding 3, and the rotor again moves through an angle of 60° (Fig. 21-28).

It is clear that we can cause the rotor to move cw or ccw in 60° steps. Thus, a pulse rate of one per second will produce a speed of 10 r/min. The successive exciting pulses are stored electronically to keep accurate track of what is going on. Knowing which coils were excited in the past, the system "knows" precisely where the rotor is at any instant. However, because we must be sure that each pulse does in

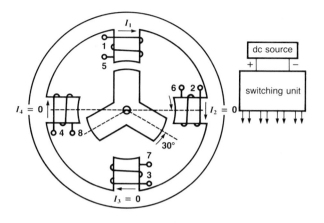

Figure 21-29
Stepper motor that advances 30° per pulse.

fact correspond to a 60° step, we must adjust the stepping rate so that the rotor has time to respond.

For example, if we apply stepping pulses at the rate of 200 per second when the rotor is at standstill, it will only vibrate in place. The rotor cannot follow these rapid signals because of its inertia and the inertia of the load. However, if the motor is already turning at 1900 r/min (corresponding to a pulse rate of 190/s), an increase in the pulse rate to 200 per second presents no problem. The rate at which the pulses may be generated depends therefore upon the speed at which the stepper motor is turning. This is true whether the motor is accelerating or decelerating. Fast response pulsing must therefore be computer-controlled to reduce the risk of missing a step.

We can diminish the angular motion per step by increasing the number of poles. For example, Fig. 21-29 shows a stepper motor that advances 30 degrees per pulse. There are many types of stepper motors, but the basic principle remains the same: to control the position of an object by digital means.

21.15 Precision motor in a watthourmeter

Watthourmeters that measure the energy consumed in domestic homes and industry contain extremely accurate single-phase motors. They rotate at a speed that is directly proportional to the active power (kilowatts) used by the consumer. Thus, it is possible to measure the energy consumed during a given period of time by recording the number of turns on a register.

The motor is similar to the shaded-pole disc motor illustrated in Fig. 21-18, except that ϕ_m is created by the line voltage and ϕ_a by the line current at the con-

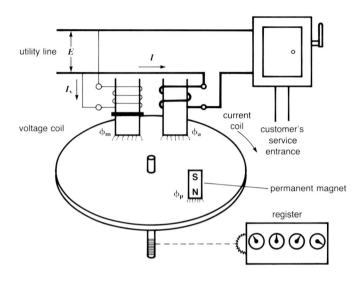

Figure 21-30
Principle of the watthourmeter.

sumer's service entrance. Figure 21-30 shows schematically how the voltage and current coils are placed above the disc and how they are connected to the consumer's line. The voltage coil is highly inductive, and so current I_v (and the corresponding flux) lags almost 90° behind the line voltage E. A small short-circuiting ring around the core adjusts the phase so that the net flux ϕ_m crossing the disc is exactly 90° behind E. The flux ϕ_a created by the current coil is in phase with the line current I.

When the line current is in phase with the line voltage, ϕ_m and ϕ_a are 90° out of phase, and so a torque is exerted on the disc. However, if the current lags 90° behind the voltage, ϕ_m and ϕ_a are in phase and no revolving field is produced. Consequently, the torque is zero. In general, the torque exerted on the disc is given by the expression $T = k_1\phi_m\phi_a\cos\theta$, which can be reduced to the form

$$T = kEI\cos\theta \qquad (21\text{-}2)$$

where

$T =$ torque, in newton meters (N.m)

$E =$ line voltage, in volts (V)

$I =$ line current, in amperes (A)

$\theta =$ phase angle between E and I, in degrees (°)

$k =$ constant that depends upon the design of the instrument motor

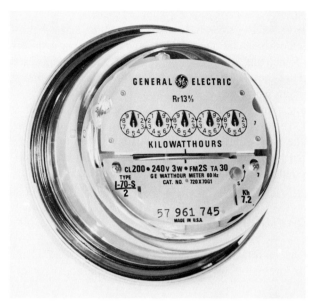

Figure 21-31
Single-phase watthourmeter for a 120/240 V, 60 Hz service entrance. The rotating aluminum disc can be observed through the narrow slot immediately below the label "kilowatthours". (*Courtesy General Electric*).

Thus, the torque is proportional to the active power $EI\cos\theta$ consumed. However, if the motor is free to rotate, it will turn at a speed that is limited only by electrical losses and mechanical friction. As a result, the speed will not be linearly related to the torque. To obtain a linear relationship (speed proportional to torque) a permanent magnet is introduced so that the disc sweeps across its flux ϕ_p.

The resulting braking torque T_b is directly proportional to the speed. Thus, $T_b = k_2 n$. But we recall from Section 1-15 that the speed of a motor is stable when the driving torque is equal to the load torque. Consequently, because $T = T_b$, the speed of rotation n is directly proportional to the active power.

These highly precise motors are calibrated so that each revolution corresponds to a definite amount of electrical energy, expressed in *watthours*. This energy is indicated by the constant K_h inscribed on the nameplate.

Example 21-3:

The disc of a watthourmeter makes 5 complete turns in 78 s. If the nameplate shows $K_h = 7.2$, calculate the active power consumed.

Solution:

a)   The energy consumed during the interval is

$$W = 5 \text{ turns} \times 7.2 \text{ W.h/turn} = 36 \text{ W.h}$$

b)   The duration of the interval is 78 s = 78/3600 h = 0.0217 h.

c)   The power is $P = W/t = 36$ W.h/0.0217 h = 1659 W.

21.16 Summary

In contrast with heavy industry, where three-phase induction motors and synchronous motors predominate, the domestic and business-machine markets use a variety of single-phase motors. We found that single-phase motors are more complicated than three-phase motors because they require a starting winding and an automatic switch to disconnect the winding when the motor is close to its rated speed.

We also discovered an interesting application of hysteresis and of reluctance torque in the design of small synchronous motors. These properties were first mentioned when we introduced the basic principles of magnetism.

Finally, we learned that synchros and stepper motors can be used in precise control systems, such as are needed in robotics and the machine-tool industry.

TEST YOUR KNOWLEDGE

21-1   If the main winding and auxiliary winding of a single-phase motor are connected in series, it is impossible to obtain a starting torque.

☐ true   ☐ false

21-2   The auxiliary winding of a split-phase motor must be disconnected quickly to

a. prevent overheating   b. produce a good starting torque
c. reduce the power drawn from the line

21-3   The rated full-load speed of a 50 Hz single-phase motor is 2850 r/min. If the motor is connected to a 60 Hz source, the speed

a. will be close to 2375 r/min   b. will be close to 3420 r/min
c. will not run on a 60 Hz line

21-4   At no-load, the line current of a single-phase motor lags 72° behind the line voltage. The no-load power factor is

a. zero   b. 30.9 %   c. 95.1 %

21-5   A capacitor-start motor develops a high torque because

a. it increases the auxiliary flux ϕ_a
b. it increases the phase angle between ϕ_a and ϕ_m
c. it absorbs less reactive power from the line

21-6   A single-phase motor is connected to a 400 Hz source. It will vibrate at a frequency of

a. 400 Hz   b. 200 Hz   c. 800 Hz

synchros

21-7   The direction of rotation of a shaded-pole motor can be reversed by interchanging the leads of the ac source.

☐ true   ☐ false

21-8   The direction of rotation of an ac series motor can be reversed by interchanging the armature leads.

☐ true   ☐ false

21-9   A hysteresis clock motor incorporated with a gear box has a no-load speed of 1.00 r/min, when connected to a 120 V, 60 Hz source. If the voltage increases to 120 V, the speed will be

a. 1.05 r/min   b. 0.952 r/min   c. 1.00 r/min

21-10   In stepper motors, the number of poles on the rotor and stator is usually different.

☐ true   ☐ false

QUESTIONS AND PROBLEMS

21-11   A 10-pole split-phase motor is connected to a 60 Hz source. What is its synchronous speed?

21-12   What is the purpose of the auxiliary winding in a single-phase induction motor? How can we change the rotation of such a motor?

21-13   State the main differences between a capacitor-start motor and a resistance split-phase motor. What are their relative advantages?

21-14   Explain the operation of a shaded-pole motor.

21-15   Why are some single-phase motors equipped with a resilient mounting? Is such a mounting necessary on three-phase motors?

21-16   What is the main advantage of a capacitor-run motor?

21-17   Which of the motors discussed in this chapter is best suited to drive the following loads:
a. small portable drill
b. 3/4 hp air compressor
c. vacuum cleaner
d. 1/100 hp blower
e. 1/3 hp centrifugal pump
f.   1/4 hp fan for use in a hospital ward
g. electric timer
h. stereo turntable

21-18   The palm of the human hand can just barely tolerate a temperature of 130°F. If the no-load temperature of the frame of a 1/4 hp motor is 64°C in an ambient temperature of 76°F

a. can a person keep his hand on the frame?
b. is the motor running too hot?

21-19   A 115 V, 60 Hz shaded-pole motor develops a rated mechanical output of 6 W at 2990 r/min. The corresponding input power is 21 W, and the line current is 0.33 A. Calculate the

a.   efficiency of the motor
b.   power factor of the motor
c.   torque developed by the motor, in millinewton meters (mN.m)
d.   slip
e.   phase angle between the line voltage and line current
f.   losses in the motor

21-20   The motor in Problem 21-19 has a locked-rotor current of 0.35 A and a locked-rotor torque or 10 mN.m. The power input is 24 W. Compare the locked-rotor losses with the full-load losses.

21-21   If the rotor of the stepper motor in Fig. 21-29 had four poles instead of three, it would be impossible to make it rotate. Explain.

21-22   If the 3-pole rotor of the stepper motor in Fig. 21-29 is replaced by a 5-pole rotor, calculate the smallest angular motion per pulse when the coils are excited in sequence. Hint: Draw a diagram of the rotor and stator.

21-23   A 4-pole, 60 Hz single-phase motor is equipped with a centrifugal switch taken from a 2-pole, 60 Hz motor. Explain why the auxiliary winding may burn out and why the motor can never operate properly.

CHAPTER 22

Industrial Motor Control

Introduction and Chapter objectives

Industrial control, in its broadest sense, consists of all the methods used to control the performance of an electrical system. When applied to machinery, it involves the starting, acceleration, reversal, deceleration, and stopping of a motor and its load. In this chapter we will study the electrical (but not electronic) control of dc motors and three-phase induction and synchronous motors. Our study will include several elementary circuits and components because they are the building blocks that apply to any system of control, no matter how complex it may be.

This chapter will also introduce the method of representing control circuits by means of schematic diagrams. Such diagrams enable us to analyze the behavior of ac and dc starters while the motor is accelerating.

22.1 Typical control devices

Every control circuit is composed of a number of basic components connected together to achieve the desired performance. The size of the components varies with the power they handle, but the principle of operation remains the same. Using only a dozen basic components, we can design control systems that are very complex. The basic components are:

1. Disconnecting switches
2. Manual circuit breakers
3. Cam switches
4. Pushbuttons
5. Relays
6. Magnetic contactors
7. Thermal relays and fuses
8. Pilot lights
9. Limit switches and other special switches
10. Resistors, reactors, transformers, capacitors

Table 22A illustrates these devices and states their main purpose and application. The symbols for these and other devices are given in Table 22B.

TABLE 22A   BASIC COMPONENTS FOR CONTROL CIRCUITS

Disconnecting switches (Fig. 22-1)

A fused disconnecting switch isolates the motor from the power source. It consists of three knife-switches and three line fuses enclosed in a metal box. An external handle opens and closes the knife-switches simultaneously. An interlocking mechanism prevents persons from opening the hinged cover when the switch is closed. Disconnecting switches are designed to carry the rated full-load current indefinitely, and to withstand short-circuit currents for brief intervals. Some disconnecting switches have no fuses: they serve only to isolate a circuit, not to protect it.

Manual circuit breakers (Fig. 22-2)

A manual circuit breaker opens and closes a circuit, like a toggle switch. It trips (opens) automatically when the current exceeds a predetermined limit. After tripping, it can be reset manually. Circuit breakers are often used instead of fused disconnecting switches because no fuses have to be replaced.

Cam switches (Fig. 22-3)

A cam switch has a group of fixed contacts and an equal number of movable contacts. The contacts can be made to open and close in a preset sequence by rotating a handle or knob. Cam switches are used to control the motion and position of hoists, cranes, and machine tools.

Pushbuttons (Fig. 22-4)

A pushbutton is a switch that is actuated by finger pressure. Two or more contacts open or close when the button is depressed. Pushbuttons are usually spring-loaded so they will return to the normal position when pressure is removed.

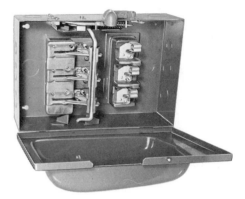

Figure 22-1
Fused disconnecting switch, rated 600 V, 30 A, 3-phase (*Courtesy Square D*)

Figure 22-2
Manual circuit breaker rated 600 V, 100 A, 3-phase. (*Courtesy Square D*)

Control relays (Figs. 22-5 and 22-6)

A control relay is an electromagnetic switch that opens and closes a set of contacts when the relay coil is energized. The coil produces a strong magnetic field which attracts a movable armature bearing the contacts. Control relays are mainly used in low-power circuits. They include time-delay relays whose contacts open or close after a definite time interval.

Thermal relays (Fig. 22-7)

A thermal relay (or overload relay) is a small protective device whose contacts open when the line current exceeds a preset limit. The current flows through a calibrated heating element which raises its temperature. When the temperature exceeds a certain limit, the relay trips and its control contacts open. Thermal relays are inherent time-delay devices because the temperature cannot follow the instantaneous changes in current.

Magnetic contactors (Figs. 22-8 to 22-10)

A magnetic contactor is basically a large control relay designed to open and close a power circuit. It has a relay coil that actuates a set of contacts. Magnetic contactors are used to control motors ranging from 0.5 hp to several hundred horsepower. The size, dimensions, and performance of contactors are standardized.

Pilot lights (Fig. 22-11)

A pilot light indicates the on/off state of a remote component in a control system.

Limit switches and special switches (Figs. 22-12 and 22-13)

A limit switch is a low-power snap-action device that opens or closes a control contact, depending upon the position of a mechanical part. Other limit switches are sensitive to pressure, temperature, liquid level, direction of rotation, and so forth.

Figure 22-3
Three-position cam switch rated 250 V, 10 A. (*Courtesy Siemens*)

Figure 22-4
Pushbutton with NO and NC contacts rated to interrupt an ac current of 6 A one million times. (*Courtesy Siemens*)

Figure 22-5
Single-phase control relay rated 25 A, 115/230 V, 60 Hz. (*Courtesy Potter and Brumfield*)

Figure 22-6
Time-delay relay continuously variable from 5 s to 1.5 min. (*Courtesy Siemens*)

Figure 22-7
Thermal overload relay rated 400 A.
(*Courtesy Siemens*)

Figure 22-8
Magnetic contactor sizes are standardized according to NEMA specifications. This picture shows the range. (*Courtesy Ward Leonard Electric*)

Figure 22-9
Magnetic contactor with the arc shield removed, exposing the three-phase contacts. (*Courtesy Siemens*)

Figure 22-10
Direct current contactor rated 65 A, 600 V, showing the holding coil on the right and two arc shields on the left. Arc suppression is more difficult in dc contactors. (*Courtesy Siemens*)

Figure 22-11
Start-stop pushbutton station with pilot light. (*Courtesy Siemens*)

Figure 22-12
Limit switch with one NC contact rated for ten million operations; position accuracy is 0.5 mm. (*Courtesy Square D*)

Figure 22-13
Liquid level switch. (*Courtesy Square D*)

TABLE 22B   GRAPHIC SYMBOLS FOR ELECTRICAL DIAGRAMS

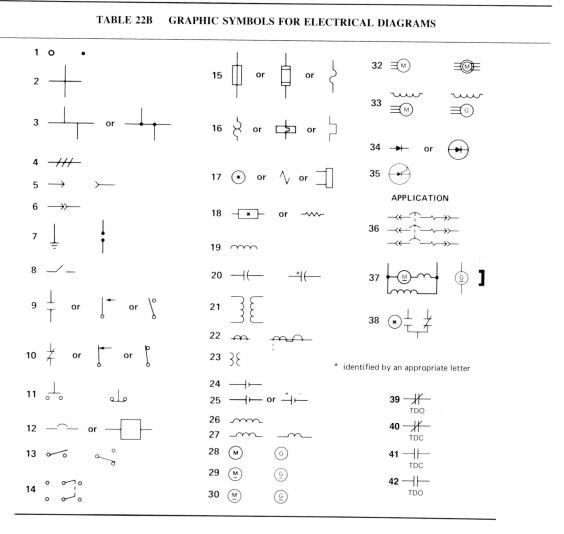

APPLICATION

* identified by an appropriate letter

1. terminal; connection   2. conductors crossing   3. conductors connected   4. three conductors   5. plug; receptacle
6. separable connector   7. ground connection; arrester   8. disconnecting switch   9. normally open contact (NO)
10. normally closed contact (NC)   11. pushbutton NO; NC   12. circuit-breaker   13. single pole switch; three-way switch
14. double pole double throw switch   15. fuse   16. thermal overload element   17. relay coil   18. resistor
19. winding, inductor or reactor   20. capacitor; electrolytic capacitor   21. transformer   22. current transformer; bushing type
23. potential transformer   24 dc source (general)   25. cell   26. shunt winding   27. series winding; commutating pole or compensating winding
28. motor; generator (general symbols)   29. dc motor; dc generator (general symbols)   30. ac motor; ac generator (general symbols)
32. 3-phase squirrel-cage induction motor; 3-phase wound-rotor motor   33. synchronous motor; 3-phase alternator   34. diode
35. thyristor or SCR   36. 3-pole circuit breaker with magnetic overload device, drawout type
37. dc shunt motor with commutating winding; permanent magnet dc generator   38. magnetic relay with one NO and one NC contact.
39. NC contact with time delay opening   40. NC contact with time delay closing   41. NO contact with time delay closing
42. NO contact with time delay opening

For a complete list of graphic symbols and references see *"IEEE Standard and American National Standard Graphic Symbols for Electrical and Electronics Diagrams"* (ANSI Y32.2/IEEE No. 315) published by the Institute of Electrical and Electronic Engineers, Inc., New York, N.Y. 10017. Essentially the same symbols are used in Canada and several other countries.

22.2 Normally open and normally closed contacts; graphic symbols

Control circuit diagrams always show components in a state of rest, that is, when they are not energized (electrically) or actuated (mechanically). In this state, some electrical contacts are open while others are closed. They are respectively called *normally open contacts* (NO) and *normally closed contacts* (NC). They are designated by the symbols given in Table 22B.

The coil of a control relay or contactor is usually represented by a small circle. The circle bears a letter, such as A, M, F, or a combination of letters and numbers, such as 1A, 2A, 1M, 2M, CR and CX, in order to identify the relay or contactor. The contacts that are associated with a given relay bear the same letter or letter-number as the relay coil. This is important because the contacts of a given relay can appear in quite different places in a schematic diagram.

Most contactors are equipped with one or more *auxiliary* contacts in addition to the main power contacts. The auxiliary contacts are small control contacts that form part of the low-power control circuit. The auxiliary contacts are sometimes identified by the letter-number of the coil and a subscript, such as A_X, or $1A_a$. The subscripts simplify the description of the control circuit.

22.3 Control diagrams

Depending upon the amount of information that is required, a control system can be represented by four types of circuit diagrams. They are listed here in order of increasing detail and completeness:

1. block diagrams

2. one-line diagrams *

3. schematic diagrams

4. wiring diagrams

A *block diagram* (Fig. 22-14) is composed of a set of rectangles, each representing a control device, together with a brief description of its function. The rectangles are connected by arrows which indicate the direction of power flow.

A *one-line diagram* (Fig. 22-15) is similar to a block diagram except that the components are shown by their symbols rather than by rectangles. The symbols give an idea of the nature of the components; consequently, one-line diagrams contain more information. The single lines connecting the various components represent one, two, or more wires.

A *wiring diagram* (Fig. 22-16) shows the connections between the components, taking into account the location of the terminals and even the color coding of the wires. These diagrams are employed when installing equipment or troubleshooting a circuit.

A *schematic diagram* (Fig. 22-17) shows all the electrical connections between components but without regard to their location or terminal arrangement. This type of diagram is indispensable when troubleshooting a circuit or analyzing its mode of operation.

* Also called single-line diagrams.

Note that the four diagrams in Figs. 22-14 to 22-17 describe the same control circuit. The symbols used to designate the various components are given in Table 22B.

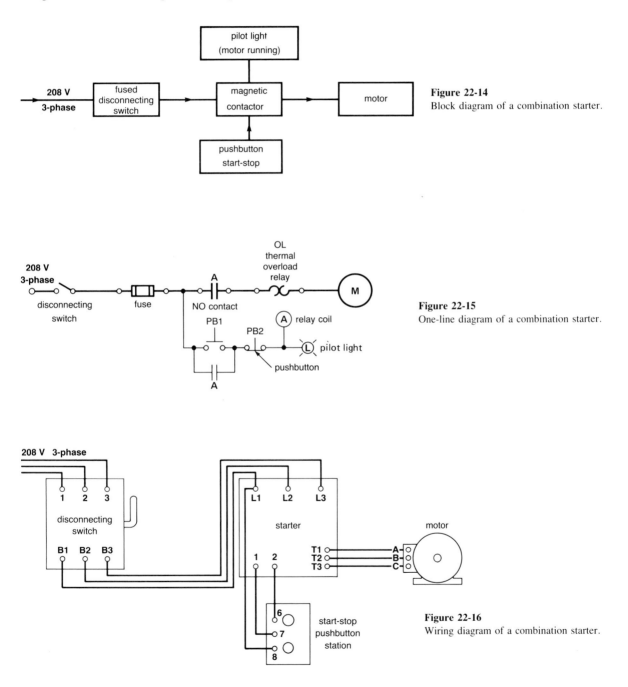

Figure 22-14
Block diagram of a combination starter.

Figure 22-15
One-line diagram of a combination starter.

Figure 22-16
Wiring diagram of a combination starter.

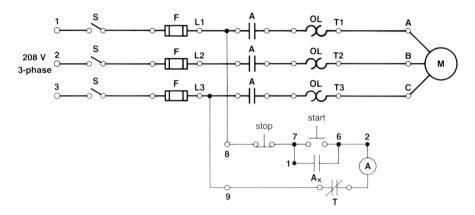

Figure 22-17
Schematic diagram of a combination starter.

22.4 Automatic dc starters

In Chapter 12 on dc motors, we learned that a resistor must be placed in series with the armature during the starting period. This can be done manually by using a face-plate starter, but in modern equipment an automatic starter is preferred. We will describe a very simple dc starter, in order to explain the basic principles that are common to all dc starters. We shall proceed in two steps. The first step will show the basic circuit that will bring the motor up to speed; the second step introduces the additional components needed to adequately protect the motor and electrical system.

Figure 22-18 is the schematic diagram of a dc starter for a shunt-wound motor. The armature terminals are labeled A1 and A2 and the field terminals F1, and F2. In addition to the start-stop pushbuttons, the circuit contains a main contactor M and a starting contactor A. The associated main contacts M and A are designed to carry the rated armature current. The auxiliary contacts M_{1X}, M_{2X}, and A_{1X} carry control currents, and so they are much smaller than the main contacts.

Contactor A is designed so that its contacts operate as soon as voltage is applied to relay coil A. However, when the coil is de-energized, it takes several seconds for the NC contact A to return to its normal closed position. This property is designated by the letters TDC (time delay closing). Similarly, it takes several seconds for the NO contact A_{1X} to return to its normal open position. This property is designated by the letters TDO (time delay opening).

The motor is brought up to speed in two steps because the starter has only one series resistor R. In larger motors, at least two resistors would be used to obtain a more gradual acceleration.

The circuit operates as follows. When the disconnecting switch is closed, the dc line voltage between L1 and L2 becomes available to drive the motor. By pressing the start pushbutton, relay coil A becomes energized, causing contact A to open and A_{1X} to close. This removes the short-circuit across R, just a fraction of a second before A_{1X} closes. The closing of A_{1X} energizes relay coil M, and main contact M

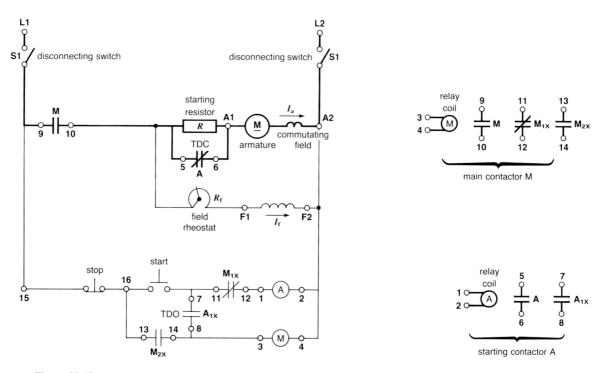

Figure 22-18
Basic control circuit of a dc starter for a shunt motor. The relay coils and contacts on the right show even more clearly how they are associated, and how they relate to the control diagram.

closes. This applies full voltage across the field circuit, as well as the armature circuit. The armature current I_a is established at once, but the field current I_f takes time to build up because of the high inductance of the shunt field.

When M is energized, contact M_{1X} opens, which removes the excitation of coil A. However, because of the TDC, TDO time-delay feature mentioned above, contact A remains open and contact A_{1X} remains closed for several more seconds.

Contact M_{2X} closes at the same time as main contact M and it plays an important role. In effect, it provides a current path to energize relay coil M, even when the start pushbutton returns to its normal position. As a result, to start the motor we have to depress the start pushbutton only momentarily. Auxiliary contact M_{2X} is sometimes called a *self-sealing* contact, because it "seals" the current path of its own relay coil.

As soon as contact A closes, full voltage appears across the armature, and the motor runs normally. To stop it, we momentarily depress the stop pushbutton, which de-energizes coil M, causing main contact M and self-sealing contact M_{2X} to open. The motor will coast to a stop, and during this interval it acts as a self-excited shunt generator.

The simple circuit of Fig. 22-18 has some serious drawbacks because it provides

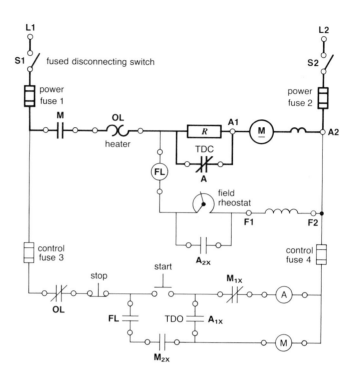

Figure 22-19
Basic control circuit of the dc starter including circuit-protective devices.

no protection against overloads, short-circuits, or overspeeding of the motor. This deficiency is overcome in Fig. 22-19. It requires the addition of fuses, a thermal overload relay OL, a field-loss relay FL, and an extra contact A_{2X} on contactor A.

The *power* fuses will blow if a short-circuit occurs in the starter box or in the armature circuit. The *control* fuses have a much smaller current rating because they protect the smaller components and smaller wiring of the control circuit. The heater of the overload relay carries the full motor current. If a sustained overload occurs, the heat given off by the heater will cause the normally closed OL contact to open, which, in turn, opens the circuit of relay coil M, causing contactor M to drop out.

The extra contact A_{2X} is closed whenever contact A is open. This means that the field rheostat is shorted-circuited during the acceleration period when resistor R is in service. As a result, the field strength is maximum, which ensures the highest possible starting torque.

The field-failure relay coil FL is energized as long as the shunt field current is above a predetermined minimum. If, for any reason, the shunt field becomes too weak, the FL contact will open, causing main contactor M to drop out. Thus, the FL relay protects the motor from overspeeding.

22.5 Starting methods for three-phase induction motors

Three-phase squirrel-cage motors are started either by connecting them directly across the line or by applying reduced voltage to the stator. The starting method (and type of starter) depends upon the power capacity of the supply line and the type of load.

Across-the-line starting is simple and inexpensive. The main disadvantage is the high starting current. It can produce a significant voltage drop, which may affect other customers connected to the line. Voltage-sensitive devices such as incandescent lamps, television sets, and high-precision machine tools respond poorly to such voltage dips. The voltage drop can also seriously reduce the torque of the motor while it is starting up.

Mechanical shock is another problem that should not be overlooked. Equipment can be damaged if full-voltage starting produces a hammer-blow impulse. Thus, printing presses, paper mills, electric trains, and elevators are always started on reduced voltage. Conveyor belts are another example in which sudden starting may not be acceptable.

In large industrial installations, we can sometimes install across-the-line starters for motors rated up to as much as 10 000 hp. The fuses and contactors must be chosen to carry the starting current during the acceleration period.

A disconnecting switch or manual circuit-breaker is always placed between the supply line and the starter. The switch and starter are sometimes mounted in the same enclosure to make what is called a *combination starter*. The line fuses (when used) are rated at about 3.5 times full-load current; consequently, they do not protect the motor against continuous overloads. Their primary function is to protect the motor and supply line against catastrophic currents resulting from a short-circuit in the motor or starter, or a failure to start up. The fuse rating, in amperes, must comply with the requirements of the National Electrical Code.

22.6 Manual across-the-line starters

Manual three-phase starters (Fig. 22-20 and 22-21) are composed of a circuit breaker, and either two or three thermal relays, all mounted in an enclosure. They are used for small motors (usually 10 hp or less) at voltages ranging from 115 V to 575 V. The thermal relays trip the circuit-breaker whenever the current in one of the phases exceeds the rated value.

22.7 Magnetic across-the-line starters

Magnetic across-the-line starters are used whenever we have to control a motor from a remote location. Figure 22-22 and 22-23 show a typical combination magnetic starter and its schematic diagram.

The starter has five main components: a disconnecting switch, a fuse block, a magnetic contactor, a thermal relay, and a control transformer. An external control station permits the motor to be started and stopped.

Figure 22-20
Manual starters for single-phase motors rated at 1 hp; left: surface mounted; center: flush mounted; right: waterproof enclosure. (*Courtesy Siemens*)

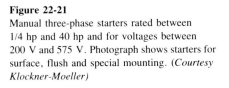

Figure 22-21
Manual three-phase starters rated between 1/4 hp and 40 hp and for voltages between 200 V and 575 V. Photograph shows starters for surface, flush and special mounting. (*Courtesy Klockner-Moeller*)

1. The *magnetic contactor* A has three main contacts A and one small auxiliary contact A_x. Contacts A must be big enough to carry the starting current and the rated full-load current without overheating. Contacts A and A_x remain closed as long as the relay coil A is energized.

2. The *thermal relay* OL protects the motor against sustained overloads. The relay is composed of three individual heater elements connected in series with the three phases. A small normally closed contact OL forms part of the relay assembly. It opens when the relay gets too hot and stays open until the relay is manually reset.

    The current rating of the thermal relay is chosen to protect the motor against sustained overloads. Contact OL opens after a period of time that depends upon the magnitude of the overload current.

3. The *control station*, composed of start-stop pushbuttons, may be located either close to, or far away from, the starter. The pilot light is optional.

4. The 600 V/120 V control transformer steps down the line voltage to a standard 120 V control voltage. The control circuit is protected by a low-caliber fuse.

5. The fused disconnecting switch serves to isolate the motor and starter during routine inspection or repairs. The fuses protect the motor and starter components against catastrophic failure.

    To start the motor, we first close the disconnecting switch and then depress the start pushbutton momentarily. Relay coil A is immediately energized, contacts A and A_x close, and full voltage appears across the motor. When the pushbutton is released, it returns to its normal position, but relay coil A remains excited because auxiliary contact A_x is now closed. Contact A_x is a self-sealing contact.

Figure 22-22
Combination across-the-line starter rated 100 hp, 575 V, 3-phase, 60 Hz. The disconnecting switch is controlled by an external handle. The three fuse supports are immediately below the switch and the magnetic contactor is mounted in the lower left-hand corner. The small 575/115 V transformer in the lower right-hand corner supplies low-voltage power for the control circuit. (*Courtesy Square D*)

To stop the motor, we push the stop button, which opens the circuit to the relay coil. In case of an overload, the opening of contact OL produces the same effect.

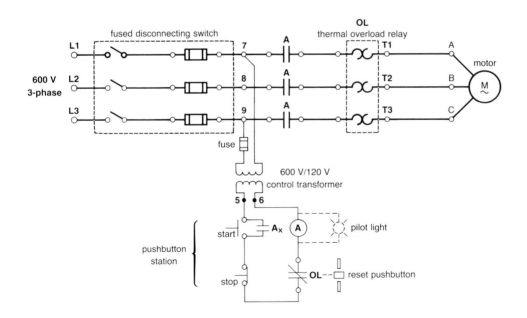

Figure 22-23
Schematic diagram of a three-phase combination starter using a low-voltage control circuit.

22.8 Inching and jogging

In some mechanical systems, we have to adjust the position of a motorized part very precisely. To accomplish this, we energize the motor in short pulses so that it barely starts before it again comes to a halt. A double-contact pushbutton J is added to the usual start-stop circuit, as shown in Fig. 22-24. This arrangement permits conventional start-stop control as well as so-called *jogging*, or *inching*.

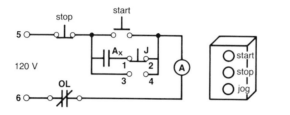

Figure 22-24
Control circuit and pushbutton station for start-stop-jog operation. Terminals 5 and 6 are connected to the secondary of the control transformer.

Jogging imposes severe duty on the power contacts because they continually make and break currents that are six times greater than normal. When jogging is required, the contactor is usually selected to be one NEMA size bigger than for normal duty.

22.9 Reduced-voltage starting

As mentioned previously, some industrial loads have to be started very gradually. Examples are coil winders, printing presses, and other machines that process fragile products. In other industrial applications, we cannot connect a motor directly to the line because the starting current may be too high.

In such cases, we have to reduce the voltage applied to the motor, either by connecting resistors (or reactors) in series with the line or by employing an autotransformer. In reducing the voltage, we should remember that:

1. The locked-rotor stator current is proportional to the stator voltage: reducing the voltage by half reduces the current by half;

2. The locked-rotor torque is proportional to the square of the voltage: reducing the stator voltage by half reduces the torque by a factor of four.

22.10 Primary resistance starting

Primary resistance starting consists of placing three resistors in series with the stator during the start-up period (Fig. 22-25)*. After a time sufficiently long so the motor is running close to rated speed, a second magnetic contactor B short-circuits the resistors. This method gives a very smooth start with negligible mechanical shock. The resistors are short-circuited after a delay that depends upon the setting of time-delay relay 1A. Note that contact 1A bears the letters TDC for time delay closing.

Figure 22-26 shows a typical primary resistance starter.

---

*   Secondary resistance starting consists of placing three resistors in the rotor circuit of a wound-rotor induction motor.

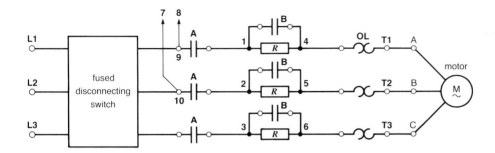

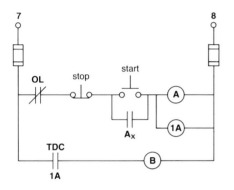

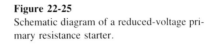

Figure 22-25
Schematic diagram of a reduced-voltage primary resistance starter.

Figure 22-26
Primary resistance starter for a 50 hp, 575 V, 3-phase, 60 Hz motor. The three resistors are in the upper left-hand corner, and the two magnetic contactors are mounted side by side below. (*Courtesy Siemens*)

22.11 Autotransformer starting

For a given locked-rotor torque, autotransformer starters draw a lower line current than do resistance starters. The disadvantages are that autotransformers cost more, and the transition from reduced voltage to full voltage is not quite as smooth.

The autotransformer is equipped with taps that may be selected to give output voltages equal to 80%, 65%, or 50% of the rated motor voltage. The corresponding starting torques are 0.64, 0.42, or 0.25 of the full-voltage starting torque. Furthermore, the starting currents on the line side are reduced to 0.64, 0.42, or 0.25 of the full voltage locked-rotor current.

Figure 22-27
Autotransformer reduced voltage starter rated 100 hp, 575 V, 3-phase, 60 Hz. (*Courtesy Square D*)

Figure 22-27 shows a starter using two autotransformers connected in open delta. Its schematic diagram is shown in Fig. 22-28, making use of the 65% tap. The time-delay relay CR has three contacts CR_1, CR_2 and CR_3. Contact CR_1 in parallel with the start pushbutton closes as soon as relay coil CR is energized. The two other contacts operate after an interval that depends upon the time-delay setting. Contactors A and B are mechanically interlocked to prevent them from closing simultaneously.

Contactor A closes as soon as the start pushbutton is depressed because contact CR_2 is closed. This excites the autotransformer, and reduced voltage (65 percent of full-line voltage) appears across the motor terminals. A few seconds later, the two CR contacts in series with coils A and B respectively open and close. Contactor A drops out, followed almost immediately by the closing of contactor B. This action applies full voltage to the motor and simultaneously disconnects the autotransformer from the line. Note that relay coil B cannot become excited before contact A_X closes. Consequently, contactor B cannot come into service before contactor A has dropped out.

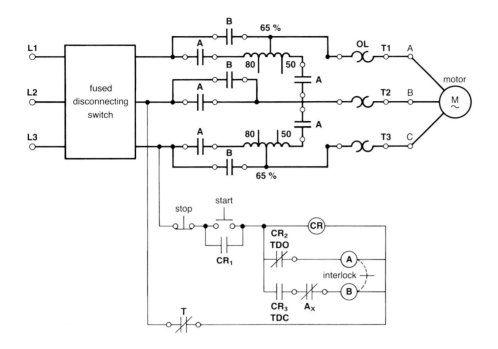

Figure 22-28
Simplified schematic diagram of an autotransformer starter.

In transferring from contactor A to contactor B, the motor is momentarily disconnected from the line. This creates a problem because when contactor B closes, a large transient current is drawn from the line. This transient surge is hard on the contacts and also produces a mechanical shock. For this reason, we sometimes employ more elaborate circuits in which the motor is never completely disconnected from the line.

Because the autotransformers are in service for only a short time (usually less than 10 s), they can be wound with much smaller conductors than continuously rated transformers. This enables manufacturers to reduce the size, weight, and cost of these starters.

Example 22-1:

A 200 hp, 460 V, three-phase, 60 Hz, 1785 r/min induction motor has a locked-rotor torque of 600 ft.lbf and a locked-rotor current of 1350 A. The motor is connected to an autotransformer starter, as shown in Fig. 22-28. Calculate:

1. the locked-rotor stator current

2. the locked-rotor torque

3. the current flowing in the 460 V supply lines L1, L2, and L3.

Solution:

1. a) the locked-rotor stator current is proportional to the stator voltage

   b) the stator voltage is   65 % x 460 V = 299 V

   c) the stator current is $I_{\text{locked-rotor}} = 1350 \times (299/460) = 877.5$ A

2. a) the locked-rotor torque is proportional to the square of the voltage across the stator. Thus, the locked-rotor torque is

$$T = 600 \times (299/460)^2 = 253 \text{ ft.lbf}$$

3. a) the apparent power delivered to the motor is

$$S = 1.732 \, EI = 1.732 \times 299 \times 877.5 = 454\,429 \text{ VA} = 454 \text{ kVA}$$

   b) the autotransformer consumes a negligible amount of apparent power. Consequently, the apparent power furnished by the 460 V lines is nearly 454 kVA.

   c) the line current in phases L1, L2 and L3 is therefore
   $$I = S/(1.732 \, E) = 454\,000/(1.732 \times 460) = 569 \text{ A}$$

22.12 Other starting methods

Several other methods are used to limit the current and torque when starting induction motors. Some require only a change in the stator winding connections. Thus, with *part-winding* starting, the motors have two identical sets of wye-connected windings, only one of which is energized during the starting period. When the motor approaches its rated speed, the second set is connected to the line. Thus, under running conditions, the two sets of windings operate in parallel. When one set of windings is energized, the locked-rotor torque is about 50 percent of what it would be if both sets were connected across the line. Furthermore, the locked-rotor current is about 65 % of its "full-winding" value.

In *wye-delta* starting, all six stator leads are brought out to the terminal box. The windings are connected in wye during start-up and in delta during normal running conditions. This starting method produces the same results as an autotransformer starter having a 58 percent tap. The reason is that during start-up, the voltage across each wye-connected winding is only 1/1.732 (=0.58) of its rated value.

Example 22-2:

A 40 hp, 575 V, three-phase, 60 Hz, 875 r/min induction motor is equipped with a delta-connected winding, and all six leads are brought out to the terminal box. The locked-rotor current is 232 A, and the locked-rotor torque is 300 ft.lbf. The motor is installed in a small plant where across-the-line starting is not feasible. If wye-delta starting is used, calculate:

1. the locked-rotor current drawn from the line

2. the locked-rotor torque

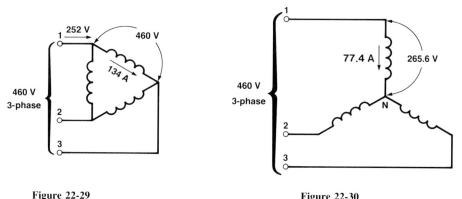

Figure 22-29
See Example 22-2.

Figure 22-30
See Example 22-2.

Solution:

1. a) the rated voltage per phase is 460 V

    b) the rated locked-rotor current in each winding when the motor is connected in delta is
        $$I = 232/1.732 = 134 \text{ A (Fig. 22-29)}$$

    c) the voltage per phase when the motor is connected in wye is
        $$E = 460/1.732 = 265.6 \text{ V}$$

    d) the current per phase is proportional to the voltage per phase. The current in each winding is therefore
        $$I = 134 \times (265.6/460) = 77.4 \text{ A}$$

    e) the locked-rotor line current is therefore 77.4 A, which is only one-third of 232 A (Fig. 22-30).

2. the locked-rotor torque is proportional to the square of the voltage per winding. The locked-rotor torque with the windings connected in wye is
    $$T = 300 \times (265.6/460)^2 = 100 \text{ ft.lbf}$$

22.13 Cam switches

Some industrial operations must remain under the continuous control of an operator. In hoists, for example, an operator has to vary the lifting and lowering rate, and the load has to be carefully set down at the proper place. Such a control sequence can be accomplished with manually-operated cam switches.

Figure 22-31 shows a 3-position cam switch designed for the forward, reverse and stop operation of a three-phase induction motor. For each position of the knob, some contacts are closed while others are open. This information is given in a table (Fig. 22-31b), usually attached to the side of the switch. An × designates a closed contact, while a blank space is an open contact. In the forward position, for exam-

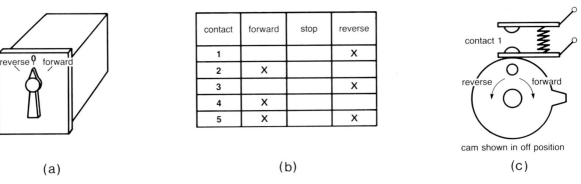

contact	forward	stop	reverse
1			X
2	X		
3			X
4	X		
5	X		X

(a)   (b)   (c)

cam shown in off position

Figure 22-31
Cam switch: (a) external appearance; (b) table listing the on-off state of the five contacts; (c) detail of the cam that controls contact 1, in the "off" position.

ple, contacts 2, 4, and 5 are closed, and contacts 1 and 3 are open. When the knob is turned to the stop position, all contacts are open. Figure 22-31c shows the shape of the cam that controls the opening and closing of contact 1.

The schematic diagram (Fig. 22-32) shows how to connect the cam switch to a three-phase motor. The state of the contacts (open or closed) is shown on the diagram for each position of the knob. For example, in the reverse position, contacts 1, 3, 5 are closed (see crosses) and contacts 2, 4 are open. The three-phase line and the motor are connected to the appropriate cam-switch terminals. Note that four jumpers J are required to complete the connections.

Some cam switches are designed to carry several hundred amperes, but magnetic contactors are often preferred to handle large currents. In such cases, a small cam switch is used to control the coils of the contactors. Very elaborate control schemes can be designed with multicontact cam switches.

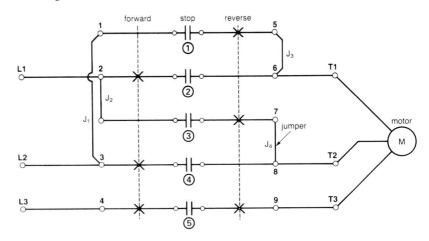

Figure 22-32
Schematic diagram of a cam switch permitting forward and reverse operation of a 3-phase motor.

22.14 Starting three-phase synchronous motors

In Section 19.9, we learned that synchronous motors are started as induction motors. Consequently, the same across-the-line and reduced voltage starting methods can be applied to synchronous motors as are used for induction motors. However, a special method is needed to control the dc field so that the synchronous motor will develop (1) a satisfactory locked-rotor torque and (2) maximum pull-in torque.

The first objective is met by connecting the field across an external "discharge" resistor, with the dc source disconnected. The second objective is met by connecting the field to the dc source only when the motor is running close to synchronous speed *and* when the N and S rotor poles are in the correct position with respect to the N and S stator poles. How can we meet this objective?

It so happens that the ac voltage E_f induced in the field winding during start-up enables us to determine both the speed of the rotor and the position of the poles. The speed is directly related to the induced frequency, and the position of the poles is related to the polarity and slope of the induced voltage. Figure 22-33 shows the voltage E_f induced in the field of a synchronous motor when it is running below synchronous speed. It is the voltage that would appear across the field terminals on open circuit. Whenever E_f is maximum, the rotor poles are midway between the stator poles. On the other hand, when E_f is zero, the rotor poles are directly facing the stator poles. Furthermore, if care is taken to properly identify the field terminals, the following relationships are true:

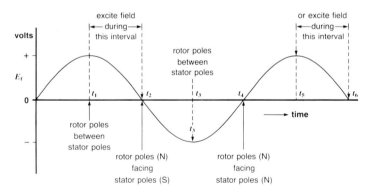

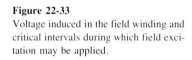

Figure 22-33
Voltage induced in the field winding and critical intervals during which field excitation may be applied.

1. When E_f is maximum positive (instant t_1), the N pole of the rotor is approaching the S pole of the stator.

2. When E_f is zero and the rate of change is negative (instant t_2), the N pole of the rotor is facing the S pole of the stator.

3. When E_f is maximum negative (instant t_3), the N pole of the rotor is approaching the N pole of the stator.

4. When E_f is zero and the rate of change is positive (instant t_4), the N pole of the rotor is facing the N pole of the stator.

Knowing that instants t_1, t_2 and t_3 are related to the position of the poles, it is possible to apply field excitation at the best moment. If possible, the field strength should be maximum when the N and S poles are facing each other (instant t_2). Consequently, field excitation should be applied slightly *before* instant t_2 so that the field current has had time to build up. However, excitation must not be applied before instant t_1. Referring to Fig. 22-33, field excitation should be applied between instants t_1 and t_2. It should be noted, however, that small, lightly loaded synchronous motors will pull into synchronism no matter when the excitation is applied.

As the motor picks up speed, the frequency and magnitude of E_f decreases progressively, just as in the rotor bars of a squirrel-cage motor. This means that the frequency of E_f can be used as an indicator of the rotor speed. If the slip is s and the stator line frequency is f, the frequency of E_f is given by Eq. 17-3:

$$f_r = sf \qquad \text{Eq. 17-3}$$

One method of detecting the speed and position of the rotor poles is shown in Fig. 22-34. The field terminals are connected to a discharge resistor R_d in series with a reactor X_m. The instantaneous magnitude of the voltage v across X_m is related to the instantaneous value of E_f. As a result, v can be used as a measure of E_f. Voltage v is detected by an electrical or electronic device D. The output of the device actuates a relay coil P at that moment when the rotor speed and position are such as to produce the highest pull-in torque. When P operates, the discharge resistor is automatically disconnected from the circuit, and the dc source is simultaneous connected to F1 and F2.

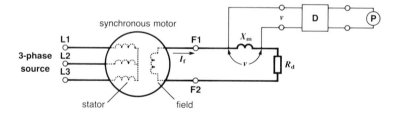

Figure 22-34
Method of detecting the position and speed of the rotor poles.

Example 22-3:

Figure 22-35 shows schematically the rotor and stator poles of a multipolar synchronous motor that has not yet come up to speed. The stator poles are moving at synchronous speed to the left. The rotor poles are also moving to the left, but at a lower speed. Although they are not yet excited, the rotor poles *will* have the polarities shown when they are connected to the dc source. What is the polarity of E_{ab} across the winding of the (future) N pole at this moment?

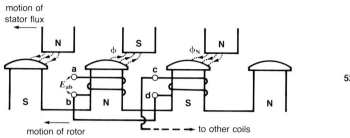

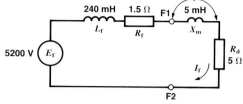

Figure 22-35
See Example 22-4.

Figure 22-36
See Example 22-5.

Solution:

From the viewpoint of the rotor, the stator flux is moving slowly toward the left. Consequently, the flux ϕ, linking coil ab is increasing. By applying Lenz's law (Section 8-5), we find that E_{ab} is positive. Thus, when the (future) N pole of the rotor is approaching the S pole of the stator E_{ab} is positive. Note that the field coils a-b, c-d, etc., are connected in series and so the induced voltage E_f is the sum of the voltages induced in each coil.

Example 22-4:

The field of a 500 hp, 2300 V, three-phase, 60 Hz, 300 r/min synchronous motor has a resistance R_f of 1.5 Ω and an inductance L_f of 240 mH. At standstill, the ac voltage E_f induced in the field is 5200 V. The field is connected to X_m and R_d as shown in Figs. 22-34 and 22-36. If X_m has an inductance L_m of 5 mH and $R_d = 5 \Omega$, calculate:

1. the voltage across F_1 and F_2 if R_d is disconnected

2. the value of I_f at standstill

3. the voltage across X_m at standstill

Solution:

1. with the field circuit open, the full value of $E_f = 5200$ V appears across the terminals

2. a) to determine the value of I_f we must first calculate the impedance of the field circuit (Fig. 22-36)

    b) the total inductance of the circuit is
        $$L = L_f + L_m = 240 + 5 = 245 \text{ mH}$$

    c) The total reactance of the circuit is
        $$X_L = 2\pi f L \qquad \qquad \text{Eq. 13-1}$$
        $$X_L = 2 \times 3.1416 \times 60 \times 0.245 = 92.4 \ \Omega$$

d)   the total resistance of the circuit is
$$R = R_d + R_f = 5 + 1.5 = 6.5 \ \Omega$$

e)   The impedance of the circuit is

$$Z = \sqrt{R^2 + X_L{}^2} \hspace{4cm} \text{Eq. 14-11}$$

$$Z = \sqrt{6.5^2 + 92.4^2} = 92.6 \ \Omega$$

f)   The current I_f is $I_f = E_f/Z = 5200/92.6 = 56$ A

3.   a)   the value of X_m is
$$X_m = 2\pi f L_m = 2 \times 3.1416 \times 0.005 = 1.88 \ \Omega$$

b)   the voltage across X_m at standstill is
$$v = I_f X_m = 56 \times 1.88 = 105.3 \ \text{V}$$

22.15   Automatic starter for a synchronous motor

Figure 22-37 shows the schematic diagram of an automatic across-the-line starter for a 500 hp, 2300 V, three-phase, 60 Hz, 300 r/min synchronous motor. The motor has a rated full-load current of 104 A, a locked-rotor current of 411 A, and the rated field current is 80 A at 125 V dc. The starter has the following notable features:

1.   The fused disconnecting switch has its blades connected to ground when it is in the open position. This is a safety feature to ensure zero voltage during maintenance and repair.

2.   A 2400 V/120 V control transformer T_1 isolates the control circuit from the high-voltage line.

3.   The main contactor relay coil M is fed by a rectifier D1 that converts the ac current into dc. The main advantage is to eliminate contactor vibration and noise.

4.   Relay A is small compared with contactor M, and so the pushbutton stations can be of conventional size.

5.   The 125 V dc source for the field is provided by a three-phase rectifier D2 that is powered by a delta-wye step-down transformer T2.

6.   Note that all circuits are protected by fuses to limit the damage if an accidental short-circuit should occur.

7.   The field discharge resistor has a value of 5 Ω, and the inductance of X_m is 5 mH.

8.   The motor can be started from two locations and can also be stopped at two locations. Note that the start pushbuttons are in parallel, while the stop pushbuttons are in series.

The circuit operates as follows. When the main disconnecting switch S is closed, the control transformer T_1 is energized, as well as the field excitation transformer

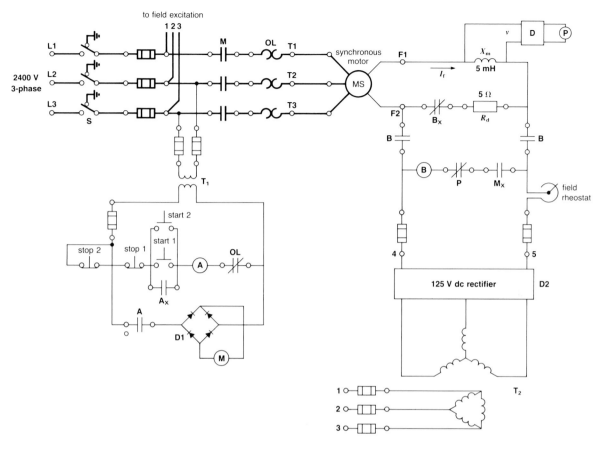

Figure 22-37
Simplified schematic diagram of an automatic synchronous motor starter.

T_2. A dc voltage appears across terminals 4 and 5 but does not reach the field because contacts B are open. When one of the start pushbuttons is depressed, the following sequence takes place.

1. relay A is excited, which causes relay coil M to close main contacts M, thus applying full voltage to the stator.

2. An ac voltage is induced in the field, causing ac current I_f to flow in reactor X_m and field discharge resistor R_d. The locked-rotor value of I_f is 56 A, and the frequency is 60 Hz. The voltage across R_d is $E_R = I_f R = 56 \times 5 = 280$ V.

The voltage across X_m is $v = I_f X_m = 56 \times 1.88 = 105.3$ V. This high input voltage v to detector D immediately actuates relay P, which keeps the circuit of relay coil B open, despite the fact that auxiliary contact M_X is now closed. With relay B out of service, the field circuit remains in the state shown in Fig. 22-37.

3. As the motor picks up speed, the induced field voltage begins to fall, which reduces the magnitude of I_f. Furthermore, the frequency of I_f decreases rapidly with increasing speed. Consequently, the voltage v across X_m falls very quickly. Its value is about 4 V when the motor turns at 285 r/min. The corresponding slip is 5 percent, and the frequency is 3 Hz.

4. When the slip is below 5 percent (speed greater than 285 r/min), v is so small that detector D is about ready to allow relay P to drop out. As already mentioned, the drop-out will occur at the precise moment that will ensure the maximum pull-in torque.

5. As soon as P is de-energized, relay coil B becomes excited because contact M_x is still closed. As a result, contact B_x opens and resistor R_d is removed from the circuit. At the same time contacts B close and dc excitation is applied to the field. Current I_f is now a dc current, and so the voltage v across X_m falls to zero, which ensures that relay P will remain de-energized.

A complete circuit diagram would show many more details, most of which relate to safety features and overload protection of the squirrel-cage and stator windings.

22.16 Summary

This chapter has shown how electrical components and devices are assembled to make automatic dc and ac starters. It has also shown how control circuit diagrams are drawn and how to interpret them.

We have also learned about the problems associated with the starting of large motors. These problems have given rise to several different starting methods such as reduced-voltage starting, part-winding starting, wye-delta starting, and so forth. The starting methods for synchronous and induction motors are very similar. However, the correct timing in applying the dc field presents a special problem in the case of synchronous motors. Thus, we found that slip, variable frequency, reactance, and even Lenz's law play an important part in the design of the field circuit. It illustrates the importance of many of the basic topics we introduced in earlier chapters. In the next chapter, we have another opportunity to see how these concepts and practical applications fit together.

TEST YOUR KNOWLEDGE

22-1   A thermal relay can be used for short-circuit protection

☐ true   ☐ false

22-2   A fused disconnecting switch can often be used instead of a circuit breaker.

☐ true   ☐ false

22-3   A normally closed contact of a relay is a contact that is closed when the relay coil is in operation.

☐ true   ☐ false

22-4   In a large three-phase starter, three fuses are sufficient to adequately protect against short-circuits.

☐ true   ☐ false

22-5   A stop pushbutton always has at least one normally closed contact

☐ true   ☐ false

22-6   A field-failure relay in a dc starter prevents the motor from overspeeding when the series field circuit is interrupted.

☐ true   ☐ false

22-7   The rated locked-rotor current of a 460 V, 3-phase squirrel-cage motor is 250 A. If the starting voltage across the motor is reduced to 230 V, the new locked-rotor stator current will be

a. 125 A   b. 250 A   c. 62.5 A

22-8   If one of the line fuses is removed in Fig. 22-23, the motor will not start.

☐ true   ☐ false

22-9   A 28-pole, 3-phase, 60 Hz synchronous motor pulls in at 247 r/min. The frequency in the field circuit at that moment is

a. 3.94 %   b. 2.37 Hz   c. 2.4638 Hz

22-10   In Fig. 22-33, if E_f has a frequency of 2 Hz, the time interval during which the field excitation can be applied is

a. 0.5 s   b. 8 s   c. 0.125 s

QUESTIONS AND PROBLEMS

22-11   Name four types of circuit diagram, and describe the purpose of each.

22-12   Without referring to the text, describe the operation of the starter shown in Fig. 22-17, and state the use of each component.

22-13   Draw the symbols for a NO contact, a NC contact, and a thermal relay.

22-14   If the start and stop pushbuttons of a starter are pushed simultaneously, what will happen?

22-15   In Fig. 22-17, if the A_x contact were removed, what effect would it have on the operation of the starter?

22-16   In Fig. 22-19, if a short-circuit occurs across the armature while the motor is running, which of the following devices will interrupt the circuit: power fuses, control fuses, or OL?

22-17   In Fig. 22-19, if the motor is overloaded by 30 percent, which device will protect the motor? Describe how this protection works.

22-18   A partial short-circuit between the turns of the stator in Fig. 22-37 causes the currents in lines L1 and L2 to be above normal. Which device will trip the motor? Describe how the protection operates.

22-19   An ac magnetic contactor can make 3 million "normal" circuit interruptions before its contacts have to be replaced. If a motor is started and stopped 15 times per day, 7 days a week, after how many years will the contacts have to be replaced?

22-20   Without referring to the text, describe the sequence of operations that occurs when the start pushbutton in Fig. 22-19 is momentarily depressed. (The disconnecting switch is closed.)

22-21   The motor in Fig. 22-28 is on the 65 % tap, and draws a stator current of 800 A when the speed is 710 r/min. Calculate the current in the three phases of lines L1, L2 and L3.

22-22   How many three-phase ac contactors are needed to start a motor using the part-winding method?

22-23   A delta-connected induction motor has a rated locked-rotor current of 360 A. Calculate the locked-rotor current if the motor is started with the windings connected in wye.

22-24   A delta-connected induction motor has a rated locked-rotor torque of 60 N.m. Calculate the locked-rotor torque if the motor is started with the windings connected in wye.

22-25   Without referring to the text, describe the sequence of operations that takes place when the start pushbutton is momentarily depressed in the starter of Fig. 22-28.

22-26   In Fig. 22-35, the field of the stator rotates to the left at 1800 r/min, and the rotor rotates to the left at 1720 r/min. Using Lenz's law, determine the polarity of E_{cd} due to flux ϕ_N.

22-27   The field voltage induced across the open-circuit terminals of a 720 r/min, 60 Hz synchronous motor is 6000 V. What is the induced voltage per pole?

22-28   The synchronous motor starter in Fig. 22-37 works well with the connections as shown. What would happen if the terminals 4 and 5 were reversed?

22-29   Without referring to the text, describe the sequence of events that takes place when the start pushbutton of Fig. 22-37 is depressed.

CHAPTER 23

Generation, Transmission, and Distribution of Electrical Energy

Introduction and Chapter Objectives

Whenever we switch on a light, start an electric motor, or turn on a stove, we seldom think of the complex system that brings this energy to our door. Whether the power drives an electric shaver or a crane, it all comes from a remote generating station where the energy is produced. This energy flows over a network of transmission lines, feeders, and cables, as well as through step-up and step-down transformers. As the energy comes closer and closer to the actual point of use, it flows over low-voltage lines and through low-voltage transformers and protective devices until it finally arrives at the outlet to which the load is connected.

This network of lines, cables, and apparatus is divided into three loosely defined sectors, called *generation, transmission* and *distribution*.

Generation relates to the generating stations that produce the electricity. It includes generators, turbines, water reservoirs, dams, nuclear reactors, and all the auxiliary equipment needed to produce the electrical energy.

Transmission relates to the lines and equipment that carry the electrical energy over considerable distances. The voltage ranges from 115 kV to 765 kV, and the equipment includes the circuit breakers, insulators, lightning arresters, and switches that are installed in high-voltage (HV) and extra-high-voltage (EHV) substations.

Distribution relates to the lines, cables, and equipment that are relatively close to the consumer. The voltage lies between 120 V and 69 kV, and the equipment includes circuit breakers, insulators, fuses, lightning arresters, and switches installed in medium-voltage (MV) and low-voltage (LV) substations. The distribution network is considered to end at the consumer's service entrance.

Figure 23A is a one-line diagram showing how the generation, transmission, and distribution sectors are interconnected to supply the energy needs of consumers, such as commercial enterprises, residences, and heavy industry. The thousands of interconnections between the generating stations and the loads are called a *network*.

Only three generating stations are shown, but there may be dozens spotted at different points throughout the system. They may be hundreds of kilometers apart, and any station may contain one or more generators. The loads are also dispersed throughout the system, often covering an area of several states or an entire country. The diagram also shows the importance of substations, where transformers, switches, and protective devices are installed.

In this chapter, we will describe the basic features of the generation, transmission, and distribution of electrical energy. We also will cover simple commercial and domestic distribution systems. Finally, we will discuss the costing of electricity and how power bills can be reduced by improving the power factor.

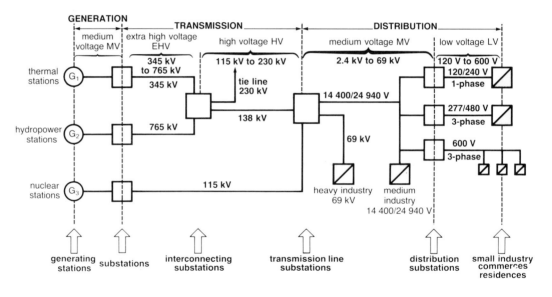

Figure 23A

One-line diagram of an electrical system. The indicated line voltages are only representative values.

23.1 Basic objectives of an electric utility system

The main purpose of an electric utility system is to deliver active power (kilowatts or megawatts) from the point where it is generated to the point where it is needed. Consequently, the first objective is to transmit this active power as efficiently as possible from the generating station to the load.

The second objective is to deliver the power at a voltage level that is reasonably

**TABLE 23A   VOLTAGE CLASSES AS APPLIED
TO INDUSTRIAL AND COMMERCIAL POWER**

voltage class	nominal system voltage		
	two-wire	three-wire	four-wire
low	120	120/240 □	—
voltage	single phase	single phase	120/208 □
		480 V □	277/480 □
LV		600 V	347/600
medium		2 400	
voltage		4 160 □	
		4 800	
MV		6 900	
		13 800 □	7 200/12 470 □
		23 000	7 620/13 200 □
		34 500	7 970/13 800
		46 000	14 400/24 940 □
		69 000 □	19 920/34 500 □
high		115 000 □	
voltage		138 000 □	
		161 000	
HV		230 000 □	
extra high		345 000 □	
voltage		500 000 □	
EHV		735 000 - 765 000 □	

All voltages are 3-phase unless indicated otherwise.
Voltages designated by the symbol □ are preferred voltages.

*Note: Voltage class designations were approved for use by IEEE Standards Board
(September 4, 1975)*

constant. The voltage may be 120 V, 440 V or 230 kV, but whatever it is, it should not vary greatly from its *nominal*, or rated, value. Typically, the electric power utility wants to keep the voltage level constant to within ±7 percent. Table 23A lists the nominal voltage levels recommended for use in North America.

A third objective is to maintain a constant frequency. The reason is that most motors are either induction or synchronous and, as we learned, their speed depends directly upon the frequency. In practice, the frequency is held to within a fraction of one percent of the nominal frequency. Nominal frequency is 60 Hz in the United States and Canada and 50 Hz in most other countries.

23.2 Reactive power and the electric utility system

Whenever we transmit ac power, we are faced with the problem that most loads operate at lagging power factor. This means that they absorb not only active power (kilowatts) but also reactive power (kilovars). The transmission and distribution network therefore has to deliver both active and reactive power to the millions of loads it serves. Unfortunately, the presence of reactive power increases the current in the transmission lines and transformers, and so they have to be bigger than if they carried only active power. To make matters worse, when reactive power flows in a network, it tends to make the line voltage fluctuate over wider limits as the load varies. In other words, reactive power produces poorer voltage regulation in the system.

For these reasons, the electric utility company tries to keep the reactive power carried by the lines as small as possible. Ideally, the transmission and distribution network should operate at a power factor of 100 percent. This objective is met reasonably well by installing capacitors throughout the system, close to the reactive loads. As we learned in Chapter 10, capacitors can generate reactive power. In a large electric system they are installed at various points (Fig. 23-1), so that substantial reactive power never flows over any great distance. Some of these capacitors are installed in substations by the electric utility, but many are installed by industrial consumers on their premises. The reason is that such consumers usually pay a premium price if their power factor falls below 90 percent.

23.3 System demand and power control

The *power demand* of an electric utility system is equal to the total active power (kilowatts) drawn by its millions of customers. The demand varies considerably throughout the day and from season to season. It is highest between noon and 6 o'clock in the evening, and lowest at around 4 o'clock in the morning. In the course of a year, the ratio between the peak demand and the lowest demand can be as much as 2.5. For example, the power demand of one large utility company varies from a minimum of 6000 MW in the summer to a peak of 16 000 MW in the winter.

To properly allocate the power among the various generating stations, *dispatch centers* are set up to monitor and control the power flow of the entire system. Dispatch centers try to minimize the cost of generating and transmitting power while maintaining system stability. For example, generators and their prime movers

Figure 23-1
This pole-mounted capacitor bank
supplies the reactive power needed
on a three-phase 24.9 kV transmission line. It is rated at 2700 kvar and
is composed of eighteen 150 kvar,
14.4 kV single-phase capacitors.
(*Courtesy General Electric*)

(turbines) operate efficiently only when they deliver power close to their full-load
rating. Thus, a turbine-generator unit never runs at half its rated output for any
length of time. As a result, generating units are periodically connected and
disconnected from the power system, depending upon whether the power demand is
high or low (Fig. 23-2).

23.4 Choice of frequency

Most large power systems generate ac voltages at a frequency of either 60 Hz or
50 Hz. The advantages of 50 Hz and 60 Hz systems are comparable, but they
outweigh those of a 25 Hz or a 100 Hz system, for example. A low frequency such
as 25 Hz improves the voltage-regulation and power-handling capacity of
transmission lines because their inductive reactance is lower. However, 25 Hz
motors and transformers tend to be bulky. Furthermore, the highest induction motor
speed is about 1500 r/min.

A 100 Hz system produces the opposite effect, and the highest induction motor
speeds tend to be greater than those required by industry. Thus, it has been found
that 60 Hz and 50 Hz systems are the most practical.

Power can also be transmitted by dc. High-voltage direct-current transmission is
attractive when large amounts of power have to be carried over distances of several
hundred kilometers. It is also useful when power has to be delivered by cable under
bodies of water. The power at the output of the dc line is converted back to ac by
means of electronic converters.

Figure 23-2
Control rooms in different generating stations are in constant communication with the dispatch center to optimize the total power flow of the system. Each control room also monitors and controls its own generating station. (*Courtesy Hydro-Québec*)

23.5 Choice of voltage level

Transmission and distribution voltage levels vary greatly throughout an electric utility system. The change from one level to another is accomplished by transformers. But what determines the voltage level at different points in the system? It depends mainly upon the amount of power and the distance over which it has to be transmitted. As a rough approximation, the voltage is given by the equation:

$$\overline{E = k \sqrt{Pl}} \qquad (23\text{-}1)$$

where

E = three-phase line voltage, in volts (V)

P = power to be transmitted, in kilowatts (kW)

l = length of the transmission line, in kilometers (km)

k = empirical coefficient, whose value ranges from 50 to 100

The approximate voltage of a transmission line is first calculated according to Eq. 23-1. A standard voltage level is then selected, using one of the preferred system voltages listed in Table 23A.

Example 23-1:

A three-phase line has to carry a load of 40 MW over a distance of 100 km. Calculate the approximate voltage of the line.

Solution:

a)   40 MW = 40 000 kW

b) $E = k \sqrt{Pl} = (50 \text{ to } 100) \sqrt{40\,000 \times 100}$ Eq. 23-1

  $E = (50 \text{ to } 100) \times 2000 = (50 \times 2000) \text{ to } (100 \times 2000)$

   $= 100 \text{ kV to } 200 \text{ kV}$

Referring to Table 23A, a voltage of 115 kV or 138 kV would probably be considered. The final choice depends upon economic and technical factors that have to be taken into account for each individual line.

Figure 23-3
The value of the transmission line voltage depends mainly upon the active power and the distance it has to be transported. This wood-pole line delivers three-phase power at 115 kV. Each ACSR conductor is supported by a string of eight suspension insulators.

23.6 System stability

In an electric utility system, the active power produced by the generators must at all times equal the active power demanded by consumers. The reason is that the system cannot store electrical energy. This means that a continuous minute-by-minute balancing act goes on, so that the power generated by the hundreds of generators is exactly equal to that demanded by the millions of consumers. As long as this state of equilibrium is maintained, the electrical system is *stable*. But it tends to become unstable as soon as the generated power is greater or less than the power being consumed. What happens when such an imbalance occurs?

When the power demand exceeds the power being generated, the generators begin to slow down and the system frequency decreases. On the other hand, if the power demand is less than the generated power, the generators begin to speed up, causing the system frequency to increase. We recall that synchronous generators are interconnected, and so they must all operate at the same frequency.

The generator speed is therefore a very precise indicator of the balance (or imbalance) between the power demanded and that being generated. A slight drop in speed of a particular generator indicates that its power input should be increased. This is done by opening the steam valve of the turbine that drives the generator. Every generator in the system is equipped with a sensitive governor that controls the steam valve and therefore the power output of the generator. In addition, auxiliary control devices ensure that proper load-sharing takes place among the various generators.

In the case of hydraulic turbines, wicket gates permit more or less water to flow through the turbine in response to the governor signal.

The electric utility system may become unstable when large loads are suddenly added to, or removed from, the system. For example, if a transformer fails, circuit breakers will immediately trip, causing a large block of demand power to be dumped. This will cause all the generators to accelerate, and the speed may increase at an alarming rate. Steam valves and wickets gates have to be closed as quickly as possible, but these adjustments cannot be done instantaneously. During the adjustment process the entire system will tend to oscillate, and overvoltages may occur. The system is said to be unstable, and quick action is required to prevent the network from failing. For example, loads of lesser importance may have to be shed (disconnected) to keep the system from losing synchronism.

GENERATING STATIONS

The three main components of an electric utility system, are generating stations, transmission lines, and distribution systems.

There are three main types of generating stations:

1. *Thermal* generating stations burn coal, oil, or gas to produce the steam that drives the steam turbine. The turbine is directly coupled to the synchronous generator.

2. *Hydropower* generating stations use the water stored behind dams, or flowing in rivers and tides to drive a hydraulic turbine. The turbine is directly coupled to the synchronous generator.

3. *Nuclear* generating stations are identical to thermal generating stations except that the source of heat is an atomic reactor rather than a fossil fuel.

23.7 Location of the generating station

The location of a thermal generating station must be carefully planned to arrive at an acceptable, economic solution. We can sometimes place it next to the primary source of energy, such as a coal mine, and use a transmission line to carry the electricity to where it is needed. When this is not practical, we can transport the primary energy (coal, oil, or gas) by ship, train, or pipeline to the generating station. The generating station may therefore be near to, or far away from, the consumers of electrical energy.

Thermal and nuclear generating stations are usually built next to a river or lake (Fig. 23-4) because when thermal energy is converted into mechanical energy, a large amount of heat is inevitably lost. This heat must be carried away by a coolant, and the cheapest one is running water. When the supply of water is limited, special

Figure 23-4
Thermal generating stations are usually situated next to rivers and lakes to obtain an adequate source of cooling water. This generating station is on the shores of the St-Lawrence River. (*Courtesy Hydro-Québec*)

cooling towers may have to be installed to carry away the heat (Fig. 23-5).

Hydropower stations must be situated wherever the waterfall happens to be. This may require transmission lines several hundred kilometers long to bring the power to the industrial site (Fig. 23-6).

Figure 23-5
When the source of cooling water is limited, cooling towers are installed. This one is situated at a nuclear generating station in Oregon. (*Courtesy Portland General Electric Company*)

Figure 23-6
View of Churchill Falls, Labrador, before the hydropower station was developed. It is 750 miles away from the nearest important consumer site. (*Courtesy Hydro-Québec*)

23.8 Makeup of a thermal generating station

Most thermal stations have ratings between 200 megawatts (MW) and 1500 MW, so as to attain the high efficiency and economy of a large installation. Such a station has to be seen to appreciate its enormous complexity and size.

Figure 23-7 is a highly-simplified schematic diagram showing its main components. They are described and further illustrated below.

● Natural gas, oil, or pulverized coal is burned inside a huge furnace called a boiler.

● Combustion is aided by means of a forced-draft fan which provides the enormous quantities of air that are needed. Approximately 10 tons of air are required for every ton of fuel that is burned (Fig. 23-8).

● The products of combustion are released in the form of hot gases that flow over water tubes and are then carried up the chimney and into the atmosphere.

- The heat causes the water in a drum to boil, releasing large quantities of steam at typical pressures and temperatures of 16 megapascals (MPa) and 550°C (1022°F).*

- The amount of steam flowing into the turbine is regulated by a steam valve. As more steam flows, the turbine delivers more power, which results in a greater power output (active power) from the generator. Figures 23-9 to 23-11 show typical turbine-generator sets.

- The steam that exhausts from the turbine is cooled and condensed in a *condenser*. A condenser is actually a huge heat exchanger, in which the heat from the steam is transferred to cool water that circulates in a separate piping system (Fig. 23-12). In modern thermal generating stations, about 50 percent of the heat released in the boiler is lost in the condenser. Together with other losses, this means that for every 1000 W of heat released in the boiler, only

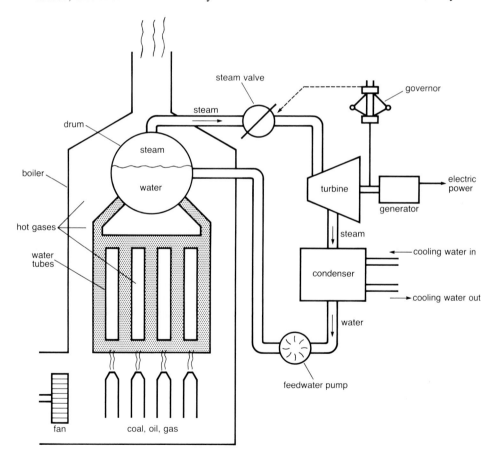

Figure 23-7
Simplified schematic diagram of a thermal generating station.

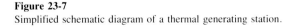

* 1 MPa is approximately equal to 145 pounds per square inch.

about 400 W is converted into electricity. The remaining power (600 W) either goes "up in smoke" by way of chimney, or "down the drain" by way of the condenser.

- The condensed steam, now simply lukewarm water at a temperature of about 30°C, is pumped back into the drum by means of a *feedwater pump*. These huge pumps, driven by motors of several thousand horsepower, force the water back into the drum against the high internal pressure of 16 MPa.

- The synchronous generator feeds its power into the utility system. As mentioned in Chapter 20, the setting of the steam valve establishes the active power output, while the dc excitation determines the reactive power delivered by the generator.

- A speed-sensitive *governor* acts upon the steam valve, so as to automatically control the active power output of the generator.

In practice, a thermal station has hundreds of other components and accessories to ensure high efficiency, safety, and economy. For example, special water purifiers maintain the required cleanliness and chemical composition of the feedwater, oil pumps keep the bearings properly lubricated, electrostatic precipitators keep fly-ash and other pollutants from reaching the atmosphere, and so on. However, the basic components we have described are sufficient to understand how a thermal station operates.

Figure 23-8
This forced-draft fan provides 455 m³/s of air for a thermal generating station. It is driven by a three-phase induction motor rated 12 000 hp, 890 r/min, 11 kV, 60 Hz. (*Courtesy Novenco Inc.*)

Figure 23-9
Exposed blades of a steam turbine in the process of being assembled. (*Courtesy Hydro-Québec*)

Figure 23-10
A 250 MVA generator driven by a steam turbine. The generator housing and brushless exciter are on the left, and the steam turbine is on the right. An identical turbine-generator unit is in the background. (*Courtesy Hydro-Québec*)

Figure 23-11
The coal-fired Cumberland generating station of the Tennessee Valley Authority consists of two 1300 MVA generating units, one of which appears in the foreground. The second unit is at the far end of the turbine hall. (*Courtesy Brown Boveri*)

Figure 23-12
Two large pipes feed cooling water into and out of this 220 MW condenser. A condenser is as necessary as a boiler is, in thermal and nuclear generating stations. (*Courtesy Foster-Wheeler Energy Corporation*)

23.9 Nuclear generating stations

A nuclear station is identical to a thermal station except that the boiler is replaced by a nuclear reactor. The reactor contains the radioactive material that generates the heat. A nuclear power plant therefore contains a generator, steam turbine, condenser, etc., similar to those found in a conventional steam station. The overall efficiency is also similar (between 30 and 40 percent) and a cooling system must be provided. Consequently, nuclear stations are also situated close to rivers and lakes. Because of these similarities, we will examine only the operating principle of the nuclear reactor itself.

23.10 Energy released by atomic fission

All nuclear reactors are based upon atomic fission, which occurs when the nucleus of an atom splits in two. The total mass of the resulting particles is less than that of the original atom. The loss in mass releases energy according to Einstein's equation:

$$E = mc^2 \qquad (23\text{-}2)$$

where

E = energy released, in joules (J)

m = loss of mass, in kilograms (kg)

c = speed of light (3×10^8 m/s)

When an atom fissions, an enormous amount of energy is released. For example, a loss in mass of only 1 gram produces 9×10^{10} joules, which is equivalent to the heat released by burning three tons of coal. Uranium is an element that fissions relatively easily, and that is why it is used in nuclear reactors.

23.11 Chain reaction

How can we provoke the fission of a uranium atom? One way is to bombard its nucleus with *neutrons*. A neutron* makes an excellent projectile because it is not repelled as it approaches the positively-charged nucleus. If the impact is strong enough, the nucleus will fission, releasing energy. Fission produces a second important result: it ejects two or three neutrons that move at high speed way from the broken nucleus. These neutrons collide with other uranium nuclei, and so a chain reaction quickly takes place, releasing a tremendous amount of heat.

In a nuclear reactor, we have to slow down the neutrons to increase their chances of striking other uranium nuclei. Towards this end, the fissionable uranium compound is immersed in a *moderator*. The moderator may be light water, heavy water, graphite, or any other material that can slow down neutrons without absorbing them. By properly distributing the uranium fuel within the moderator, we can reduce the speed of the neutrons so they have the required velocity to initiate other fissions. Only then will a chain reaction take place, causing the reactor to "go critical".

---

*   A neutron has the same mass as a proton, but has no electric charge.

23.12 Makeup of a nuclear generating station

Figure 23-13 is a highly simplified schematic diagram of a nuclear generating station using a so-called *pressurized water reactor* (PWR).

● An enclosed nuclear reactor contains the fissionable atomic fuel and a moderator. In this diagram, the moderator is light (ordinary) water, and it also serves as a coolant.

● The coolant transfers heat from the nuclear reactor to the heat exchanger, where the steam is produced. (In other types of reactors, the coolant is a fluid, such as heavy water or liquid sodium, or a gas, such as helium or carbon dioxide.)

● The coolant circulating pump drives the coolant around the loop so that efficient heat exchange takes place. The pumps are driven by motors of several thousand horsepower.

● The heat exchanger in a PWR isolates the radioactive coolant from the steam that flows through the steam turbine.

A single reactor can produce typical thermal outputs of 1800 MW and corresponding electrical outputs of 600 MW. The overall efficiency is therefore about 30 percent.

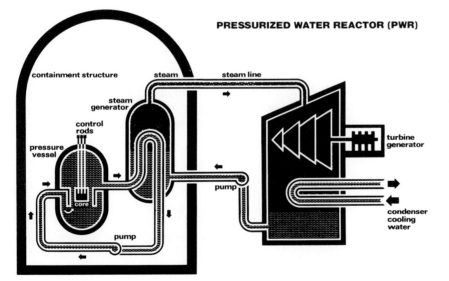

Figure 23-13

Highly simplified diagram of one type of nuclear generating station. The PWR shown here is a type of reactor fueled by slightly enriched uranium in the form of oxide pellets held in zirconium alloy tubes in the core. Water is pumped through the core to transfer heat to the heat exchanger (steam generator). This coolant water is kept under pressure to prevent boiling and transfers heat to the water in the steam generator to make the steam. The arrangement of the steam turbine, turbine generator, condenser and feedwater pump is the same as in Figure 23-7. (*Courtesy Portland General Electric Company*)

Figure 23-14
Aerial view of a light-water nuclear generating station. The large rectangular building houses a 667 MVA, 90 percent power factor, 19 kV, 60 Hz, 1800 r/min synchronous generator and steam turbine; the circular concrete building contains the reactor. (*Courtesy Connecticut Yankee Atomic Power Company; photo by George Betancourt*)

23.13 Hydropower generating stations

Hydropower generating stations convert the energy of moving water into electrical energy by means of a hydraulic turbine that is coupled to a synchronous generator.

The power that can be extracted from water depends upon the height through which it falls and its rate of flow. The available power can be calculated by the equation:

$$P = 9.8qh \qquad (23\text{-}3)$$

where

P = available waterpower, in kilowatts (kW)

q = water flow, in cubic meters per second (m³/s)

h = head of water, in meters (m)

9.8 = constant which takes account of units and the density of water

Because of friction losses in the water conduits, turbine casing, and the turbine itself, the mechanical power output of a hydraulic turbine is less than that calculated by Eq. 23-3. However, the efficiency of large turbines is between 90 and 94 percent.

Example 23-2:

A waterfall is found to have a height of 20 meters, and delivers a steady flow of 300 cubic meters of water per second. Calculate the electric power that can be harnessed, assuming the turbine and piping have an efficiency of 90 percent, and the generator has an efficiency of 97 percent.

Solution:

The available water power is:

$$P = 9.8 \; qh = 9.8 \times 300 \times 20 = 58 \; 800 \; kW \qquad \text{Eq. 23-3}$$

The mechanical power output of the turbine is:

$$P_m = 58 \; 800 \times eff = 58 \; 800 \times 0.90 = 52 \; 920 \; kW \qquad \text{Eq. 1-11}$$

The electrical power output of the generator is:

$$P = 52 \; 920 \times eff = 52 \; 920 \times 0.97 = 51 \; 332 \; kW = 51.3 \; MW$$

23.14 Makeup of a hydropower station

A hydropower installation consists of dams, waterways, and conduits that impound and channel the water toward the turbines. These, and other items described below enable us to understand the basic features and components of a hydropower generating station (Fig. 23-15).

- A dam made of earth or concrete is built across a riverbed to create a huge reservoir of water. Dams enable us to regulate the water flow so that the generating station may run at close to full capacity throughout the year. Dams and reservoirs often serve a dual purpose, providing irrigation, recreational, flood control, and navigation facilities, in addition to their power-generating role. Some reservoirs extend over thousands of acres, in order to create the required stock of water.

- *Spillways* are provided next to the dam to discharge water whenever the reservoir level is too high. Spillways are opened during the spring and rainy seasons when the water supply is excessive.

- The water behind the dam is channeled through *penstocks* (huge steel pipes), which bring the water to *spiral cases* that surround the individual turbines.

- The amount of water flowing into the turbine is controlled by means of *wicket gates* that open and close in response to signals given by the turbine governor. The wicket gates play the same role as the steam valve in a steam turbine.

- After the water has passed through the turbine, it moves through a vertical channel called a *draft tube*. The draft tube draws the water downward and improves the efficiency of the system. The water then flows out to the *tailrace* and downstream in the original riverbed.

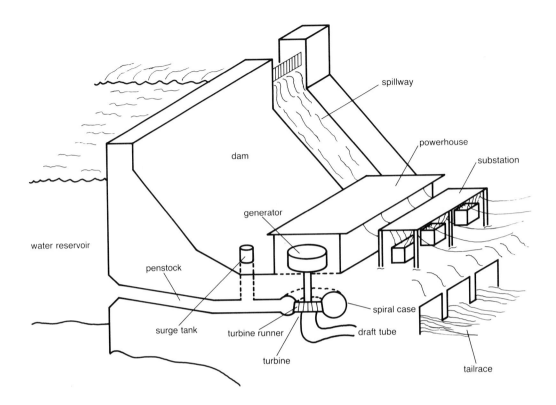

Figure 23-15
Highly simplified diagram of one type of hydropower station.

- The powerhouse contains the synchronous generators, transformers, circuit breakers, and associated control equipment. Instruments, relays, and meters are contained in a central control room where the entire generating station can be monitored and controlled.

  Figure 23-16 to 23-19 show construction details of a large hydropower generating station.

Figure 23-16
Grand Coulee Dam on the Columbia River, in the state of Washington is 108 m high and
1270 m wide. It is the largest hydropower plant in the world, having 18 generating units of
125 MW, and 12 generating units of 600 MW, for a total of 9450 MW of installed capacity.
The spillway is in the middle of the dam. (*Courtesy U.S. Department of the Interior, water
and Power Resources Service; photo by Bob Isom*)

Figure 23-17
The spiral case will surround the turbine runner, and the water flow will be controlled by wicket gates that are mounted around the inner circle. (*Courtesy U.S. Department of the Interior, water and Power Resources Service; photo by H.S. Holmes*)

Figure 23-18
Turbine runner, 10 m in diameter, being lowered inside the scroll case. The wicket gates can be seen side by side at the bottom of the pit. The turbine is rated 620 MW, 72 r/min and operates at a nominal head of 87 m. (*Courtesy U.S. Department of the Interior, water and Power Resources Service; photo by H.S. Holmes*)

Figure 23-19
Salient-pole rotor having a diameter of 60 ft, a height of 11 ft, and weighing 1972 tons is lowered inside the stator. The generator has a rating of 600 MW. (*Courtesy U.S. Department of the Interior, water and Power Resources Service; photo by H.S. Holmes*)

Figure 23-20
These transformers, located in a generating station, step up the generator voltage to the level needed for transmission. (*Courtesy Kearney-National*)

TRANSMISSION OF ELECTRICAL ENERGY

The medium voltages generated by synchronous generators are stepped up to much higher values by means of transformers located in the substation which is part of the generating station (Figure 23-20). The resulting transmission-line voltages may range from 115 kV to 765 kV.

23.15 Components of a transmission line

A transmission line is composed of conductors, insulators, and supporting structures, such as wooden poles or steel towers.

1. *Conductors.* Conductors for high-voltage lines are always bare. Stranded copper conductors or steel-reinforced aluminum cables (ACSR) are used. ACSR conductors are preferred because they result in a lighter and more economical line.

2. *Insulators.* Insulators serve to support the conductors and insulate them from each other, and from ground (Fig. 23-3). They are usually made of porcelain, but synthetic materials are also used.

   From an electrical standpoint, insulators must offer a high resistance to surface leakage currents; consequently, they are molded and wrinkled to increase the leakage path. (Fig. 23-21).

   The insulators must also be strong enough to withstand the tremendous pull from the weight of the conductors.

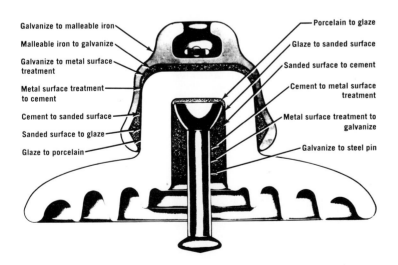

Galvanize to malleable iron
Malleable iron to galvanize
Galvanize to metal surface treatment
Metal surface treatment to cement
Cement to sanded surface
Sanded surface to glaze
Glaze to porcelain

Porcelain to glaze
Glaze to sanded surface
Sanded surface to cement
Cement to metal surface treatment
Metal surface treatment to galvanize
Galvanize to steel pin

Figure 23-21

Cross section of a suspension insulator, showing detailed construction and wrinkled lower surface. (*Courtesy Canadian Ohio Brass Co. Ltd.*)

Figure 23-22
String of suspension insulators able to support a load of 25 000 lb. (*Courtesy Canadian Porcelain*)

Figure 23-23
Twelve parallel strings of 33 insulators are used to support the extra long span of conductors crossing the Saguenay River. (*Courtesy Hydro-Québec*)

Figure 23-24
Pin-type insulator with several skirts to increase the leakage path. The conductor is fixed at the top and a steel pin screws into the base to support the insulator. (*Courtesy Canadian Ohio Brass Co. Ltd.*)

For voltages above 70 kV, suspension-type insulators, strung together pin to cap (Fig. 23-22) are always used. The number of insulators in the string depends upon the voltage. For example, 13 to 16 insulators are used on a 230 kV line. Figure 23-23 shows an insulator arrangement for a 735 kV line. It is composed of 12 parallel strings of 33 insulators to provide both mechanical and electrical strength.

For voltages below 70 kV, pin-type insulators are often used.

3. *Supporting structures.* Steel towers or wooden poles are used to keep the conductors at a safe height from the ground and at an adequate distance from each other. At voltages below 70 kV we can use single wooden poles equipped with cross arms. However, for EHV lines the towers are made of galvanized angle iron, bolted together (Fig. 23-25).

The spacing between conductors must be sufficient to prevent arc-over under gusty wind conditions. The spacing has to be increased as we increase the distance between towers and as the line voltage becomes higher.

23.16 Corona effect and bundled conductors

High line voltages produce a continual electrical discharge around the conductors because of local ionization of the air. This discharge, or *corona effect*, produces losses over the entire length of the transmission line. In addition, corona emits high-frequency noise that interferes with nearby radio receivers and TV sets. We can decrease corona by increasing the diameter of the conductors. The easiest way to achieve this is to use two, three, or more conductors per phase (Fig. 23-26). Such

Figure 23-25
Two transmission-line towers made of angle-iron support two three-phase lines. The ice covering the conductors increases the strain on the insulators. The two ground wires leaving the top of the nearby tower are clearly visible.
(*Courtesy Hydro-Québec*)

Figure 23-26
This uncompleted 735 kV transmission line shows one phase composed of four bundled conductors. The conductors have a diameter of 35 mm and are held 457 mm apart by steel spacers. Each ACSR conductor is composed of 42 aluminum strands 4.6 mm in diameter, and 7 steel strands 2.5 mm in diameter. (*Courtesy Hydro-Québec*)

bundling increases the effective electrical diameter of the conductor. Bundled conductors also reduce the inductive reactance of the line, which increases its power-handling capacity.

23.17 Grounding

Transmission line towers are always connected to ground by a heavy copper conductor. A low ground resistance is required so that lightning strokes will be less likely to produce a flashover across the insulators and a consequent line outage.

Bare conductors are also strung from tower to tower at the very top of the towers (Fig. 23-25). These conductors, called *ground wires*, shield the transmission line by intercepting lightning strokes before they hit the current-carrying conductors below. Grounding wires do not normally carry current; consequently they are often made of steel. They are connected to ground at each tower.

23.18 Electrical properties of a transmission line

We have already learned that a single-phase transmission line has resistance, inductance, and capacitance.* Such lines can be represented by the equivalent circuit of Fig. 23-27. The values of R, X_L, and X_C depend upon the conductor size, the length of the line, the spacing between conductors, and the frequency. The same circuit can be used for three-phase transmission lines, in which case E is the line-to-neutral voltage and X_C is the line-to-neutral capacitance.

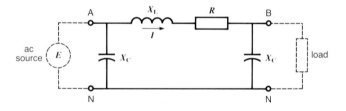

Figure 23-27
Equivalent circuit of a single-phase transmission line. The same circuit can be used to represent one phase of a three-phase transmission line.

The conductor size increases with increasing line current I, which tends to reduce R. Furthermore, as the line voltage E increases, the spacing between the conductors has to be increased. This tends to increase the inductance and reduce the capacitance of the line. As a result, X_L and X_C both tend to increase with increasing voltage.

---

*   See Sections 3-12, 9-9, and 10-14.

To illustrate, consider an ordinary 750 V, 60 Hz, No. 10 gauge, 3-phase cable 50 km long (Fig. 23-28). For such a line, the values of R, X_L, and X_C per phase are respectively 340 Ω, 12 Ω, and 6000 Ω. By comparison, a 69 kV, 60 Hz overhead transmission line of equal length but using 800 mm^2 ACSR conductors has corresponding values of 1.8 Ω, 25 Ω, and 25 000 Ω (Fig. 23-29).

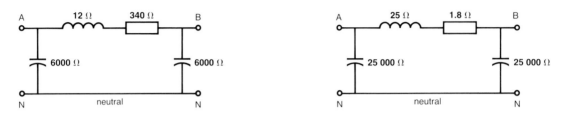

Figure 23-28
Equivalent circuit and relative impedance values of a LV three-phase cable having small conductors spaced close together. The line is 50 km long.

Figure 23-29
Equivalent circuit and relative impedance values of a HV three-phase aerial line having large conductors spaced far apart. The line is 50 km long.

Thus, on LV lines of low power, the conductor resistance tends to predominate, and so we can represent such lines by a simple resistance R. (Fig. 23-30). The inductive and capacitive reactances can be neglected.

On HV lines of high power, the inductive reactance predominates, therefore they can be represented by a simple reactance X_L (Fig. 23-31).

Figure 23-30
Simplified equivalent circuit of a LV and low power transmission line. The resistance of the line predominates.

Figure 23-31
Simplified equivalent circuit of a HV and high power transmission line. The inductive reactance predominates.

On EHV lines, the capacitive reactance draws a large capacitive current because the voltage is so high. Consequently, such lines can be represented by the circuit of Fig. 23-32.

23.19 Voltage regulation of a transmission line

Consider the HV line of Fig. 23-31 feeding a unity power factor load. Assume the voltage E_s of the source is fixed. As the load increases, line current I will increase

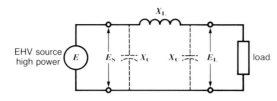

Figure 23-32
Equivalent circuit of an EHV, high power transmission line.
All the line impedances must be retained.

and the voltage E_L across the load will drop. However, we want to keep E_L as constant as possible and close to its rated value. Ideally, E_S and E_L should have the same magnitude. We could install a tap-changing transformer between the line and the load to keep the load voltage constant.

Another solution is to connect a variable capacitor bank X_{CR} at the load end of the line (Fig. 23-33). By adjusting the value of X_{CR}, we can attain a partial series resonance effect with X_L, which will keep E_L at its rated value. When the load is very small, X_{CR} can be removed, but as the load increases, more and more capacitance has to be added.

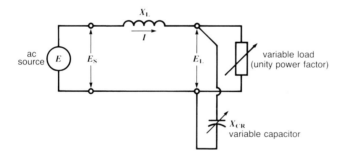

Figure 23-33
The voltage at the load end of the transmission line can be regulated by varying the capacitance X_{CR}.

What happens if the load is suddenly reduced and X_{CR} is kept unchanged? The load voltage E_L will rise considerably above its rated value because of series resonance between X_L and X_{CR}. This overvoltage condition is to be avoided because it may cause overheating of the loads that are still connected to the line.

Let us now turn back to the EHV line (Fig. 23-32). The presence of the line capacitance X_C at the end of the line means that we have to add less capacitance X_{CR} as the load increases. However, at light loads, E_L will rise above its nominal value because of partial series resonance between X_L and X_C. To prevent this from happening, we have to add an appropriate *inductive* reactance X_{LR} at light loads

(Fig. 23-34). Thus, on long HV and EHV transmission lines, we must install a variable inductive/capacitive reactance at the end of the line, in order to keep the voltage stable as the load varies. This can be done by using either a synchronous capacitor or a *static var compensator*.

The synchronous capacitor is a large synchronous motor that operates at no-load. The dc excitation is varied so as to vary the amount of reactive power that is delivered (or absorbed) by the machine. The principle of var control was covered in Section 20-7.

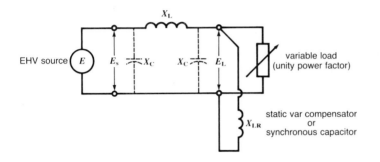

Figure 23-34
When EHV lines operate at light load, an inductive reactance must be added to absorb the reactive power generated by the capacitance of the transmission line. Otherwise E_L will rise above its rated value.

A static var compensator (Fig. 23-35) does the same job. However, it consists of a large three-phase capacitor bank and a somewhat smaller inductor bank. The amount of capacitive reactance (or inductive reactance) is varied electronically, by means of power thyristors. Static var compensators are preferred to synchronous capacitors because they can respond more quickly to voltage changes. This is particularly important when sudden overloads or short-circuits occur on the line. The very fast response of the static var compensator prevents the voltage from collapsing and tends to maintain stability of the system.

23.20 Power-handling capacity

We recall that the reactance X_L of a HV or EHV line is much greater than its resistance. Consequently, it is the *reactance* that sets an upper limit to the amount of power that a three-phase HV line can deliver. The maximum power is given by the equation

$$P_{max} = \frac{E^2}{X_L} \qquad (23\text{-}4)$$

where

P_{max} = maximum active power the three-phase line can deliver, in megawatts (MW)

E = line voltage, in kilovolts (kV)

X_L = inductive reactance per phase, in ohms (Ω)

Figure 23-35
Static var compensator for a HV transmission line. (*Courtesy General Electric*)

Example 23-3:

Calculate the maximum active power that the three-phase line of Fig. 23-29 can deliver, assuming a line voltage is 115 kV.

Solution:

$$P_{max} = E^2/X_L = 115^2/25 = 529 \text{ MW} \qquad\qquad \text{Eq. 23-4}$$

This is a large amount of power, enough to supply the needs of a modern city of 250 000 people.

The power that has to be delivered often exceeds the capacity of a three-phase line. Thus, in the previous example, if we had to deliver 700 MW, we would have to use two three-phase lines, running side by side between the generating station and the load. The second line is often supported by the same towers, but sometimes this is not possible. We must then run two transmission lines in parallel which is costly because the towers, insulators, and conductors have to be doubled (Fig. 23-36). Furthermore, a wider right-of-way has to be secured, and such rights are not easily obtained from property owners.

One solution is to use a higher transmission line voltage. For example, if the line voltage were increased from 115 kV to 230 kV, the power-handling capacity of the line would become

$$P = E^2/X_L = 230^2/25 = 2116 \text{ MW}$$

which is four times the previous value.

Figure 23-36
Two 735 kV transmission lines are needed to carry the power from a large generating site in northern Quebec. The clearing through the forest indicates the importance of securing rights-of-way. (*Courtesy Hydro-Québec*)

Raising the line voltage requires more insulators, and increases the spacing between lines and to ground. Furthermore, the size of the conductors has to be increased because the current is greater. However, this is still cheaper than running lines in parallel.

But there is always a limit to everything. Thus, the highest practical transmission line voltage is currently about 765 kV. If we have to transmit, say, 5000 MW over a distance of 1000 km, we still need *three* 765 kV transmission lines, running in parallel, to carry the load.

SUBSTATION EQUIPMENT

There is a need throughout the electrical system to raise and lower voltages by means of transformers, and to energize and de-energize electrical lines by means of switches and circuit breakers. There is a further need to provide lightning arresters, fuses, overvoltage and overcurrent relays, and other protective devices. This equipment is assembled in a *substation* (Fig. 23-37 and 23-38). Depending upon the power it has to handle and the number of incoming and outgoing transmission and

Figure 23-37
Small substation to meet the needs of a rural community.

Figure 23-38
Architecturally pleasing substation in the center of a large city. (*Courtesy Hydro-Québec*)

feeder lines it has to serve, the substation may cover an area of several acres or it may be a small metal-clad unit sitting next to a factory.

In the description that follows, we will examine some of the major pieces of equipment, or switchgear, used in a substation. We will conclude this overview with a typical substation that provides power to a small suburb.

23.21 Circuit breakers

Circuit breakers are designed to interrupt either normal or short-circuit currents. They behave like big switches that may be opened or closed by local pushbuttons or distant radio signals. Furthermore, circuit breakers will automatically open a circuit whenever the line current exceeds a preset limit. They can be set more accurately than fuses can be, and do not have to be replaced after each fault.

The most important types of circuit-breakers are:

1. oil circuit breakers (OCBs),

2. air-blast circuit breakers,

3. SF_6 circuit breakers.

The nameplate on a circuit breaker usually indicates (1) the maximum steady-state current it can carry, (2) the maximum current it can interrupt, (3) the maximum line voltage, and (4) the interrupting time in cycles. The interrupting time may last from 2 to 8 cycles on a 60 Hz system. High-speed interruption limits the damage to transmission lines and equipment and, equally important, it improves the stability of the system.

The tripping action is usually initiated by an overload relay that detects abnormal line conditions.

1. *Oil circuit breakers*. Oil circuit breakers are composed of a steel tank filled with insulating oil. In one version (Fig. 23-39), a group of porcelain bushings channels the three-phase line currents to a set of fixed contacts. A group of movable contacts opens and closes the circuit.

    When the circuit breaker is closed, the line current of each phase penetrates the tank through one porcelain bushing, flows through the fixed and movable contacts, and then flows out by a second bushing.

    If an overload occurs, the tripping coil releases a powerful spring which causes the contacts to open. As soon as the contacts separate, a violent arc is produced, which volatilizes the surrounding oil. The pressure of the hot gases creates turbulence around the contacts which causes cool oil to swirl around the arc, thus extinguishing it.

2. *Air-blast circuit breakers*. Air-blast circuit breakers interrupt the circuit by blowing compressed air at supersonic speed across the opening contacts. Compressed air is stored in reservoirs and is replenished by a compressor in the substation. The most powerful circuit breakers can typically interrupt short-circuit currents of 40 kA at a line voltage of 765 kV in a matter of 2 cycles, on a 60 Hz line. The noise accompanying the air blast is so loud that noise-

suppression methods must be used when the circuit breakers are installed near residential areas. Figure 23-40 shows a typical air-blast circuit breaker.

3.   SF$_6$ *circuit breakers*. These totally enclosed circuit breakers, insulated with SF$_6$ gas,* are used whenever space is at a premium, such as in downtown sub-stations. They are much smaller than any other type of circuit breaker of equiv-alent power, and are far less noisy than air-blast circuit breakers.

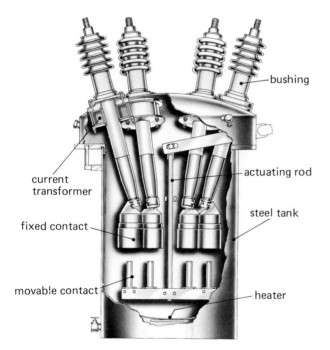

current
transformer

fixed contact

movable contact

bushing

actuating rod

steel tank

heater

Figure 23-40
Air-blast circuit breaker composed of three identical modules connected in series. Compressed air stored in the cylindrical tank is channeled upwards in the three porcelain columns and blown across the open-ing (enclosed) contacts above. One such circuit breaker is needed for each phase on a three-phase line. (*Courtesy Brown Boveri*)

Figure 23-39
Cross section of a three-phase oil circuit breaker, showing four of the six brushings. The heater keeps the oil at a satisfactory temperature during cold peri-ods. (*Courtesy Canadian General Electric*)

*   Sulfur hexafluoride

23.22 Air-break switches

Air-break switches can interrupt only the exciting currents of transformers or the moderate capacitive currents of unloaded transmission lines. They cannot interrupt normal load currents.

Air-break switches are composed of a movable contact that engages a fixed contact; both contacts are mounted on insulating supports (Fig. 23-41). Arcing horns are attached to the fixed and movable contacts so that when the main contact is broken, an arc is set up between them. As the arc moves upward it becomes progressively longer until it eventually blows out (Fig. 23-42).

Figure 23-41
Air-break disconnecting switch rated at 3000 A to open one phase of a three-phase, 735 kV, 60 Hz line. The long horizontal arm at the top swings upward when the motorized control box is energized. The control box is situated at the base of the column on the left. The arcing horn is the vertical rod with the sphere on top. (*Courtesy Kearney-National*)

Figure 23-42
The arc produced by the opening air-break switch provides the light for this night picture. The arc is due to interrupting the exciting current of a HV transformer (*Courtesy Hydro-Québec*)

23.23 Disconnecting switches

Unlike air-break switches, disconnecting switches are unable to interrupt any current at all. They must be opened and closed only when the current is zero. They are basically isolating switches, enabling us to isolate circuit breakers, transformers, transmission lines, and so forth, from a live network. Disconnecting switches are essential to carry out maintenance work and reroute power flow (Figure 23-43 and 23-44).

23.24 Grounding switches

Grounding switches are safety switches; they ensure that a transmission line is definitely grounded while repairs are being made. To short-circuit the line to ground, three grounded blades swing up to engage the stationary contact connected to each phase. Grounding switches are opened and closed only when the line is de-energized.

Figure 23-43
Disconnecting switch rated 2000 A, 230 kV, 3-phase to reroute power in a substation. (*Courtesy Kearny-National; photo by Stan Walter*)

Figure 23-44
Three-pole disconnecting switch permits the isolation of an oil circuit breaker whose bushings are seen immediately below. (*Courtesy Hydro-Québec*)

Some disconnecting switches also act as grounding switches: when they are in the open position, the switch blades are connected to ground.

23.25 Lightning arresters

The purpose of a lightning arrester* is to limit the overvoltages that may occur across transformers and other electrical apparatus due to lightning or switching surges. The upper end of the arrester is connected to the line that has to be protected, and the lower end is connected to ground (Fig. 23-45).

---

*   Also called surge arrester or surge diverter.

The arrester is composed of an external porcelain tube containing an ingenious arrangement of stacked discs, air gaps, and blow-out coils. The discs are usually composed of a silicon carbide material whose resistance decreases drastically when the voltage exceeds a certain level.

Figure 23-45
Surge arrester connected between a 735 kV line and ground. The nominal voltage across the arrester is 424 kV. The tank in the background is a reactor to compensate the capacitive reactance of the transmission line. (*Courtesy Hydro-Québec*)

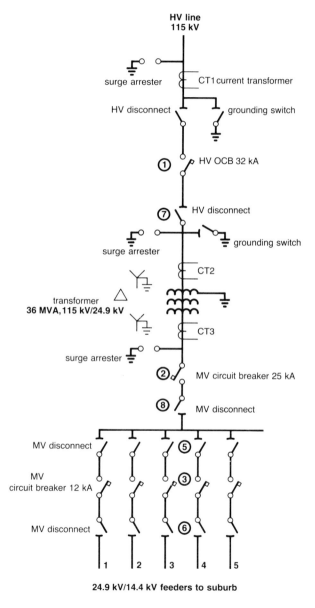

Figure 23-46
One-line diagram of a small substation feeding a suburb.

Under normal voltage conditions, the air gaps prevent current from flowing through the tubular column. However, if an overvoltage occurs, the air gaps break down and the surge discharges to ground. The arc is quickly snuffed out, making the arrester ready to protect the apparatus against the next voltage surge. The discharge period is very short, rarely lasting more than a fraction of a millisecond.

23.26 Example of a substation

Figure 23-46 shows the principal components of a small substation providing power to a suburb. Three-phase power is fed into the substation at 115 kV and is redistributed by five feeders at 24.9 kV to various load centers within the suburb.

The substation contains a three-phase transformer rated at 36 MVA, 115 kV/24.9 kV. The windings are connected in wye-wye with a tertiary winding. Automatic tap-changers inside the transformer regulate the secondary voltage.

An oil circuit breaker (1) having an interrupting capacity of 32 kA protects the HV side. Another OCB (2) having an interrupting capacity of 25 kA protects the MV side. In addition, the five outgoing feeders are protected by OCBs, each having an interrupting capacity of 12 kA.

To understand how the substation operates, suppose that it is functioning normally, with all disconnect switches and circuit breakers closed. Let us consider two cases.

Case 1. Severe short-circuit occurs on MV feeder 3

This will cause the 12 kA circuit breaker (3) protecting feeder 3 to trip. If repairs have to be made on the feeder, the corresponding MV disconnects (5) and (6) will be opened to isolate the feeder from the live MV bus.

Case 2. 36 MVA transformer develops an internal short-circuit

Under normal conditions, the ratio between the primary and secondary currents is constant, and this is reflected in the "balanced" condition of current transformers CT2 and CT3. But when an internal short-circuit occurs, the primary current suddenly becomes much greater. This unbalanced condition is detected by the two CTs, causing the HV oil circuit breaker (1) to trip.

During the repair stage, the motorized disconnect (7) is opened, which also closes the associated grounding switch. The MV circuit breaker (2) is also opened, as well as the MV disconnect (8). The transformer is now isolated, and so it may be removed and replaced by a spare.

LOW-VOLTAGE DISTRIBUTION IN BUILDINGS

The electrical distribution system in a building is the final link between the consumer and the original source of electrical energy. Such "in-house" distribution systems must meet certain basic requirements. They relate to:

1. *Safety*:
    a. protection against electric shock
    b. protection of conductors against physical damage
    c. protection against overloads
    d. protection against hostile environments

2. *Conductor voltage drop*: It should not exceed 1 or 2 percent of the nominal line voltage.

3. *Life expectancy*: The distribution system should last a minimum of 50 years.

4. *Economy*: The cost of the installation should be minimized while observing the pertinent standards.

Standards are set by the National Electrical Code*, and every electrical installation must be approved by an inspector before it can be put into service.

23.27 Principal components of an electrical installation

Many components are used in the makeup of an electrical installation. The block diagrams of Figs. 23-47 and 23-48, together with the following definitions†, will show the purpose of the major items.

1. **Service conductors**. These are the conductors that extend from the street main or from transformers to the service equipment of the premises supplied.

2. **Service equipment**. The necessary equipment, usually consisting of a circuit breaker or switch and fuses, and their accessories, located near the point of entrance of supply conductors to a building or other structure, or an otherwise defined area, and intended to constitute the main control and means of cutoff of the supply.

3. **Metering equipment**. Various meters and recorders to indicate the electrical energy consumed on the premises.

4. **Panelboard**. A single panel or group of panel units designed for assembly in the form of a single panel; including buses, automatic overcurrent devices, and with or without switches for the control of light, heat or power circuits; designed to be placed in a cabinet or cutout box placed in or against a wall or partition and accessible only from the front.

5. **Switchboard**. A large single panel, frame, or assembly of panels on which are

---

*   In Canada, by the Canadian Electrical Code.

†   Some taken from the National Electrical Code; See Article 100 — Definitions.

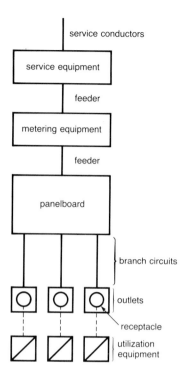

Figure 23-47
Block diagram of the electrical distribution system in a residence.

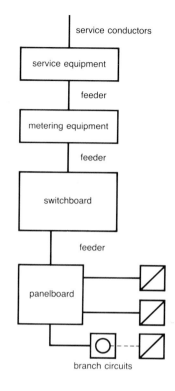

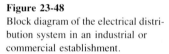

Figure 23-48
Block diagram of the electrical distribution system in an industrial or commercial establishment.

mounted, on the face, or back, or both, switches, overcurrent and other protective devices, buses and usually instruments. Switchboards are generally accessible from the rear as well as from the front and are not intended to be installed in cabinets.

6. **Feeder**. All circuit conductors between the service equipment, or the generator switchboard of an isolated plant, and the final branch-circuit overcurrent device.

7. **Branch circuit**. The circuit conductors between the final overcurrent device protecting the circuit and the outlet(s).

8. **Outlet**. A point on the wiring circuit at which current is taken to supply utilization equipment.

9. **Receptacle**. A contact device installed at the outlet for the connection of a single attachment plug.

10. **Utilization equipment**. Equipment which utilizes electrical energy for mechanical, chemical, heating, lighting, or similar services.

The greatly simplified diagrams of Figs. 23-47 and 23-48 indicate the type of distribution systems used respectively in a home and in an industrial or commercial establishment. Figure 23-49 shows in greater detail some of the components found in a typical home. The metering consists of a watthourmeter and the service equipment is a fused disconnecting switch with the neutral grounded to a water-pipe. The panelboard consists of several circuit breakers (or fuses) that protect each of the branch circuits. The branch circuits consist of two- or three-conductor cables, all single-phase.

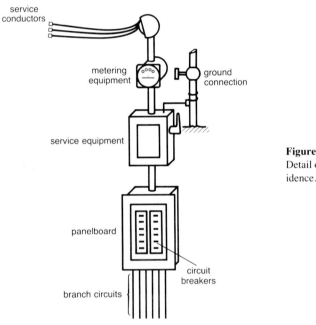

Figure 23-49
Detail of the service entrance in a residence.

23.28 LV distribution systems

The most common LV systems used in the United States and Canada are:

1.   Single-phase, 2-wire, 120 V
2.   Single-phase, 3-wire, 120/240 V
3.   Three-phase, 4-wire, 120/208 V
4.   Three-phase, 3-wire, 480 V
5.   Three-phase, 4-wire, 277/480 V
6.   Three-phase, 3-wire, 600 V
7.   Three-phase, 4-wire, 347/600 V

In some countries, three-phase, 220/380 V, 50 Hz systems are used. Despite the different voltages employed, the basic principles of LV distribution are the same.

Single-phase, 2-wire, 120 V system. This simple distribution system is used only for very small loads. When heavier loads have to be serviced, the 120 V system is not satisfactory because large conductors are required. Such systems are now absolete.

Single-phase, 3-wire, 120/240 V system. In order to reduce the current and conductor size, the voltage is raised to 240 V. However, because the 120 V level is still very useful, the 120 V/240 V 3-wire system was developed (Fig. 23-50). This type of distribution system is widely used. It is produced by a distribution

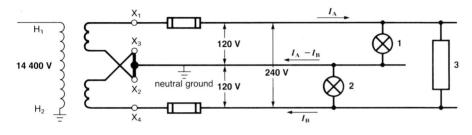

Figure 23-50
Single-phase, 120/240 V distribution system.

transformer having a double secondary winding (Section 16-13). The common wire, called neutral, is connected to ground. When the "live" lines A and B are unequally loaded, the current in the neutral current is equal to the difference between the line currents ($I_A - I_B$). We try to distribute the 120 V loads as equally as possible between the two live lines. In effect, the current in the neutral wire is then zero, which reduces the I^2R losses in the conductors and the secondary winding of the distribution transformer.

What are the advantages of such a 3-wire system?

1. The line to ground voltage is only 120 V, which is reasonably safe for use in a home

2. Lighting and small motor loads can be energized at 120 V, while larger loads, such as electric stoves, motors, and so forth, can be fed from the 240 V line.

Both live lines are protected by fuses or circuit breakers. The neutral wire is *never* protected.

Three-phase, 4-wire, 120/208 V system. We can create a 3-phase, 4-wire system by using three single-phase transformers connected in delta-wye. The neutral of the secondary is grounded, as shown in Fig. 23-51.

This system is used in commercial buildings and small industries because the 208 V line voltage can be used for electric motors or other large loads, while the 120 V lines can be used for lighting circuits and convenience outlets. The loads between the three live lines and neutral are balanced as best as possible. When the loads are perfectly balanced, the current in the neutral wire is zero.

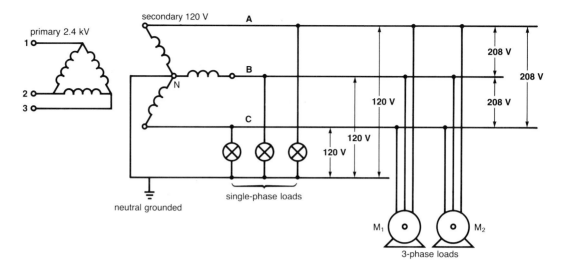

Figure 23-51
Three-phase, 4-wire 120/208 V distribution system.

Three-phase, 3-wire, 600 V system. A 600 V, 3-phase, 3-wire system is used for power circuits where fairly large motors ranging up to several hundred horsepower are installed (Fig. 23-52). Separate 600 V/240-120 V step-down transformers, spotted throughout the premises, are used to service the lighting loads and convenience outlets. (Similar remarks apply to a three-phase, 3-wire, 480 V system.)

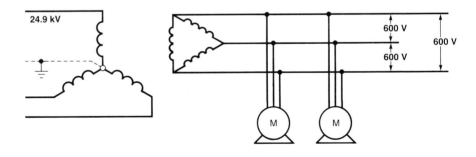

Figure 23-52
Three-phase, 3-wire 600 V distribution system.

Three-phase, 4-wire, 277/480 V system. In large buildings and commercial centers, a 480 V, 4-wire distribution system is used because motors can be operated at 480 V and fluorescent lights at 277 V. For 120 V convenience outlets, separate transformers are required, usually fed from the 480 V line. Similar remarks apply to a three-phase, 347/600 V, 4-wire system.)

23.29 Grounding electrical installations

The grounding of electrical systems is an effective way of preventing accidents. As we have seen, most electrical distribution systems in buildings are grounded, usually by connecting the neutral to a water pipe or to the massive steel structure.

The exposed metallic parts of electrical equipment, such as motors, electric stoves, hand tools, distribution panels, conduits, and so forth must *always* be grounded. Such grounding is used to reduce the danger of electric shock.

23.30 Electric shock

It is difficult to specify whether a voltage is dangerous or not because electric shock is actually caused by *the current* that flows through the body. The current depends mainly upon the skin contact resistance which varies with the thickness and wetness of the skin.

It is generally claimed that currents below 5 mA are not dangerous. Between 10 mA and 20 mA, the current is potentially dangerous because the victim loses muscular control and so may not be able to let go. Above 50 mA, the consequences may be fatal.

The resistance of the human body, measured between two hands, or between one hand and a leg, ranges from 500 Ω to 50 kΩ. If the resistance of a dry hand is 50 kΩ, then momentary contact with a 600 V line may not be fatal (I = 600 V/50 kΩ = 12 mA). But the resistance of a damp hand is much lower, therefore a voltage as low as 25 V may prove fatal (I = 25 V/500 Ω = 50 mA).

When current flows through the body, the muscular contractions prevent the victim from letting go. The current is particularly dangerous when it flows in the region of the heart. It induces temporary paralysis, and the person may have to be revived by applying artificial respiration.

The duration of current flow is equally important. For example, a current of 50 mA flowing through the chest for 10 s is far more dangerous than a current of 500 mA that flows for 20 ms.

THE COST OF ELECTRICITY

Electric utility companies have established billing rates, called tariffs, that are designed to be as equitable as possible for their customers. The rates are based upon three factors:

1. the amount of energy consumed, expressed in kilowatthours or megajoules.

2. the maximum power demanded during the billing period, expressed in kilowatts. The maximum power is called the *demand*. It is the highest average power required during a specified interval of time, such as 10, 15, or 30 minutes.

3. the power factor of the load.
    Electricity used for domestic consumption is usually billed on an energy basis

alone. Demand and power factor are taken into account only for commercial and industrial consumers.

23.31 Tariff based upon energy

The cost of electricity depends first upon the amount of energy consumed. However, even a customer using no energy has to pay a minimum service charge because it costs money to keep him connected to the line.

As energy consumption increases, the cost per kilowatthour drops, usually on a sliding scale. Thus, the domestic tariff may start at 10 cents per kilowatthour for the first one hundred kW.h, then fall to 5 cent/kW.h for the next two hundred kW.h, and bottom out at 4 cent/kW.h for the rest of the energy consumed. The same general principle applies to medium-power and large-power users of electrical energy.

The energy consumption is measured by a watthourmeter, shown in Fig. 21-31. It includes a high-precision motor (Section 21.15) whose speed of rotation is directly proportional to the active power consumed by the customer. The number of turns is therefore proportional to the energy. The number of turns (energy) is displayed on a register, which is read at monthly intervals.

23.32 Tariff based upon demand

The monthly cost of electricity for a large customer depends not only upon the energy he consumes, but also upon the rate at which he uses it. Thus, the cost also depends upon the active power drawn from the line.

The reason is that the size and investment cost of line conductors, switches, and transformers depends upon the maximum current they have to carry. Because a higher power requires a higher current, a customer drawing 500 kW for 100 hours during the month should pay more for electricity than a customer drawing 100 kW for 500 hours. The amount of energy consumed is the same in each case (500 000 kW.h), but it takes much larger electric utility equipment to service the customer that draws 500 kW.

It is advantageous, both to the customer and the utility company that energy be consumed at a constant rate during the month. The more regular the power flow, the lower the cost will be.

However, a power surge, such as that drawn by a motor during start-up, does not last long enough to warrant the installation of large equipment by the utility company. How long does the power flow have to last, to be considered significant? The answer depends upon several factors, but the period is usually taken to be 10, 15, or 30 minutes. For very large power users, such as municipalities, the averaging period may be as long as 60 minutes. It is called the *demand interval*.

23.33 Demand meter

To monitor the power flowing into the plant, a special meter is installed at the customer's service entrance. It automatically measures the average power during successive demand intervals (15 minutes, for example). The average power measured during each interval is called the *demand*. As time goes by, the demand meter records the demand every 15 minutes and a pointer moves up and down a calibrated scale as the demand changes. In order to record the maximum demand, the meter carries a second pointer that is pushed upscale by the first.

The second pointer simply rests at the highest position to which it is pushed. At the end of the month, a meter reader takes the maximum demand reading and resets the pointer to zero. Demand meters are installed at the service entrance of most industries and commercial establishments. Figure 23-53 shows a combination demand meter and watthourmeter in which the maximum demand in kilowatts is read out directly, rather than using a two-pointer system.

Figure 23-53

In this combination watthourmeter / demand meter, the maximum demand in kilowatts is registered by the three lower dials. These dials are reset to zero at the end of each billing period. (*Courtesy General Electric*)

23.34 Tariff based upon power factor

Alternating current machines, such as induction motors and transformers, absorb reactive power to produce their magnetic fields. The power factor of these machines is therefore less than unity, and so is the power factor of the factory where they are installed. A low power factor increases the cost of electrical energy because the line current is higher for a given active load. For example, a consumer that takes 830 kW from a 480 V, 3-phase line will draw a line current of 1000 A, if the power factor is 100 percent. But if the power factor is 50 percent, the current will be 2000 A. Thus, even though the active power and energy consumed are the same, a power factor of 50 percent requires much larger utility company equipment to service the load.

For this reason, most electric utilities stipulate that the power factor of their industrial clients be 90 percent or more, in order to benefit from the minimum rate. To determine the power factor, the electric utility company records the apparent

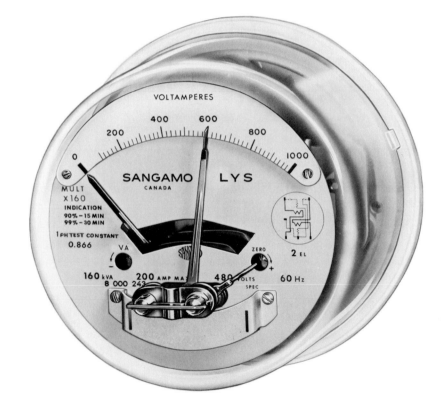

Figure 23-54
Demand meter that records the maximum apparent power consumed during the billing period. If the apparent power is greater than the active power demand, the power factor is less than unity. (*Courtesy Sangamo*)

power absorbed by the plant by means of a voltamperemeter (Fig. 23-54).

When the power factor is too low, it is usually to the customer's advantage to improve it, rather than pay the higher monthly bill. This is usually accomplished by installing capacitors at the service entrance to the plant on the load side of the metering equipment. These capacitors may supply part, or all, of the reactive power required by the plant. Industrial capacitors for power factor correction are made in single-phase and three-phase units rated from 5 kvar to 200 kvar.

23.35 Typical rate structures

Electrical power utility rates vary greatly from one region to another, therefore we can give only a general overview of the subject. Most companies divide their customers into categories, according to their power demand. For example, one utility company uses four power categories:

1. *Domestic power* — power corresponding to the needs of houses and rented apartments;

2. *Small power* — power of less than 100 kw;

3. *Medium power* — power of 100 kw and more, but less than 5000 kw;

4. *Large power* — power in excess of 5000 kw.

Table 23B shows typical rate structures that apply to residential and medium-power categories.

TABLE 23B   TYPICAL RATE STRUCTURES

Residential Rate Structure

Typical contract clauses:

1. "... This rate shall apply to electric service in a single private dwelling

2. ... This rate applies to single-phase alternating current at 60 Hz..."

Rate schedule

Minimum monthly charge: $10.00 plus
Energy charge:
    First 100 kW.h per month at 10 cent/kW.h
    Next 200 kW.h per month at 8 cent/kW.h
    Excess over 300 kW.h per month at 6 cent/kW.h

General Power Rate Structure (medium power)

Typical contract clauses:

1. "... The customer's maximum demand for the month, or its contract demand, is at least 50 kW but not more than 5000 kW ...

2. ... The power shall be delivered at a nominal 3-phase line voltage of 480 V, 60 Hz ...

3. ... The power taken under this contract shall not be used to cause unusual disturbances on the Utility Company's system. The customer shall, at its expense, correct such disturbances ...

4. ... Voltage variations shall not exceed 7 percent higher or lower from the nominal line voltage ...

5. ... Utility Company shall make periodic tests of its metering equipment to maintain a high standard of accuracy ...

6. ... Customer shall use power so that current is reasonably balanced on the three phases. Customer agrees to take corrective measures if the current on the more heavily loaded phase exceeds the current in either of the two other phases by more than 20 percent. If said unbalance is not corrected, Utility Company may meter the load on individual phases and compute the billing demand as being equal to 3 times the maximum demand on any phase ...

7. ... The maximum demand for any month shall be the greatest of the demands measured in kilowatts during any 30-minute period of the month ...

8. "... If 90 percent of the highest average kVA measured during any 30 minute period is higher than the maximum demand, such amount will be used as the billing demand ..."

Rate Schedule

Demand charge: $5.00 per month per kW of billing demand
Energy charge: 6 cent/kW.h for the first 100 hours of billing demand
                4 cent/kW.h for the next 50 000 kW.h per month
                3 cent/kW.h for the remaining energy

Example 23-4:

A homeowner consumes 900 kW.h during the month of August. Calculate his electricity bill using the residential rate schedule given in Table 23B.

Solution:

Minimum charge	=	$ 10.00
First 100 kW.h @ 10 cent/kW.h	=	10.00
Next 200 kW.h @ 8 cent/kW.h	=	16.00
Remaining energy consumed		
(900 − 300) = 600 kW.h @ 6 cent/kW.h	=	36.00

Total bill for the month: $ 72.00

This represents an average cost of 7200/900 = 8 cent/kW.h.

Example 23-5:

A small industry operating 24 hours a day, 7 days a week consumes 260 000 kW.h per month. The maximum demand is 1200 kW, and the maximum kVA demand is 1700 kVA. Calculate the electricity bill using the medium-power rate schedule in Table 23B.

Solution:

1. Clause 8 is important here because 0.90 x 1700 kVA = 1530 kVA which is greater than the maximum demand of 1200 kW. Consequently, the demand for billing purposes (1530 kVA) is higher than the metered demand (1200 kW).

2. Applying the rate schedule, the demand charge is:
    1530 kW at $5.00/kW   = $ 7 650

3. The energy charge is:
    a) 1530 kW × 100 hours = 153 000 kW.h @ 6 cent/kW.h   = $ 9 180

    b) 50 000 kW.h at 4 cent/kW.h =   = $ 2 000

    c) The remainder of the energy is
        (260 000 − 153 000 − 50 000) = 57 000 kW.h

    d) 57 000 kW.h at 3 cent/kW.h   = $ 1 710

    Total bill for the month: = $20 540

    The average cost of energy = 20 540/260 000 = 7.9 cent/kW.h

Management could reduce the monthly bill by installing capacitors. The kvar capac/304/ity should be such that the maximum kVA demand is equal to, or less than, 1/0.9 = 1.1 times the maximum demand (kW).

23.36 Power factor correction

Power factor correction (or improvement) is economically feasible when the monthly decrease in the cost of electric power exceeds the amortized cost of installing the capacitors. In some cases, the customer has no choice, but most comply with the minimum power factor specified by the utility company.

The power factor may be improved by installing capacitors at the service entrance to the plant or commercial enterprise.

Example 23-6:

A factory draws an apparent power of 3000 kVA at a power factor of 65% (lagging). What capacity in kvar must be installed at the service entrance to bring the overall power factor to (a) unity, and (b) 90 percent lagging.

Solution:

1. a) Apparent power absorbed by the plant is:
        $S = 3000 \text{ kVA}$

    b) Active power absorbed by the plant is:
        $P = S \cos \theta = 3000 \times 0.65 = 1950 \text{ kW}$ Eq. 14-23

    c) Reactive power absorbed by the plant is:

    $$Q = \sqrt{S^2 - P^2} = \sqrt{3000^2 - 1950^2} = 2280 \text{ kvar}$$ Eq. 14-20

    The respective power flows are shown in Fig. 23-55.

    To raise the power factor to unity, we have to supply all the reactive power absorbed by the load (2280 kvar). The capacitors must therefore have a capacity of 2280 kvar, or about 760 kvar per phase on a three-phase line. Figure 23-56 shows the active and reactive power flow after the capacitors are installed.

2. a) The factory still draws the same amount of active power (1950 kW) because the mechanical and thermal loads are fixed. If the new overall power factor is to be 0.90 lagging, the apparent power drawn from the line must be:

        $S = P/\cos \theta = 1950/0.90 = 2167 \text{ kVA}$

    b) The new reactive power supplied by the line is:

        $Q = \sqrt{2167^2 - 1950^2} = 945 \text{ kvar}$

    c) Because the plant still needs 2280 kvar, the difference must come from the capacitors. The capacity of these units is:

        $Q = (2280 - 945) = 1335 \text{ kvar}$

Thus, if a power factor of 0.90 (instead of unity) is acceptable, we can reduce the size of the capacitor bank from 2280 kvar to 1335 kvar. Figure 23-57 shows the power flow in the transmission line and the factory. Note that the factory draws the same active and reactive power, regardless of the size of the capacitor bank.

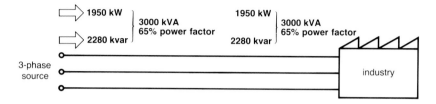

Figure 23-55
Active and reactive power before capacitors are installed (see Example 23-6)

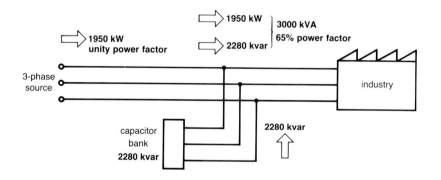

Figure 23-56
Active and reactive power after capacitors are installed and power factor of the line is unity.

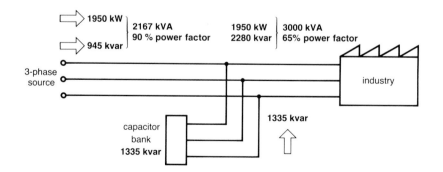

Figure 23-57
Active and reactive power when the power factor of the line is raised to 90 percent.

23.37 Summary

This chapter has given a broad overview of how a modern electrical system is integrated. Most electric power is generated by either thermal, hydropower or nuclear plants. The location of these plants is determined by the energy source, but a reliable source of water is also important.

When large quantities of energy have to be transported, we learned that the reactance of the transmission line plays a predominant role. The resistance of the line determines the efficiency and heating effects, but the amount of power that can be transmitted depends upon X_L. We also learned that a HV transmission line can produce resonance effects, which can be used to regulate the voltage near the load. Thus, voltage control by static var compensators and synchronous capacitors is made possible.

This chapter also showed how important are the concepts of active, reactive and apparent power. The cost of electricity depends upon the power factor at the consumer's premises and the nature of his demand. Thus, a large consumer can save thousands of dollar per month by correcting the power factor or by reducing his peak demand.

We also learned that power can be turned on and off by fast-acting circuit breakers, and that disconnecting switches are needed to isolate equipment and to reroute power. At the low-voltage end of the spectrum we examined some of the more common distribution systems that are in use. They permit a combination of single-phase loads and three-phase loads to be connected to the line.

Finally, the illustrations have given us an opportunity to see the size of the machines and components that make up an electrical system. They indirectly point out the importance of the mechanical and thermal aspects in the generation, transmission and distribution of electrical energy.

TEST YOUR KNOWLEDGE

23-1   The designation HV is the voltage range

    a. 2.4 kV to 69 kV   b. 115 kV to 230 kV   c. 69 kV to 230 kV

23-2   The voltage drop over the entire length of a 200 km transmission line is

    a. relatively small   b. relatively large   c. large and variable

23-3   On a medium-voltage system, one of the preferred voltages is

    a. 13.2 kV   b. 13.8 kV   c. 14 kV

23-4   The voltage at the load end of a 115 kV transmission line varies from 115 kV at no-load to 109 kV. The voltage drop is

    a. 600 V   b. 6000 V   c. 60 000 V

23-5   In Problem 23-4, the voltage drop in percent is

a. 5.2 %   b. 5.5 %   c. 5.357 %

23-6   A load having a lagging power factor

a. absorbs reactive power   b. generates reactive power
c. absorbs no active power

23-7   If the reactive power carried by a transmission line is reduced, the line current always decreases.

☐   true   ☐   false

23-8   A transmission operates at unity power factor, and the line current is 260 A. If capacitors are installed at the end of the line, the line current will

a. increase   b. decrease   c. remain the same

23-9   In Problem 23-8, the power factor of the line will

a. improve   b. worsen   c. remain the same

23-10   The power demand of an electric utility system is usually measured in

a. MVA   b. kW   c. MW

23-11   If we double the voltage of a HV transmission line, it can carry

a. four times as much power   b. twice as much power
c. six times as much power

23-12   The governor of a turbine-generator set attempts to automatically keep the frequency constant as the generator load varies.

☐   true   ☐   false

23-13   Each ton of coal in a generating station releases 20 GJ of heat. If 10 tons are burned per hour, the approximate electric power output of the station is

a. 17 MW   b. 55 MW   c. 55 MVA

23-14   The steam that emerges from a turbine escapes up the chimney.

☐   true   ☐   false

23-15   The energy released when one ounce of matter is converted into energy is about

a. 8.5 MJ   b. 2.5 PJ   c. 2.5 PW

23-16   The energy delivered in one year by a 200 MW generating station that operates night and day is about

a. 6.3 PJ   1.7 TJ   c. 1 752 000 kW.h

23-17   A synchronous capacitor operates at times as a three-phase inductor

☐   true   ☐   false

23-18   A surge arrester is used to limit switching surges as well as lightning surges.

☐   true   ☐   false

23-19   The three conductors leading into the service entrance of a home are the respective phases of a three-phase system

☐   true   ☐   false

23-20   The intensity of an electric shock is due to

a. the current   b. the voltage   c. the power of the source

23-21   A plant draws 600 kW at a lagging power factor of 82 percent. To bring the power factor up to 100 percent, the capacitor bank rating should be about

a. 345 kvar   b. 420 kvar   c. 345 kVA

23-22   An OCB opens a 60 Hz line in 2.4 cycles. The time to interrupt the current is

a. 144 ms   b. 40 ms   c. 20 ms

23-23   The condenser in a thermal generating station is also a heat exchanger

☐   true   ☐   false

QUESTIONS AND PROBLEMS

23-24   Name the three main sectors of an electric utility system.

23-25   Name the three basic objectives of an electric utility system.

23-26   State the meaning of the abbreviations LV, MV, HV and EHV.

23-27   Name two undesirable effects that result when a transmission line carries a substantial amount of fluctuating reactive power.

23-28   What is the approximate transmission-line voltage range that would be considered for a line that has to carry 5 000 kW over a distance of 30 km?

23-29   Name four types of generating stations.

23-30   Why are thermal and nuclear generating stations usually located near rivers and lakes?

23-31   Explain how a large electric system may become unstable.

23-32   Why are forced-draft fans of several thousand horsepower needed in a generating station?

23-33   Calculate the theoretical energy that is released when 1 lb of matter is converted into energy. For how long could this energy drive a 200 mW generator, assuming an efficiency of 30 percent?

23-34   What is the purpose of a moderator in a nuclear reactor?

23-35   A large hydropower generating station is proposed at a site where the water falls through a height of 1060 feet. If the steady water flow during the year is 1500 cubic yards per second, calculate the available water power, in megawatts.

23-36   Why are bundled conductors always used on EHV lines?

23-37   Explain the purpose of the overhead ground wire on a HV line.

23-38   Explain why synchronous capacitors are sometimes used on HV and EHV lines.

23-39   What is the purpose of a static var compensator?

23-40   A 69 kV, 60 Hz transmission line that is 30 km long has an inductive reactance of 0.5 Ω/km. Calculate its approximate power-handling capacity and the corresponding line current.

23-41   In some transmission lines the power-handling capacity cannot be increased by increasing the conductor size. Explain why.

23-42   Name three types of circuit breakers used in substations.

23-43   State the advantages of a 120/208 V, three-phase distribution system over a 208 V, three-phase system.

23-44   When a person walks across a room, the static electricity may sometimes rise to more than 10 000 V. Why does this not kill the person when discharge occurs?

23-45   State the factors that determine the monthly electricity bill of a manufacturing plant.

23-46   State the factors that determine the monthly electricity bill of a private home.

23-47   Explain the purpose and operation of a demand meter.

23-48   A homeowner consumes 7 kW.h during the month of June. Using the rate structure given in Table 23B, calculate the average cost per kW.h.

23-49   The motor of an electric clock consumes 6 W. Calculate the yearly cost to the homeowner if the average cost is 8 cents per kW.h.

23-50   The average 60 W incandescent lamp has a lifetime of 900 H and costs 40 cents. If the average cost of electricity is 6 cent/kW.h, what is the *total* expenditure on a 60 W bulb over its lifetime?

23-51   A manufacturing plant contains the following major loads
         incandescent lighting — 200 kW.
         induction motors — 1200 kVA at a power factor of 80 percent.
         induction motors — 650 kVA at a power factor of 70 percent.

   Calculate:

   a)   the total active power absorbed by the plant

   b)   the total reactive power supplied by the electric utility company

   c)   the average power factor of the plant.

23-52   In Problem 23-52 calculate the capacitor bank rating needed to bring the power factor of the plant up to

   a)   100 percent

   b)   95 percent

23-53   A large paper mill contains six 4000 hp synchronous motors having an efficiency of 0.96 and a power factor rating of 0.85 (leading). Calculate

   a)   the active power supplied to the motors

   b)   the reactive power this group of motors can generate.

Conversion Charts for Units

The relative size of a unit is the key factor that gives us a "feel" for a unit.

The conversion charts in this appendix show the relative size of a unit by the position it occupies on the chart. The largest unit is at the top, the smallest at the bottom, and intermediate units are ranked in between.

The rectangles bearing SI units extend slightly towards the left, to distinguish them from other units.

The units are connected by straight lines, each of which bears an arrow and a number. The number is the ratio of the larger to the smaller of the units so connected, and so its value is always greater than one. The arrow always points toward the smaller of the two units.

In Fig. A-1, for example, five units of length — the mile, meter, yard, inch and millimeter — are listed in descending order of size. The numbers show the relative size of the connected units; the yard is 36 times larger than the inch, the inch is 25.4 times larger than the millimeter, and so on. With this arrangement, we can easily convert from one unit to any other.

Suppose we want to convert from yards to millimeters. Starting from **yard** in Fig. A-1, we have to move downward and in the direction of the arrows for both lines (36 and 25.4) until we reach **millimeter**. Conversely, if we want to convert from millimeters to yards, we start at **millimeter** and move upward against the direction of the arrows until we reach **yard**. To convert from one unit to another, we apply the following rule:

1. *If, in going from one unit to another, we move in the direction of the arrow, we multiply by the associated number.*

2. *Conversely, if we move against the arrow, we divide.*

In moving from one unit to another, we can follow any path we please; the conversion result is always the same.

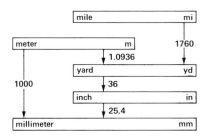

Figure A-1
Units of length.

Example A-1:

Convert 2.5 yards to millimeters.

Solution:

Starting from **yard** and moving toward **millimeter**, we always move in the direction of the arrows. We must therefore multiply the numbers associated with each line:

2.5 yd = 2.5 × 36 × 25.4 millimeters
   = 2286 mm.

Example A-2:

Convert 2000 meters into miles.

Solution:

Starting from **meter** and moving toward **mile**, we move once with, and once against the direction of the arrows. Consequently, we have:

$$2000 \text{ meters} = 2000 \times 1.0936 \div 1760 \text{ miles}$$

$$= \frac{2000 \times 1.0936}{1760}$$

$$= 1.24 \text{ miles}$$

Example A-3:

Convert 75 100 Btu to kilowatthours.

Solution:

Referring to the ENERGY chart, we now move past three arrows, two of which point opposite to the path we are following. Consequently:

75 100 Btu = 75 000 × 1.055 ÷ 1000 ÷ 3.6

   = 22 kW.h

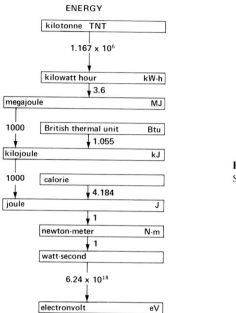

Figure A-2
See Example A-3

Quantities such as AREA, MASS, VOLUME, and so forth, are listed in alphabetical order for quick reference. Multipliers between units are either exact, or accurate to ± 0.1 percent.

Examples at the bottom of each page are intended to assist the reader in applying the conversion rule which basically states:

WITH THE ARROW — MULTIPLY

AGAINST THE ARROW — DIVIDE

MULTIPLES AND SUB-MULTIPLES OF SI UNITS

exa	E
↓ 1000	
peta	P
↓ 1000	
tera	T
↓ 1000	
giga	G
↓ 1000	
mega	M
↓ 1000	
kilo	k
↓ 10	
hecto	h
↓ 10	
deca	da
↓ 10	
UNIT	1
↓ 10	
deci	d
↓ 10	
centi	c
↓ 10	
milli	m
↓ 1000	
micro	μ
↓ 1000	
nano	n
↓ 1000	
pico	p
↓ 1000	
femto	f
↓ 1000	
atto	a

(1000 links: kilo → milli, and 1000 links UNIT region: kilo→UNIT 1000, UNIT→milli 1000)

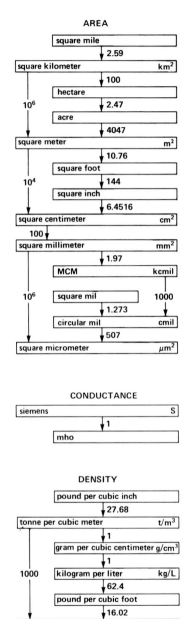

AREA

- square mile
- ↓ 2.59
- square kilometer — km²
- ↓ 100
- hectare
- ↓ 2.47
- acre
- ↓ 4047
- square meter — m² (10⁶)
- ↓ 10.76
- square foot
- ↓ 144
- square inch (10⁴)
- ↓ 6.4516
- square centimeter — cm²
- ↓ 100
- square millimeter — mm²
- ↓ 1.97
- MCM — kcmil (10⁶)
- square mil — 1000
- ↓ 1.273
- circular mil — cmil
- ↓ 507
- square micrometer — μm²

CONDUCTANCE

- siemens — S
- ↓ 1
- mho

DENSITY

- pound per cubic inch
- ↓ 27.68
- tonne per cubic meter — t/m³
- ↓ 1
- gram per cubic centimeter g/cm³
- ↓ 1
- kilogram per liter — kg/L (1000)
- ↓ 62.4
- pound per cubic foot
- ↓ 16.02
- kilogram per cubic meter — kg/m³

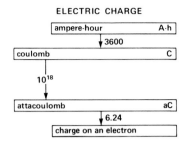

ELECTRIC CHARGE

- ampere-hour — A·h
- ↓ 3600
- coulomb — C
- ↓ 10¹⁸
- attacoulomb — aC
- ↓ 6.24
- charge on an electron

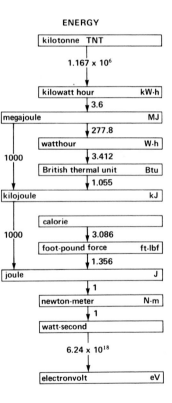

ENERGY

- kilotonne TNT
- ↓ 1.167 x 10⁶
- kilowatt hour — kW·h
- ↓ 3.6
- megajoule — MJ (1000)
- ↓ 277.8
- watthour — W·h
- ↓ 3.412
- British thermal unit — Btu
- ↓ 1.055
- kilojoule — kJ (1000)
- calorie
- ↓ 3.086
- foot-pound force — ft·lbf
- ↓ 1.356
- joule — J
- ↓ 1
- newton-meter — N·m
- ↓ 1
- watt-second
- ↓ 6.24 x 10¹⁸
- electronvolt — eV

Example: Convert 1590 MCM to square inches.

Solution: 1590 MCM = 1590 (÷ 1.97) (÷ 100) (÷ 6.4516) in² = 1.25 in².

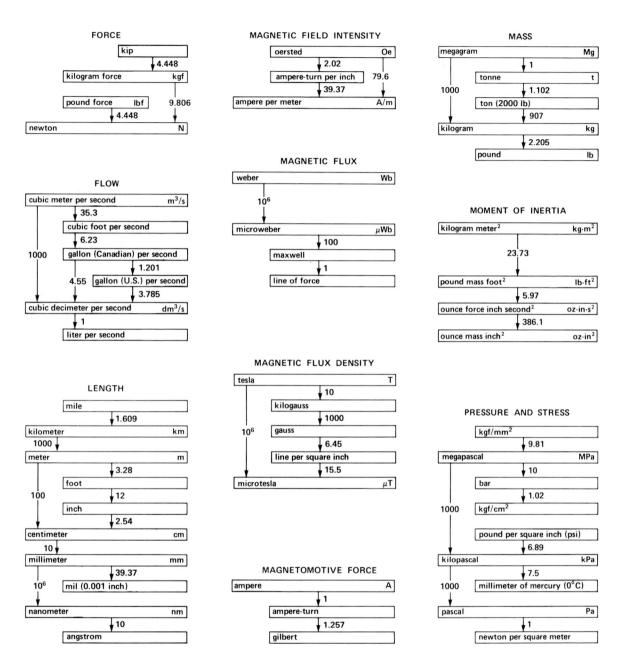

FORCE

kip

↓ 4.448

kilogram force   kgf

pound force   lbf   9.806

↓ 4.448

newton   N

MAGNETIC FIELD INTENSITY

oersted   Oe

↓ 2.02

ampere-turn per inch   79.6

↓ 39.37

ampere per meter   A/m

MASS

megagram   Mg

↓ 1

tonne   t

1000   ↓ 1.102

ton (2000 lb)

↓ 907

kilogram   kg

↓ 2.205

pound   lb

FLOW

cubic meter per second   m³/s

↓ 35.3

cubic foot per second

↓ 6.23

1000   gallon (Canadian) per second

↓ 1.201

4.55   gallon (U.S.) per second

↓ 3.785

cubic decimeter per second   dm³/s

↓ 1

liter per second

MAGNETIC FLUX

weber   Wb

10⁶ ↓

microweber   μWb

↓ 100

maxwell

↓ 1

line of force

MOMENT OF INERTIA

kilogram meter²   kg·m²

23.73

pound mass foot²   lb·ft²

↓ 5.97

ounce force inch second²   oz·in·s²

↓ 386.1

ounce mass inch²   oz·in²

MAGNETIC FLUX DENSITY

tesla   T

↓ 10

kilogauss

↓ 1000

10⁶   gauss

↓ 6.45

line per square inch

↓ 15.5

microtesla   μT

LENGTH

mile

↓ 1.609

kilometer   km

1000 ↓

meter   m

↓ 3.28

foot

100   ↓ 12

inch

↓ 2.54

centimeter   cm

10 ↓

millimeter   mm

↓ 39.37

10⁶   mil (0.001 inch)

↓

nanometer   nm

↓ 10

angstrom

PRESSURE AND STRESS

kgf/mm²

↓ 9.81

megapascal   MPa

↓ 10

bar

↓ 1.02

1000   kgf/cm²

pound per square inch (psi)

↓ 6.89

kilopascal   kPa

↓ 7.5

1000   millimeter of mercury (0°C)

pascal   Pa

↓ 1

newton per square meter

MAGNETOMOTIVE FORCE

ampere   A

↓ 1

ampere-turn

↓ 1.257

gilbert

Example: Convert 580 psi to megapascals.
Solution: 580 psi = 580 (x 6.89) (÷ 1000) MPa = 4 MPa.

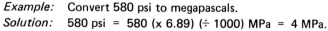

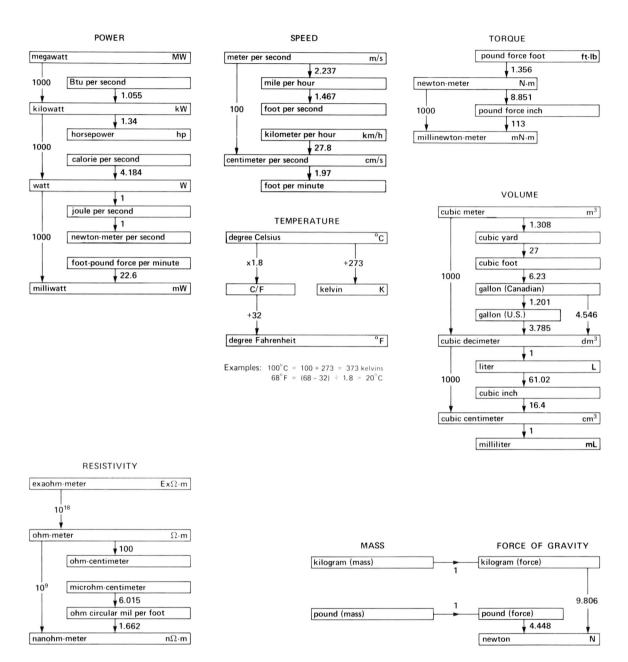

POWER

megawatt	MW
1000	
Btu per second	
1.055	
kilowatt	kW
1.34	
horsepower	hp
1000	
calorie per second	
4.184	
watt	W
1	
joule per second	
1	
1000 newton-meter per second	
foot-pound force per minute	
22.6	
milliwatt	mW

SPEED

meter per second	m/s
2.237	
mile per hour	
1.467	
100 foot per second	
kilometer per hour	km/h
27.8	
centimeter per second	cm/s
1.97	
foot per minute	

TEMPERATURE

degree Celsius °C

×1.8   +273

C/F   kelvin K

+32

degree Fahrenheit °F

Examples: 100°C = 100 + 273 = 373 kelvins
68°F = (68 − 32) ÷ 1.8 = 20°C

TORQUE

pound force foot	ft·lb
1.356	
newton-meter	N·m
8.851	
pound force inch	
1000	
113	
millinewton-meter	mN·m

VOLUME

cubic meter	m³
1.308	
cubic yard	
27	
cubic foot	
6.23	
1000 gallon (Canadian)	
1.201	
gallon (U.S.)	4.546
3.785	
cubic decimeter	dm³
1	
liter	L
1000 61.02	
cubic inch	
16.4	
cubic centimeter	cm³
1	
milliliter	mL

RESISTIVITY

exaohm-meter	Ex Ω·m
10^{18}	
ohm-meter	Ω·m
100	
ohm-centimeter	
10^9 microhm-centimeter	
6.015	
ohm circular mil per foot	
1.662	
nanohm-meter	nΩ·m

MASS   FORCE OF GRAVITY

kilogram (mass)	→ 1	kilogram (force)	
			9.806
pound (mass)	→ 1	pound (force)	
		4.448	
		newton	N

Example: Calculate the force of gravity (in newtons) that acts on a mass of 9 lb.
Solution: 9 lbm → 9 (× 1) (× 4.448) N = 40 N.

Properties of Round Copper Conductors

The American wire Gauge system follows a geometric progression in which the daameters of wire gauges No. 0000 and No. 36 are fixed at 460 mils and 5 mils respectively. Because there are 39 wire sizes between No. 0000 and No. 36, the ratio between two successive diameters is $\sqrt[39]{460/5} = 1.1229\ldots$. The corresponding ratio between the cross sections of two consecutive wire sizes is $(1.1229\ldots)^2 = 1.261$.

For example, the theoretical diameter of a No. 35 wire is $5 \times 1.1229\ldots = 5.614$ mils. The theoretical cross section of a No. 35 wire is $5^2 \times 1.261 = 31.52$ circular mils.

In practice, the diameters and cross sections of the conductors are rounded off according to the values given in the following table.

PROPERTIES OF ROUND COPPER CONDUCTORS

Gauge number AWG/ B & S	Diameter of bare conductor		Cross section		Resistance mΩ/m or Ω/km		Weight g/m or kg/km	Typical diameter of *insulated*
	mm	mils	mm²	cmils	25°C	105°C		
250MCM	12.7	500	126.6	250 000	0.138	0.181	1126	magnet
4/0	11.7	460	107.4	212 000	0.164	0.214	953	wire
2/0	9.27	365	67.4	133 000	0.261	0.341	600	used in
1/0	8.26	325	53.5	105 600	0.328	0.429	475	relays
1	7.35	289	42.4	87 700	0.415	0.542	377	magnets
2	6.54	258	33.6	66 400	0.522	0.683	300	motors
3	5.83	229	26.6	52 600	0.659	0.862	237	transformers
4	5.18	204	21.1	41 600	0.833	1.09	187	etc.
5	4.62	182	16.8	33 120	1.05	1.37	149	
6	4.11	162	13.30	26 240	1.32	1.73	118	
7	3.66	144	10.5	20 740	1.67	2.19	93.4	**mm**
8	3.25	128	8.30	16 380	2.12	2.90	73.8	
9	2.89	114	6.59	13 000	2.67	3.48	58.6	3.00
10	2.59	102	5.27	10 400	3.35	4.36	46.9	2.68
11	2.30	90.7	4.17	8 230	4.23	5.54	37.1	2.39
12	2.05	80.8	3.31	6 530	5.31	6.95	29.5	2.14
13	1.83	72.0	2.63	5 180	6.69	8.76	25.4	1.91
14	1.63	64.1	2.08	4 110	8.43	11.0	18.5	1.71
15	1.45	57.1	1.65	3 260	10.6	13.9	14.7	1.53
16	1.29	50.8	1.31	2 580	13.4	17.6	11.6	1.37
17	1.15	45.3	1.04	2 060	16.9	22.1	9.24	1.22
18	1.02	40.3	0.821	1 620	21.4	27.9	7.31	1.10
19	0.91	35.9	0.654	1 290	26.9	35.1	5.80	0.98
20	0.81	32.0	0.517	1 020	33.8	44.3	4.61	0.88
21	0.72	28.5	0.411	812	42.6	55.8	3.66	0.79
22	0.64	25.3	0.324	640	54.1	70.9	2.89	0.70
23	0.57	22.6	0.259	511	67.9	88.9	2.31	0.63
24	0.51	20.1	0.205	404	86.0	112	1.81	0.57
25	0.45	17.9	0.162	320	108	142	1.44	0.51
26	0.40	15.9	0.128	253	137	179	1.14	0.46
27	0.36	14.2	0.102	202	172	225	0.908	0.41
28	0.32	12.6	0.080	159	218	286	0.716	0.37
29	0.29	11.3	0.065	128	272	354	0.576	0.33
30	0.25	10.0	0.0507	100	348	456	0.451	0.29
31	0.23	8.9	0.0401	79.2	440	574	0.357	0.27
32	0.20	8.0	0.0324	64.0	541	709	0.289	0.24
33	0.18	7.1	0.0255	50.4	689	902	0.228	0.21
34	0.16	6.3	0.0201	39.7	873	1140	0.179	0.19
35	0.14	5.6	0.0159	31.4	1110	1450	0.141	0.17
36	0.13	5.0	0.0127	25.0	1390	1810	0.113	0.15
37	0.11	4.5	0.0103	20.3	1710	2230	0.091	0.14
38	0.10	4.0	0.0081	16.0	2170	2840	0.072	0.12
39	0.09	3.5	0.0062	12.3	2820	3690	0.055	0.11
40	0.08	3.1	0.0049	9.6	3610	4720	0.043	0.1

References

BOOKS

ANSI, 1981. *National Electrical Safety Code*. New York: Institute of Electrical and Electronic Engineers.

————, 1976. *Metric Practice* (ANSI Z210.1-1976)/(IEEE Std 268-1976)/(ASTM E380-76). New York: Institute of Electrical and Electronic Engineers.

ASME, 1968. *Letter symbols for quantities used in electrical science and electrical engineering*. New York: American Soc. of Mech. Eng.

Bean, R.L.; Chackan, N.; Moore, H.R. and Wentz, E.C. 1959. *Transformers for the electric power industry*. New York: McGraw.

Beeman, D. 1955. *Industrial power systems handbook*. New York: McGraw.

Biddle, 1967. *"Getting down to earth . . ." A manual on earth resistance testing for the practical man*. Plymouth Meeting: James G. Biddle Co.

Bloomquist, W.C. (ed). 1950. *Capacitors for industry*. New York: Wiley.

Bonneville Power Administration (1976). *Transmission line reference book HVDC to ± 600 kV. Palo Alto: Electric Power Research Institute*.

Boylestad, R.L. 1977. *Introductory circuit analysis*. Columbus: Merrill.

Canadian Standards Association, 1979. *Canadian metric practice guide Z 234.1-79*. Toronto: CSA.

Considine, D.M. (ed). 1977. *Energy technology handbook*. New York: McGraw.

Edison Electric Institute, 1973. *Questions and answers about the electric utility industry*. New York: Edison Electric Institute.

Federal Power Commission, 1978. *Electric power statistics*. Washington: Superintendent of Documents.

Fink, D.G. and Beaty, H.W. (eds). 1978. *Standard Handbook for electrical engineers*. New York: McGraw.

General Electric, 1975. *Transmission line reference book 345 kV and above*. New York: Fred Weidner and Son Printers Inc.

Gingrich, H.W. 1979. *Electrical machinery, transformers and controls.* Englewood Cliffs: Prentice-Hall.

IEEE, 1972. *IEEE standard dictionary of electrical and electronic terms.* New York: Wiley.

————, 1979. *IEEE standard metric practice.* New York: Inst. Electrical and Electronics Eng.

————, 1971. *Graphic symbols for electrical and electronics diagrams.* New York: Inst. Electrical and Electronics Eng.

Jackson, H.W. 1976. *Introduction to electric circuits.* Englewood Cliffs: Prentice-Hall.

Johnson, R.C. 1971. *Electrical wiring.* Englewood Cliffs: Prentice-Hall.

Kosow, I.L. 1972. *Electric machinery and transformers.* Englewood Cliffs: Prentice-Hall.

————, 1973. *Control of electric machines.* Englewood Cliffs: Prentice-Hall.

Kurtz, E.B. and Showmaker, T.J. 1976. *The lineman's and cableman's handbook.* New York: McGraw.

National Fire Protection Association, 1978. *National Electrical code.* Boston: NFPA.

NEMA, 1970. *Electrical insulation terms and definitions.* New York: Insulating materials division, National Elect. Manufacturers Ass'n.

————, 1978. *Motors and Generators* (ANSI/NEMA No MG1-1978). Washington: National Electrical Manufacturers Association.

Nesbitt, E.A. 1962. *Ferromagnetic domains, a basic approach to the study of magnetism.* Baltimore: Bell Telephone Labs, Inc.

Oklahoma State University, 1977. *Power factor improvement.* Stillwater: OSU Elect. Power Technology Dept.

Richardson, D.V. 1978. *Rotating electric machinery and transformer technology.* Reston: Reston Publ. Co.

Stevenson, W.D. Jr. 1975. *Elements of power system analysis.* New York: McGraw.

Stigant, S.A. and Franklin, A.C. 1973. *The J & P transformer book.* Boston: Newnes-Butterworths.

TVA, 1975. *Annual report of the Tennessee Valley Authority, Volume II - appendices.* Knoxville: Tennesse Valley Authority.

Underwriters Laboratories, 1978. *Systems of insulating materials* (UL 1446). Northbrook, Ill: Underwriters Laboratories Inc.

————, 1977. *Ground-fault circuit interrupters* (UL 943). Northbrook, Ill: Underwriters Laboratories Inc.

Werninck, E.H. (ed). 1978. *Electric motor handbook*. New York: McGraw.

Westinghouse, 1964. *Electrical transmission and distribution reference book.* Pittsburgh: Westinghouse Electric Corp.

Wildi, T. 1973. *Understanding Units.* Québec: Sperika Enterprises Ltd.

————, 1981. Electrical Power Technology. New York: Wiley.

REFERENCES

ARTICLES AND TECHNICAL REPORTS

Apprill, M.R. 1978. Capacitors reduce voltage flicker. *Electrical World* 189, No. 4: 55-56.

Beaty, H.W. (ed). 1978. Underground distribution. *Electrical World* 189, No. 9: 51-66.

Creek, F.R.L. 1976. Large 1200 MW four-pole generators for nuclear power station in the USA. *GEC J. Sc. and Tech.* 43, No. 2: 68-76.

Dalziel, C.F. 1968. Reevaluation of lethal electric currents. *IEEE Trans. Ind. and Gen. Appl.* 1GA-4, No. 5: 467-475.

Elliot, T.C. (ed). 1976. Demand control of industry power cuts utility bills, points to energy savings. *Power* 120, No. 6: 19-26.

Harris, S.W. 1960. Compensating dc motors for fast response. *Control Engineering* Oct. 1960: 115-118.

Hoffmann, A.H. 1969. Brushless synchronous motors for large industrial drives. *IEEE Trans. Ind. and Gen. Appl. 1GA-5, No. 2:* 158-162.

Hossli, W. (ed). 1976. Large steam turbines. *Brown Boveri Rev.* 63: 84-147.

Kolm, H.H. and Thornton, R.D. 1973. Electromagnetic flight. *Scientific American.* 229, No. 4: pp. 17-25.

Krick, N. and Noser, R. 1976. The growth of turbo-generators. *Brown Boveri Rev.* 63: 148-155.

Methé, M. 1971. The distribution of electrical power in large cities. *Canadian Engineering Journal.* October: 15-18.

Moor, J.C. 1977. Electric drives for large compressors. *IEEE Trans. Ind. Gen. Appl.* 1A-13, No. 5: pp. 441-449.

Quinn, G.C. 1977. Plant primary substation trends. *Power* 121, No. 3: 29-35.

Rieder, W. 1971. Circuit breakers. *Scientific American* 224, No. 1: 76-84.

Sebesta, D. 1978. Responsive ties avert system breakup. *Electrical World* 180, No. 7: 54-55.

Summers, C.M. 1971. The conversion of energy *Scientific American* 224, No. 3: 148-160.

Woll, R.F. 1964. High temperature insulation in ac motors. *Westinghouse Engineer* Mar. 1964: 1-5.

Woll, R.F. 1975. Effect of unbalanced voltage on operation of polyphase induction motors. *IEEE Trans. Ind. Gen. Appl.* 1A-11, No. 1: 38-42.

Woll, R.F. 1977. Electric motor drives for oil well walking beam pumps. *IEEE Trans. Ind. Gen. Appl.* 1A-13, No. 5: 437-441.

Zimmermann, J.A. 1969. Starting requirements and effects of large synchronous motors. *IEEE Trans. Ind. and Gen. Appl.* 1GA-5, No. 2: 169-175.

Answers to Problems

The following numerical answers are usually rounded off to an accuracy of ±1%.

Chapter 1
1) false; **2)** c; **3)** b; **4)** b; **5)** a; **6)** d; **7)** c;
8) false; **9)** d; **10)** c; **11)** c; **12)** a; **13)** c; **14)** c;
15) a; **16)** a; **17)** c; **18)** d; **19)** c; **20)** a; **21)** b;
22) c; **23)** b; **24)** c; **25)** c; **26)** c; **27)** a; **28)** a;
29) a; **30)** c; **31)** true; **32)** false; **33)** false;
34) b; **35)** c; **36)** a; **37)** b; **38)** c; **39)** c; **40)** c

Chapter 2
1) c; **2)** true; **3)** true; **4)** d; **5)** c; **6)** c; **7)** true;
8) b; **9)** c; **10)** a; **11)** c; **12)** c; **13)** third orbit has
16 electrons; **14a)** 28; **14b)** 4; **16a)** 960 W;
16b) 15 Ω; **17)** 240 W; 15 Ω; **18)** 2.4 A;
19a) 3 200 000 MW; **19b)** 12.5 kΩ; **19c)** 26.67
kW.h; **20)** 254 V

Chapter 3
1) false; **2)** a; **3)** b; **4)** c; **5)** a; **6)** false; **7)** false;
8) false; **9)** c; **10)** true; **11)** true; **12a)** 1.5 A;
12b) 3 W; **13a)** 1 A; **13b)** 0.67 W; **14a)** 5.1 V;
14b) 1.7 V; **14c)** 20 h; **15a)** 1.4985 V;
15b) 133 h; **16)** six groups in parallel; each
group consisting of 20 cells in series;
17) 7.5 A; **8a)** 0.4 A; **18b)** 4 W; **19a)** 900 V;
19b) 45 kW; **20a)** 125 V; **20b)** 93.75 W;
21) case 1: 10 V; 1 Ω; case 2: 0.1 W; 10 mA;
case 3: 400 V; 0.5 A; case 4: 10 Ω; 100 A;
case 5: 2000 V; 2 mA; **22)** 5 A; **23)** 5 A flowing
away from the common terminal.

Chapter 4
1) b; **2)** b; **3)** b; **4)** c; **5)** b; **6)** a; **7)** c; **8)** c; **9)** b;
11) 180°C; **12)** 160 months; **13a)** 317.5 kV;
13b) 76.2 kV; **14)** 253 mm^2; **16)** No. 1;

17) 0.04 mm; **18)** No. 34; **19a)** 7.39 km;
19b) 21.36 kg; **20)** 0.001 59 mm^2;
21a) 358.6 Ω; **21b)** 519.4 Ω; **22a)** 1.333 Ω;
22b) 160 V; 19.2 kW; **23a)** 2200 Ω ±10%;
23b) 12 MΩ ±5%; **24)** 2.21 mΩ

Chapter 5
1) b; **2)** c; **3)** false; **4)** false; **5)** a; **6)** false; **7)** c;
8) false; **12)** S pole; **13)** 1250 At; **14)** positive;
15a) repulsion; **15b)** attraction;
15c) attraction; **16a)** 10 000 At; 4000 At;
16b) 1250 W; 1750 W; **16c)** B; **16d)** repel;
6e) repel; **18)** 0.4 T

Chapter 6
1) true; **2)** true; **3)** b; **4)** true; **5)** false; **6)** true;
7) c; **9)** downward; **10)** A is a S pole;
11) toward the reader; **12)** downward;
13a) cw; **13b)** cw; **14)** 25.4 N; **15a)** 12 N;
15b) 72 N.m; 53.1 lbf.ft.

Chapter 7
1) c; **2)** b; **3)** c; **4)** a; **5)** true; **6)** a; **7)** c; **8)** false;
9) c; **10)** b; **11)** a; **12)** c; **13)** false; **14)** no;
15) negative; **21)** 6.18 A; **22)** 324°

Chapter 8
1) b; **2)** a; **3)** b; **4)** true; **5)** b; **6)** b; **7)** c; **8)** 20 V;
9a) increasing; **9b)** (+); **10)** 28.3 V;
13) 25 mV; E_{AB} is negative; the voltage is dc.

Chapter 9
1) true; **2)** true; **3)** a; **4)** true; **5)** b; **6)** b; **7)** b;
8) true; **9)** a; **10)** b; **11)** 4 H; **12a)** 30 V;

12b) 20 H; **13)** 40 V; **14)** (−); **15)** 2400 J;
16a) 180 kJ; **16b)** 24 V; **16 c)** 2.5 s; **17a)** 6 A;
17b) 10 V; **17c)** 2 V; **17d)** 0.5 A/s; **17e)** 50 J;
18) 0.6 A; 48 A; 360 A; 21 600 A; **19)** 500 V;
20) 10.7 kV

Chapter 10
1) true; **2)** a; **3)** false; **4)** b; **5)** a; **6)** b; **7)** a;
8) true; **9)** a; **10)** true; **11)** true; **13a)** 5 min;
13b) 36 kJ; **14a)** 64 056 v/s; **14b)** 6.4 A;
15a) 3000 V; **15b)** 270 J; **15c)** 5 μF; **16)** 24 s;
17) 40 V; **19)** 6 groups in parallel; each group
consisting of two capacitors in series;
20a) 0.248 μF; **20b)** 40 kV; **20c)** 198 J;
21a) 1.3 MJ; **21b)** 2250 J; **21c)** 578

Chapter 11
1) true; **2)** b; **3)** b; **4)** a; **5)** b; **6)** a; **7)** c;
11a) 96 V; **11b)** 288 W; **13a)** 4014 At;
13b) 6814 At

Chapter 12
1) a; **2)** b; **3)** a; **4)** c; **5)** b; **6)** c; **7)** a; **8)** a; **9)** a;
10) a; **20a)** 490.4 V; **20b)** 789 hp;
20c) 11.5 kW; **21)** 62.5 kA; **22a)** 600 At;
22b) 800 At

Chapter 13
1) b; **2)** c; **3)** c; **4)** a; **5)** a; **6)** c; **7)** c; **8)** true;
9) c; **10)** 120 W; **11b)** +200 V 1 positive with
respect to 2; **12a)** 12 A; **12b)** 169.7 V;
13a) 1136 A; **13b)** 0.3872 Ω; **13c)** 6850 μF;
13d) 622 V; **13e)** 1325 J; **14a)** 259 A;
14b) 1642 Ω; **14c)** 4.35 H; **14d)** 366 A;
14e) 146 kJ

Chapter 14
1) b; **2)** c; **3)** b; **4)** a; **5)** b; **6)** a; **7)** b; **8)** b; **9)** a;
10) a; **11)** 0.64 lagging; **12a)** 681.8 kVA;
12b) 323.8 kvar; **13)** 4.3 kvar; **14)** 20 Ω;
15a) 5 A; 2 A; **15b)** 5.38 A; **15c)** 92.8 %;
15d) 18.57 Ω; **16a)** 6 A; **4 A; 16b)** 7.21 A;
16c) 83.2 % leading; **16d)** 16.64 Ω; **17a)** 6 A;
4 A; **17b)** 2 A; **17c)** 0% lagging; **17d)** 60 Ω;
18a) 25 Ω; **18b)** 28 V; 96 V; **18c)** 112 W;

19a) 5 Ω; **19b)** 80 V; 90 V; **19c)** 0; **20a)** 37 Ω;
20b) 140 V; 48 V; **20c)** 192 W; **21a)** 3 A;
21b) 5 A; **21c)** 5.83 A; **21d)** 2110 W;
21e) 450 var; **21f)** 2157 VA; **21g)** 5.83 A;
21h) 370 V; **21i)** 97.8 %; **21j)** 12°;
22a) 18.97 Ω; **22b)** 30 V; **22c)** 50 V;
22d) 20 V; **22e)** 0.333 A; **22f)** 6.67 W;
22g) 30 var; **22h)** 50 var; **22i)** 20 var;
22j) 21.08 VA; **22k)** 31.6 % leading;
22l) 71.5°; **23)** 0.571 A

Chapter 15
1) b; **2)** c; **3)** true; **4)** a; **5)** b; **6)** c; **7)** b; **8)** b;
9) false; **10a)** 358 V; **10b)** 23.9 A;
10c) 25.6 kW; **11a)** 694 A; **11b)** 13.2 kV;
11c) 9161 kW; **11d)** 100 %; **12)** dim;
13) 23.5 kW; **14)** 0.9 Ω; **16a)** 103.6 kVA;
16b) 80.4 %; **16c)** 61.6 kvar

Chapter 16
1) c; **2)** b; **3)** c; **4)** true; **5)** c; **6)** a; **7)** c; **8)** a;
9) true; **10)** b; **11)** b; **f; 12)** a; **13)** c; **14)** b;
15) c; **16)** a; **17)** b; **21)** higher; **30)** 120 V;
31) 60 kV; **32a)** 110.4 V; **32b)** 22.08 A;
33a) 2420 W; **33b)** 22 A; **11 A; 34)** 50 A;
1250 A; 35a) zero; **35b)** 520 V; **36a)** short-
circuit; **36b)** no change; **38a)** 129 V;
38b) 93 A; 5.22 A; **39a)** 470 kW;
39b) 99.12 %

Chapter 17
1) false; **2)** b; **3)** c; **4)** b; **5)** b; **6)** true; **7)** b; **8)** a;
9) b; **15)** 69 N; 15.5 lbf; **16a)** 514.286 r/min;
16b) 483.43 r/min; **16c)** 3.6 Hz

Chapter 18
1) false; **2)** b; **3)** c; **4)** false; **5)** c; **6)** b; **7)** a;
8) c; **9)** b; **10)** a; **16a)** 371 N.m; **16b)** 556 N.m;
16c) 742 N.m; 960 r/min; **16d)** 464 N.m;
240 r/min; **16e)** 18.75 hp; **17)** locked rotor
torque = 229 N.m; **18)** 3.88 Ω; **19a)** 1453 W;
19b) 2344 VA; **19c)** 2.94; **20)** 90.3 Ω

Chapter 19
1) true; **2)** true; **3)** b; **4)** b; **5)** c; **6)** b; **7)** a;

13a) line current increases; **13b)** temperature increases; **14)** 168 kW; **15a)** 0.08; **15b)** 0.0167; **15c)** 0.004 48

Chapter 20
1) a; **2)** b; **3)** c; **4)** a; **5)** true; **6)** false; **7)** false; **8)** true; **9)** true; **10)** true; **11)** false; **12)** b; **13)** false; **14)** true; **17a)** 20 poles; **17b)** 360 r/min; **18)** Excitation must be increased; **21a)** 575 V; **21b)** 28.75 V

Chapter 21
1) true; **2)** a; **3)** b; **4)** b; **5)** b; **6)** c; **7)** false; **8)** true; **9)** c; **10)** true; **11)** 720 r/min; **17a)** series; **17b)** capacitor-start; **17c)** series; **17d)** shaded-pole; **17e)** resistance split-phase; **17f)** capacitor-run; **17g)** reluctance or hysteresis; **17h)** hysteresis; **18a)** too hot; **18b)** temperature rise is not defined at no-load; **19a)** 28.6 %; **19b)** 55.3 %; **19c)** 19.2 mN.m; **19d)** 16.9 %; **19e)** 56.4°; **19f)** 15 W; **20)** 24 W vs 15 W; **22)** 18°

Chapter 22
1) false; **2)** true; **3)** false; **4)** true; **5)** true; **6)** false; **7)** a; **8)** true; **9)** b; **10)** c; **19)** 548 years; **21)** 520 A; **23)** 120 A; **24)** 20 N.m; **26)** E_{cd} is (−); **27)** 600 V

Chapter 23
1) b; **2)** a; **3)** b; **4)** b; **5)** a; **6)** a; **7)** true; **8)** a; **9)** b; **10)** c; **11)** a; **12)** true; **13)** a; **14)** false; **15)** b; **16)** a; **17)** true; **18)** true; **19)** false; **20)** a; **21)** b; **22)** b; **23)** true; **28)** 19.3 kV to 38.7 kV; **33)** 198 PJ; **35)** 3630 MW; **40)** 317.4 MW; 2656 A; **48)** $1.53; **49)** $4.20; **50)** $3.64; **51a)** 1615 kW; **51b)** 1184 kvar; **51c)** 80.6 %; **52a)** 1184 kvar; **52b)** 653 kvar; **53a)** 18 650 kW; **53b)** 11.5 Mvar

Index

-A-

Acceleration, of a drive system, 18
Active power, 246
ACSR cable, 492
Air, 71, 73
Air gap, 99, 162
Alnico, 93
Alternator (see Generator, synchronous)
Aluminum, 34
Ambient temperature (see Temperature)
Ammeter, 38, 39, 50
Ampacity, 77, 78
Ampere, 38
Ampere hour, 59
Angle,
    electrical, 128
    torque, 400, 401
Apparent power (see Power)
Arc suppression, 161
Arcing horns, 504
Arcos, 22
Arcsin, 22
Arctan, 22
Armature,
    of a dc generator, 190-192, 205
    of a dc motor, 214
    reaction, 195, 209
Asynchronous motor, 342
Atom, 33, 34
Autotransformer, 315, 318
    variable, 318
Auxiliary winding, 419
AWG gauge, 74
Axes, of a graph, 20

-B-

Battery, 58
    charging of, 61
Billing demand, 518, 519
Blow-out coil, 114
Boiler, 481
Braking,
    of a dc motor, 230
    of an induction motor, 362, 363
Branch circuit, 509
Breakdown torque, 345
Brown and Sharp gauge (see AWG gauge)
Brush, 188, 190, 194
Brushless excitation, 384
Bundled conductors, 494
Burden, 312
Bus bar, 77
Bushing, 155, 313, 502

-C-

Capacitive reactance, 243, 245
Cam switch (see Switch)
Capacitance, 171, 172, 181
Capacitor, 44, 172, 176, 474
    discharge of, 176, 179
    energy in a, 174
    in parallel, 177
    in series, 177
    size of, 177
Cell,
    carbon-zinc, 58
    equivalent circuit of, 59
    internal resistance of, 59
    parallel connection of, 60

primary, 37, 58
secondary, 58
series connection of, 60
series-parallel connection of, 60
Centrifugal switch, 420, 422
Celsius, degree, 14
Ceramic permanent magnet, 93
Chain reaction, 484
Charge, 34, 36
transfer of, 169
Circuit, dc
complex, 62-63
parallel, 52
series, 49
series-parallel, 55
Circuit breaker
air-blast, 502
manual, 442
oil, 502
sulfur hexafluoride, 503
Circuits, ac
capacitive, 243
inductive, 239
resistive, 235
solution of, 259, 260, 288
Circular mil, 74
Clock motor, 432
Coil, 44, 159
Commutating poles, 201, 209
Commutation, 208
Commutator, 189, 208
Condenser, 481
synchronous (see Synchronous capacitor)
Conductors,
electrical properties, 35, 76, 80
gauge number, 75
resistance of, 57, 70
round, 76
shapes, 77
thermal properties of, 81
Contact,
normally closed, 446
normally open, 446
resistance, 83
self-sealing, 449, 452

Contactor,
magnetic, 97, 443
Control diagram, 446
Cooling tower, 480
Copper, 34
Corona effect, 494
Cosine,
of an angle, 19, 20
law, 22
Coulomb, 36, 170
Counter emf, 221
Current, unit of, 36, 38
Current flow, conventional, 38
Current transformer, 313
Cutting of flux lines, 144
Cycle, 27, 124

-D-

Dam, 487
Damper winding, 396, 399
Degree, electrical, 128
Delta connection,
voltage and current in, 285-287
Demand, 474, 513
meter, 514
Diagram (Control), 446
Dielectric, 173
constant, 173
strength, 72
Diode, 161
Dispatch center, 474
Distribution systems, 471, 472
low voltage, 508
standard voltage classes, 473
three-phase 3-wire, 511
three-phase 4-wire, 511, 512
Domains (see Magnetic domains)
Draft tube, 487
Drip-proof (motor), 354
Dynamo (see Generator, dc)

-E-

Eddy currents, 146
Effective value, 237
Efficiency,

of a machine, 13
of dc machines, 216
of induction motors, 344
of transformers, 300
Electromagnet, 96
    typical application of, 96
Electromagnetic force, 109
    magnitude of, 111
Electromagnetic induction, 135
Electromagnetic torque, 115
Electron, 34
Energy,
    measurement of, 514
    storage of, 44, 157, 171
    transformation of, 12
    unit of, 12, 43
Exciter,
    brushless, 385, 398
    pilot, 383
Explosion-proof (motor), 355

-F-

Fahrenheit, degree, 14
Farad, 171
Faraday, law of electromagnetic induction, 135
Feeder, 509
Field,
    of a dc machine, 192
    revolving, 334
Fission, 484
Fleming's right-hand rule, 146
Flux (see Magnetic flux)
    distorsion, 199
Flux density (see Magnetic flux density)
Force,
    of gravity, 4
    on a coil, 114
    on a conductor, 112
    unit of, 3
Foucault currents (see Eddy currents)
Frequency, 27, 125
    choice of, 475
Fuse, 58, 83

-G-

Gauss, 95
Gear motor, 360
Generating stations
    hydropower, 486
    location of, 479
    nuclear, 484
    thermal, 480
Generation, 471
Generator (see source)
Generator, direct-current, 187
    compound, 200
    construction of, 190-194
    equivalent circuit of, 194
    flux distorsion in, 199
    induced voltage, 188, 197
    neutral position, 194
    over compound, 202
    rating, 204
    separately excited, 194
    shunt, 198
    under load, 199, 203
Generator (synchronous), 391, 399
    3-phase, 281, 391
    Brushless excitation of, 398
    construction of, 391-395, 398
    equivalent circuit, 406
    excitation of, 396, 398, 404
    frequency of, 394
    mechanical pole shift, 408
    power output, 399, 403
    stationary-field, 391
    synchronization of, 404
    synchronous reactance, 406
    synchronous speed, 394
    torque angle, 400, 401
    under load, 397, 410
    voltage regulation, 412
Governor, 482
Ground,
    electrode, 84
    resistance of, 84
    wire, 495

Grounding,
    of electrical systems, 513
    of equipment, 513

-H-

Heating,
    by induction, 148
    of electrical machines, 12, 14
Henry, 154, 157
Hertz, 3, 27, 125
Horsepower, 3, 9
Hot spot temperature, 72
Hunting, 385
Hydrogen, 71, 380
Hydropower station, 486
    power of, 486
Hysteresis,
    loss, 103, 104
    motor, 431
    torque, 104

-I-

Impedance,
    of a transformer, 309
    of ac circuits, 258, 261-265
Inching, (see Jogging)
Induced current, 137, 146
Induced voltage, 135, 137
    examples of, 139
    Faraday's law, 135
    polarity of, 139, 145
Inductance, 154, 156
    current in, 160
    energy in, 157
    unit of, 154
    voltage induced in, 159, 163
Induction,
    mutual, 153
    self, 156
Inductive reactance, 241
Inductor (see Inductance)
Inertia, 18, 363
Infinite bus, 399
Instability, 478
Instantaneous values (voltage, current), 130

Insulation classes, 72
    life expectancy, 77, 78
Insulators,
    breakdown of, 72
    classification of, 71
    deterioration of, 71
    gaseus, 71
    liquid, 71
    pin-type, 493
    properties of, 35, 73
    resistivity of, 70
    solid, 71
    suspension-type, 492
    thermal classification of, 71
Interpole (see Commutating poles)
Ionization, 71
Iron losses, 300

-J - K - L-

Jogging, 454
Joule, 11, 43
Kelvin,
Kilovoltampere, 259
Kilowatt, 43
Kilowatthour, 43
Kinetic energy, 15
Kirchhoff's laws, 62
KVA, 259
Lagging, (phase angle), 128
    power factor, 272
Laminations, 149, 190
Lap winding,
    of a dc generator, 191, 207
    of an induction motor, 337, 338
    of a synchronous machine, 372
Leading, (phase angle), 128
    power factor, 272
Lenz's law, 139
Lightning, arresters, 505
Line voltage, 282
Line of force, properties of, 91
Linear induction motor, 348
Limit switch, 443
Load, definition of, 40
    active, 246

balanced, 285, 288
    reactive, 246
    unbalanced, 289, 290
Locked-rotor current, 224, 338
Locked-rotor torque, 343
Losses, in electrical machines, 12, 216
    in transmission lines, 58, 295, 494
    in transformers, 300, 311

-M-

Magnetic field,
    of a conductor, 93
    of a coil, 94, 95
Magnetic,
    circuit, (properties of), 97
    domains, 92
    energy, 157
    field, 90
    flux, 90
    flux density, 95
    flux lines, 90
    force of attraction, 89, 101
    permeability, 100
    torque, 101
Magnetism,
    induced, 92
    remanent, 92
    residual, 198
Magnetizing current (see Transformer)
Magnetomotive force,
    of a coil, 95
    of a permanent magnet, 95
Mass, unit of, 3
Maxwell, 91
MCM, 75
Mil, 74
    circular, 74
Moderator, 484
Molecule, 33
Motor, direct current, 213
    dynamic braking, 230
    compound, 215
    losses in, 216
    mechanical power, 220
    plugging, 230

    series, 215
    shunt, 214
    speed control, 228
    speed of, 222
    stabilized shunt, 226
    starting of, 218, 223
    theory of, 219
    torque of, 221
    torque-speed characteristic, 214, 215, 225-228
Motor, single-phase induction,
    capacitor-run, 427
    capacitor-start, 423
    choice of, 429
    construction of, 420-422
    hysteresis, 431
    power factor and efficiency of, 424
    principle of, 418
    reluctance torque, 433
    series, 428
    shaded-pole, 427
    slip, 426
    resistance split-phase, 423
    starting means, 419
    synchronous speed, 417
    synchro, 430
    vibration of, 425
    watthourmeter, 435
Motor, three-phase synchronous,
    construction of, 371-375
    equivalent circuit, 406
    excitation of, 376, 383, 387
    power factor of, 377, 378
    rating of, 387
    reactive power of, 377
    starting a, 373, 388
    synchronous speed, 371
    torque, 376
    under load, 381
    V-curve, 378
    versus induction motor, 389

Motor, three-phase induction,
    abnormal operating conditions, 364
    application of, 366
    braking of, 362, 363
    classification of, 353-359
    construction of, 337
    direction of rotation, 362
    efficiency, 344
    enclosures, 353-355
    linear type, 348
    mechanical power, 17
    plugging of, 362
    power factor, 344
    principle of, 329-333
    rotating field, 334
    rotor voltage and frequency, 341, 342, 343, 358
    sector type, 347
    slip, 333, 342, 344
    speed of, 336
    squirrel cage, 333, 337
    standardization of, 353
    synchronous speed, 333, 335, 349
    torque, 341
    torque-speed characteristic, 343, 345, 357
    two speed, 361, 362
    typical characteristics of, 356-359
    windings, 334, 346
    wound rotor, 346
Mutual induction, 153
    inductance, 154

-N - O-

National Electrical Code, 78
Network, 472
Neutral,
    of single-phase system, 311, 511
    of three-phase system, 282, 511
    position, 194
Neutron, 484
Newton, 3, 4
Nuclear,
    generating stations, 484
    reactor, 484, 485
Nucleus, 34
Ohm, 41

Ohm's law, 41
Ohmmeter, 39
Oil,
    as coolant, 71, 308
    as insulator, 71, 308
    as fuel, 478
Outlet, 509
Overcompound generator, 202

-P - Q-

Panelboard, 508
Pascal, 3
Peak demand, 474
Penstock, 487
Percent impedance, 310
Permanent magnet, 93
Permeability, 100
Phase,
    angle, 128
    meaning of, 128, 129
    sequence, 292, 406
Phasor(s), 127
    diagram, 130, 248, 254
    addition of, 253
    components of, 255
Pilot exciter, 383
Pilot light, 443
Plugging, 230, 362
Polarity,
    additive, subtractive, 304
    of a magnet, 89
    of a terminal, 119
    of a transformer, 303
    of a voltage, 119
    negative, 36
    positive, 36
Pole,
    geographic, 89
    magnetic, 89
    salient, 192, 392, 396
Polyphase systems, 279
Positive and negative
    voltage, 119
    current, 123
Potential,

difference of, 36, 170
level, 52
Potential transformer (see Voltage transformer)
Power, 8
    active, 266, 270
    angle (see Torque angle)
    apparent, 259, 272
    electric 39, 42
    factor (see Power factor)
    in ac circuits, 236, 240, 244, 266
    in 3-phase circuits, 287, 289
    instantaneous, 237
    of a motor, 8, 10
    measurement of, 10, 267, 289
    mechanical, 8
    reactive, 266, 270
    unit of, 9
Power factor,
    correction, 520
    in rate structures, 516
    of a circuit, 272, 287
Power triangle, 272
Pressurized water reactor, 485
Pull-in torque, 376, 388, 461
Pull-out torque, 382, 402
Pushbutton, 442
Primary, 296
Prime mover, 393
Prony brake, 10
Protective devices, 58, 83, 450, 500
Pull-up torque, 343

-R-

Rate structure, 514, 516, 517
Rate of change,
    of a voltage, 120, 123, 126, 180
    of a current, 124, 126, 154, 156
Rating,
    of a dc generator, 204
    of a transformer, 300,303
    of a synchronous generator, 412
    of a synchronous motor, 387
Reactance, 241, 243
Reactive power, 240, 244, 246, 247, 474

Reactor,
    line compensating, 242, 497
    nuclear, 484, 485
Real power (see Active power)
Receptacle, 509
Rectifier, 384, 464
Regulating transformer (see Tap-changing transformer)
Relay, 97
    control, 443
    overload, 443
    thermal, 443
    time delay, 443, 448
Reluctance, 100
    motor, 433
    torque, 102, 376, 433
Remanent magnetism, 92
Residual magnetism, 198
Resilient mounting, 425
Resistance,
    calculation of, 79
    equivalent, 51, 53
    ground, 84
    unit of, 41
Resistivity,
    of conductors, 70
    of insulators, 70
    of the earth,
    unit of, 70
Resistors, 50, 69
    classification of, 81
    color coding of, 82
Resonance, 265
Reversal
    dc motor, 229
    induction motor, 334
Rheostat, 82
    field, 198
    wound rotor, 346
Ripple, 191
Rotating field,
    in a three-phase machine, 334
    in a single-phase motor, 418
    synchronous speed of, 336

-S-

Salient pole, 192, 392, 396
Saturation curve, 99
    of a dc generator, 196
    of a vacuum, 98
    of a soft magnetic material, 98
Saturation flux density, 99
Scroll case (see Spiral case)
Secondary, 296
Sector motor, 347
Segment, commutator, 188, 190
Self-inductance, 156, 208
Selsyn (see Synchro)
Series motor,
    dc, 215
    single-phase, 428
Service conductors, 508
Service entrance, 510, 515
Service factor, 364
Servo (see Synchro)
Shaded-pole, 427
Shock (electric), 513
Short-circuit,
    of a transmission line, 58
    protection (see Protective devices)
Shunt field, 192
Sign notation,
    positive and negative current, 123
    positive and negative voltage, 120
Sine, of an angle, 19, 20
Sine wave 23, 125
Single-phasing, 365
Source,
    definition of, 36, 40
    reactive, 245
Slip, 333, 342, 344, 462
Slip ring, 188, 383
Speed, of a drive system, 18
Spillway, 487
Spiral case, 487
Splash-proof (motor), 354
Squirrel cage, 333
Stability, 478
Starter,
    across-the-line, 451
    autotranformer, 456
    combination, 451
    dc motor, 231, 448
    part winding, 458
    primary resistance, 454
    reduced voltage, 454
    synchronous motor, 461-466
    wound rotor, 346
    wye-delta, 458
Static var compensator, 498
Stator, 334, 337, 338, 372
Star connection (see Wye connection)
Steam-turbine generator, 396
Stepper motor, 433
Substation, 471, 500, 507
Sulfur hexafluoride, 71, 73
Surge diverter (see Lightning arrester)
SVC (see Static var compensator)
Switch,
    air break, 504
    cam, 442, 459
    centrifugal, 420, 422
    disconnecting, 442, 504
    grounding, 504
    limit, 443
Switchboard, 508
Symbols, electrical diagram, 445
Synchro, 430
Synchronization, 404
Synchronous capacitor, 379, 380
Synchronous motor (see Motor, three-phase synchronous)
Synchronous reactance, 406
Synchronous speed,
    of single-phase motors, 417
    of synchronous motors, 371
    of 3-phase generators, 394
    of 3-phase induction motors, 335
Synchroscope, 405

-T-

Tailrace, 487
Tangent, of an angle, 20
Tap, 305

Tap-changing transformer, 306
Tariff (see Rate structure)
Temperature,
    ambient, 14, 81
    coefficient of resistance, 81
    hot spot, 72
    rise (see Temperature rise)
    scales, 14
Temperature rise, 14
    by resistance method, 81, 362, 363
    of electrical wiring, 78
    of insulation classes, 71
Tertiary winding, 322
Tesla, 95
Thermal generating stations, 480
Three-phase,
    voltage (generation of), 280, 283
    neutral, 282
    currents, 283
    phasor diagrams, 282-284
    circuits (solution of), 288
    power (measurement of), 289
Time constant,
    of a coil, 160
    of a capacitor, 178
Torque, 6
    angle, 376, 382
    breakdown, 345
    of a drive system, 15-17
    locked rotor, 218, 338, 343
    measurement of, 10
    pull-in, 376, 388, 461
    pull-out, 382, 402
    pull-up, 343
    unit of, 6
Totally enclosed fan-cooled (motor), 354
Transformers, 295-324
    autotransformer, 315, 318
    construction of, 297
    cooling of, 307
    distribution, 296, 311
    equivalent circuit, 309
    flux in, 296
    ideal, 309

impedance of, 309
induced voltage, 296
in parallel, 307
iron losses in, 300
losses in, 300
magnetizing current, 296, 299
parallel operation of, 307
polarity of, 303
primary winding, 296
rating of, 300, 303
ratio, 296, 297, 300
saturation curve, 303
secondary winding, 296
single-phase, 296
special transformers, 311
taps, 305
temperature rise, 301
three-phase (see Transformers, three-phase)
toroidal, 315
Transformers (three-phase), 322
    delta-delta, 319
    delta-wye 320
    open-delta, 322
    polarity of, 319
    tertiary winding, 322
    wye-delta, 321
    wye-wye, 321
Transmission, 471, 472
Transmission lines,
    capacitance of, 181
    components of, 492
    dc, 57, 475
    equivalent circuit, 495
    inductance of, 162
    interconnection of, 472
    power of, 498
    selection of line voltage, 476
    standard voltages of, 473
    towers, 494
    voltage regulation, 496
Triangles, solution of, 22
Trigonometry, 19-22
Turbine, 481, 486
    runner, 490

-U - V-

Unbalanced loads,
    single-phase, 511
    three-phase, 289, 290
Units,
    commonly-used, 3
    conversion of, (see Appendix A), 527
    multiples and submultiples, 4, 5, 530
    SI, 2, 527
    U.S. Customary, 2, 527
Universal motor, 428
Uranium, 484
Var, 240, 245
Variac (see autotransformer, variable)
Varmeter, 268, 270, 404
V-curve, 378
Volt, 36
Voltage,
    ac, 121
    choice of transmission line, 476
    drop, 57, 63
    effective value, 252, 254
    induced, 135, 137
    rise, 63
    peak, 235
Voltage regulation,
    of a dc generator, 204
    of a synchronous generator, 412
    of a transmission line, 496
Voltage transformer, 312
Voltampere, 241, 259
Voltmeter, 37, 39, 50

-W - X - Y - Z-

Ward-Leonard system, 229
Watt, 39
Watthourmeter, 435-437
Wattmeter, 267, 268, 270, 290
Weber, 91
Weight (see Force of gravity)
Wicket gates, 487
Wire gauge, 74, 75
Work, 7
    unit of, 7

Wound-rotor motor, 346
    starting of, 347
Wye connection, 285
    voltage and current in, 285-287
Zero-speed switch, 230